Advances in Industrial Control

Advances in Industrial Control is a series of monographs and contributed titles focusing on the applications of advanced and novel control methods within applied settings. This series has worldwide distribution to engineers, researchers and libraries.

The series promotes the exchange of information between academia and industry, to which end the books all demonstrate some theoretical aspect of an advanced or new control method and show how it can be applied either in a pilot plant or in some real industrial situation. The books are distinguished by the combination of the type of theory used and the type of application exemplified. Note that "industrial" here has a very broad interpretation; it applies not merely to the processes employed in industrial plants but to systems such as avionics and automotive brakes and drivetrain. This series complements the theoretical and more mathematical approach of Communications and Control Engineering.

Indexed by SCOPUS and Engineering Index.

Proposals for this series, composed of a proposal form downloaded from this page, a draft Contents, at least two sample chapters and an author cv (with a synopsis of the whole project, if possible) can be submitted to either of the:

Series Editors

Professor **Michael J. Grimble**
Department of Electronic and Electrical Engineering, Royal College Building, 204 George Street, Glasgow G1 1XW, United Kingdom
e-mail: m.j.grimble@strath.ac.uk

Professor **Antonella Ferrara**
Department of Electrical, Computer and Biomedical Engineering, University of Pavia, Via Ferrata 1, 27100 Pavia, Italy
e-mail: antonella.ferrara@unipv.it

or the

In-house Editor

Mr. **Oliver Jackson**
Springer London, 4 Crinan Street, London, N1 9XW, United Kingdom
e-mail: oliver.jackson@springer.com

Proposals are peer-reviewed.

Publishing Ethics

Researchers should conduct their research from research proposal to publication in line with best practices and codes of conduct of relevant professional bodies and/or national and international regulatory bodies. For more details on individual ethics matters please see: https://www.springer.com/gp/authors-editors/journal-author/journal-author-helpdesk/publishing-ethics/14214

More information about this series at https://link.springer.com/bookseries/1412

Ling Zhao · Yuanqing Xia ·
Hongjiu Yang · Jinhui Zhang

Pneumatic Servo Systems Analysis

Control and Application in Robotic Systems

Springer

Ling Zhao
State Key Laboratory of Precision
Measuring Technology and Instruments
Tianjin University
Tianjin, China

Hongjiu Yang
School of Electrical
and Information Engineering
Tianjin University
Tianjin, China

Yuanqing Xia
School of Automation
Beijing Institute of Technology
Beijing, China

Jinhui Zhang
School of Automation
Beijing Institute of Technology
Beijing, China

ISSN 1430-9491 ISSN 2193-1577 (electronic)
Advances in Industrial Control
ISBN 978-981-16-9514-8 ISBN 978-981-16-9515-5 (eBook)
https://doi.org/10.1007/978-981-16-9515-5

This Springer imprint is published by the registered company Springer Nature Singapore Pte Ltd.
The registered company address is: 152 Beach Road, #21-01/04 Gateway East, Singapore 189721, Singapore

Series Editor's Foreword

Pneumatic Servo Systems Analysis, Control and Application in Robotic Systems is a new monograph, which comes to enrich the series *Advances in Industrial Control*. This volume is especially timely because pneumatic systems and pneumatic robots, in particular, are becoming increasingly important in the contexts of Industry 4.0, and of surgical and biomedical applications.

The first examples of pneumatic technology date back to the first century. Yet the first air pump, used to study the phenomenon of vacuum, was only invented by the German physicist Otto von Guericke in 1650, and it is necessary to wait until 1800 to see the great impact of pneumatics on society. In the first three decades of the nineteenth century the first air compressors were developed. The water-jacketed cylinders, so important for the automotive industry, were invented in 1872. In 1867, the first example of a pneumatic subway train was showcased at the American Institute Exhibition of New York, demonstrating how passengers could be transported along a sort of pipe just by exploiting pneumatic power.

Afterward, Pneumatic pipes were used in post offices to transport letters from one part of the building to others, as well as in department stores to transfer cash from cashiers to the safe-box or vice versa. The pneumatic drill and the pneumatic-powered hammer were invented in the 1870s and 1890s, respectively. Then in the twentieth century pneumatics reached full maturity with plenty of pneumatic components employed in a variety of sectors. They were mainly used to replace manpower in heavy or repetitive tasks, creating the conditions for the diffusion of automatic control devices in industry, and for the establishment of industrial automation as a reality.

Nowadays, digitally controlled pneumatic devices are used in numerous industrial fields, from manufacturing to the food industry, from the production of medical devices to the avionics sector. The amazing thing is that research on pneumatic systems and efficient control systems based on them has not yet stopped, as evidenced by the fact that innovative solutions, more and more efficient, less expensive, and significantly more precise, are still being put on the market with a certain regularity.

Pneumatic actuators are often among the essential parts of industrial automation systems. In fact, by virtue of their better cost–benefit ratio, pneumatic systems are replacing servo systems based on hydraulic devices in many industrial applications. Apart from being economical, using pneumatic systems is also environmentally friendly, since air is available at no cost and can be easily stored in tanks after being compressed.

In contrast to the advantages just mentioned, however, there is a major drawback which cannot be ignored when considering the use of pneumatic devices for industrial applications, i.e., the difficulty of controlling them and reaching the high precision normally required by industrial production standards. This is mainly due to the fact that pneumatic components are inherently nonlinear and often do not allow the formulation of sufficiently accurate linearized models of their behaviors.

The present book is aimed at tackling the issues related to the design and developing of pneumatic servo systems in a clear but rigorous manner. It is a book very rich in terms of contents. It starts from an overview of pneumatic servo system applications, also accurately describing the control methods that have been applied so far to address the nonlinear control problems associated with them. Then, it discusses, in a very detailed way, different plants that rely on pneumatic servo systems: a manipulator driven by pneumatic artificial muscles, a pneumatic dexterous hand, a pneumatic motion simulation platform, a pneumatic rod–cylinder servo system, and a pneumatic right-angle composite motion platform. All the examples are extremely interesting and useful for learning the essential elements of nonlinear control theory applied to this class of systems.

In the book, the authors address feedback control systems based on observers, then they deal with backstepping-based controllers of adaptive type and finally they also discuss sliding mode control. The control algorithms considered are of advanced type, but their presentation is very readable, so that even a non-control-expert reader can understand the fundamental concepts and appreciate the distinguishing characteristics of the various set-ups by following step-by-step the developments reported in the chapters.

With this volume, the series *Advances in Industrial Control* is enlarged with a new monograph that helps to make the series more varied, but, at the same time, confirms the mission of the series itself to be a tool for people who deal with industrial automation to keep up-to-date and learn about the various facets that industrial automation presents.

Antonella Ferrara
University of Pavia
Pavia, Italy

Preface

The pneumatic servo system is an automatic control system which can make the output change with the input target arbitrarily. It takes a air compressor as driving source, and the compressed air is used as working medium for energy transmission. Various pneumatic servo systems have been investigated widely because of their advantages, such as clean, low cost, simple structure, and convenient maintenance. Pneumatic servo systems are used extensively in mechanical operations, medical areas, automation systems, robots, and many other fields. However, it is difficult to achieve higher position control precision for the pneumatic servo systems because of strong nonlinearity which is caused by air compressibility, large friction forces and nonlinear properties of servo valves. Therefore, it has great significance to conduct further research on position control for the pneumatic servo systems. The active disturbance rejection control (ADRC) technique is a new practical technology which absorbs achievements of modern control theory and develops essences of PID. The ADRC does not depend on an accuracy mathematical model which are hardly obtained because of strong nonlinearity, unmodeled dynamics and external disturbances for the pneumatic servo systems. Therefore, it is a key idea to solve these uncertain factors effectively by using the ADRC for the pneumatic servo systems. This book aims to analyze the position control of the pneumatic servo systems via the improved ADRC.

In Chap. 1, an overview on application and a motivation on the research are provided for typical pneumatic servo systems. In Chap. 2, an overview on control methods is introduced to deal with nonlinear problems in pneumatic servo systems.

Then this book is divided into five parts:

Part I: Three kinds of controllers are designed for a pneumatic manipulator system with an one degree of freedom (DOF) joint driven by pneumatic artificial muscles (PAMs). In Chap. 3, a pneumatic manipulator platform is designed and introduced. In Chap. 4, a linear error feedback controller combining with compensation of disturbances is designed to ensure response speed for the pneumatic manipulator system. In Chap. 5, a nonlinear error feedback controller is designed to control the pneumatic manipulator system via a reduced-order extended state observer. In Chap. 6, a sliding mode controller based on an extended state observer is designed to guarantee stability of the pneumatic manipulator system.

Part II: Some results on a pneumatic dexterous hand system are presented. In Chap. 7, an experimental platform of the pneumatic dexterous hand system is designed and introduced. In Chap. 8, a backstepping nonlinear error feedback controller is designed to study the position problem of the pneumatic dexterous hand system via an extended state observer. In Chap. 9, a sliding mode controller with an extended state observer is proposed to cope with nonlinearities and disturbances for the pneumatic dexterous hand system. In Chap. 10, an ADRC technique is proposed to improve precision for the pneumatic dexterous hand system.

Part III: Three kinds of controllers are designed for a pneumatic motion simulation system driven by PAMs. In Chap. 11, an experimental platform for the pneumatic motion simulation system is designed and introduced. In Chap. 12, a nonlinear error feedback controller based on a linear extended state observer is applied in the pneumatic motion simulation system under varying load conditions. In Chap. 13, a linear error feedback controller based on a nonlinear extended state observer is introduced for the pneumatic motion simulation system. In Chap. 14, an adaptive backstepping controller based on an adaptive extended state observer (AESO) is presented for the pneumatic motion simulation system.

Part IV: Some results on position control are given for a pneumatic rod cylinder servo system. In Chap. 15, an experimental platform for the pneumatic rod cylinder servo system is designed and introduced. In Chap. 16, a finite-time controller is designed to improve response rapidity and anti-interference ability of the pneumatic rod cylinder servo system based on super-twisting extended state observer. In Chap. 17, a nonsingular fast terminal sliding mode (NFTSM) finite-time tracking control strategy is presented for the pneumatic rod cylinder servo system via an extended state observer. In Chap. 18, a generalized nonlinear extended state observer (GNESO) is applied in the pneumatic rod cylinder servo system with varying loads.

Part V: Some developments of position control are introduced for a pneumatic right-angle composite motion system. In Chap. 19, an experimental platform for the pneumatic right-angle composite motion system is introduced. In Chap. 20, an ADRC technique is given for the pneumatic right-angle composite motion system. In Chap. 21, a multi-point position is study for the pneumatic right-angle composite motion system by using the ADRC. In Chap. 22, a linear error feedback controller is applied to improve response speed of the pneumatic right-angle composite motion system.

We would like to acknowledge the collaborations for their contributions in this monograph with Tao Wang in Beijing Institute of Technology, Zhixin Liu in Yanshan University, Yafei Yang and Jiahui Sun in SMC (China) Company Limited, Bin Zhang in Qinhuangdao WKW Automotive Parts Co.,Ltd., Haiyan Cheng of doctoral candidate in Xi'an Jiaotong University, Xin Liu and Shaomeng Gu of doctoral candidates in Beijing Institute of Technology, Yang Yu of doctoral candidate in Northwestern Polytechnical University, Qi Li, Linlin Ge and Chengfei Zheng of master degree candidates in Yanshan University. In addition, sincerely thanks to the other members in our research group who have provided their help for this book. Special thanks go to the series Editors Antonella Ferrara and Michael J. Grimble, the Editor Oliver Jackson, the Editorial Assistant Sridevi, Mengchu Huang and Bhagyalakkshme for their help during the preparation of the manuscript. This work was also supported by the National Natural Science Foundation of China under Grant (62073238, 61733012, 61973230).

Tianjin, China — Ling Zhao
Beijing, China — Yuanqing Xia
Tianjin, China — Hongjiu Yang
Beijing, China — Jinhui Zhang
February 2021

Acknowledgements

This monograph gives a self-contained presentation of our recent work in design, analysis and control for pneumatic servo systems. The materials of the monograph have been adapted from a number of our recent publications. We acknowledge the following publishers for granting us the permission to reuse materials from our publications copyrighted by these publishers in this monograph.

Acknowledgement is given to IEEE for reproducing materials from

© 2015 IEEE. Reprinted, with permission, from Ling Zhao, Qi Li, Hongjiu Yang, Hongbo Li, "Positioning control of a one-DOF manipulator driven by pneumatic artificial muscles based on active disturbance rejection control," *Proceedings of the 34th Chinese Control Conference*, 2015. 7, Hangzhou, China, pp. 841–845 (material found in Chap. 4).

© 2019 IEEE. Reprinted, with permission, from Ling Zhao, Qi Li, Bo Liu, Haiyan Cheng, "Trajectory tracking control of a one degree of freedom manipulator based on a switched sliding mode controller with a novel extended state observer framework," *IEEE Transactions on Systems, Man, and Cybernetics: Systems*, vol. 49, no. 6, pp. 1110–1118, 2019 (material found in Chap. 6).

© 2019 IEEE. Reprinted, with permission, from Ling Zhao, Haiyan Cheng, Yuanqing Xia, Bo Liu, "Angle tracking adaptive backstepping control for a mechanism of pneumatic muscle actuators via an AESO," *IEEE Transactions on Industrial Electronics*, vol. 66, no. 6, pp. 4566–4576, 2019. (material found in Chap. 14).

© 2021 IEEE. Reprinted, with permission, from Ling Zhao, Chengfei Zheng, Yingjie Wang, Bo Liu, "A finite-time control for a pneumatic cylinder servo system based on a super-twisting extended state observer," *IEEE Transactions on Systems, Man, and Cybernetics: Systems*, vol. 51, no. 2. pp. 1164–1173, 2021 (material found in Chap. 16).

© 2017 IEEE. Reprinted, with permission, from Ling Zhao, Bin Zhang, Hongjiu Yang, Yingjie Wang, "Finite-time tracking control for pneumatic servo system via extended state observer," *IET Control Theory and Applications*, vol. 11, no. 16, pp. 2808–2816, 2017 (material found in Chap. 17).

© 2020 IEEE. Reprinted, with permission, from Ling Zhao, Bin Zhang, Hongjiu Yang, Yingjie Wang, "Observer-based integral sliding mode tracking control for a pneumatic cylinder with varying loads," *IEEE Transactions on Systems, Man, and Cybernetics: Systems*, vol. 50, no. 7, pp. 2650–2658, 2020 (material found in Chap. 18).

© 2015 IEEE. Reprinted, with permission, from Ling Zhao, Yafei Yang, Yuanqing Xia, Zhixin Liu, "Active disturbance rejection position control for a magnetic rodless pneumatic cylinder," *IEEE Transactions on Industrial Electronics*, vol. 62, no. 9, pp. 5838–5846, 2015 (material found in Chap. 20).

© 2017 IEEE. Reprinted, with permission, from Ling Zhao, Yuanqing Xia, Yafei Yang, Zhixin Liu, "Multicontroller positioning strategy for a pneumatic servo system via pressure feedback," *IEEE Transactions on Industrial Electronics*, vol. 64, no. 6, pp. 4800–4809, 2017 (material found in Chap. 21).

Acknowledgement is given to Elsevier for reproducing materials from

© 2018 Elsevier. Reprinted, with permission, from Ling Zhao, Haiyan Cheng, Tao Wang, "Sliding mode control for a two-joint coupling nonlinear system based on extended state observer," *ISA Transactions*, vol. 73, pp. 130–140, 2018 (material found in Chap. 9).

© 2017 Elsevier. Reprinted, with permission, from Hongjiu Yang, Yang Yu, Jinhui Zhang, "Angle tracking of a pneumatic muscle actuator mechanism under varying load conditions," *Control Engineering Practice*, vol. 61, pp. 1–10, 2017 (material found in Chap. 12).

© 2019 Elsevier. Reprinted, with permission, from Ling Zhao, Jiahui Sun, Hongjiu Yang, Tao Wang, "Position control of a rodless cylinder in pneumatic servo with actuator saturation," *ISA Transactions*, vol. 90, pp. 235–243, 2019 (material found in Chap. 22).

Acknowledgement is given to AMSE for reproducing materials from

© 2020 AMSE. Reprinted, with permission, from Ling Zhao, Xin Liu, Tao Wang. "Observer-based nonlinear decoupling control for two-joint manipulator systems driven by pneumatic artificial Muscles," *Journal of Dynamic Systems, Measurement, and Control-Transactions of the ASME*, vol. 142, no. 4, p. 041001, 2020 (material found in Chap. 10).

Acknowledgement is given to SAGE for reproducing materials from

© 2018 SAGE. Reprinted, with permission, from Ling Zhao, Xin Liu, Tao Wang, Bo Liu, "A reduced-order ESO based trajectory tracking control for one-DoF pneumatic manipulator," *Advances in Mechanical Engineering*, vol. 10, no. 4, pp. 1–9, 2018 (material found in Chap. 5).

© 2018 SAGE. Reprinted, with permission, from Ling Zhao, Linlin Ge, Wang Tao, "Position control for a two-joint robot finger system driven by pneumatic artificial muscles," *Transactions of the Institute of Measurement and Control*, vol. 40, no. 4, pp. 1328–1339, 2018 (material found in Chap. 8).

Acknowledgement is given to Springer for reproducing materials from

Contents

Symbols and Acronyms

R	Field of real numbers
R^n	n-dimensional real Euclidean space
$R^{n \times m}$	Space of $n \times m$ real matrices
C	Field of complex numbers
I	Identity matrix
A	System matrix
A^{-1}	Inverse of matrix A
A^T	Transpose of matrix A
$A \geq 0$	Symmetric positive semi-definite
$A > 0$	Symmetric positive definite
$A \leq 0$	Symmetric negative semi-definite
$A < 0$	Symmetric negative definite
lim	Limit
min	Minimum
max	Maximum
SMC	Sliding mode control
PAM	Pneumatic artificial muscle
PAMs	Pneumatic artificial muscles
DOF	Degree of freedom
ADRC	Active disturbance rejection control
TD	Tracking differentiator
ESO	Extended state observer
AESO	Adaptive extended state observer
GNESO	Generalized nonlinear extended state observer
NFTSM	Nonsingular fast terminal sliding mode
$\text{rank}(A)$	Rank of matrix A
$\lambda(A)$	Eigenvalue of matrix A
$\lambda_{\min}(A)$	Minimum eigenvalue of matrix A
$\lambda_{\max}(A)$	Maximum eigenvalue of matrix A
$\sigma(A)$	Singular value of matrix A

$\sigma_{\min}(A)$	Minimum singular value of matrix A
$\sigma_{\max}(A)$	Maximum singular value of matrix A
$\mid x \mid$	Absolute value (or modulus) of x
$\| x \|$	Euclidean norm
$\| x \|_2$	The second norm for vectors
$\| x \|_{m\infty}$	The infinite norm for vectors
$\forall$	For all
$\in$	Belong to
$\subseteq$	Inclusion sign
$\rightarrow$	Tend to, or mapping to (case sensitive)
$\sum$	Sum
$G(\cdot)$	Characteristic equation
$\exp(x)$	e^x
$\mathrm{sign}(x)$	The sign function of x
$\mathrm{sat}(x)$	The saturation function of x
sup	Supremum
inf	Infimum

Chapter 1
Introduction

1.1 Background

With the implementation of national strategies, such as German industry 4.0, advanced manufacturing in America, made in China 2025, and so on, great changes in the manufacturing industry have been promoted by the new industrial revolution, which brings important opportunities and challenges for robots. Pneumatic robots have received more and more attention for manufacturing, in which pneumatic systems play important roles in driving mechanical devices [1]. For various advantages such as clean, safety, low cost, convenient maintenance, and good flexibility [2, 3], the pneumatic systems are extensively used in textile industry, chip manufacturing, automobile manufacturing, pharmaceutical manufacturer, food packing, and so on [4–6]. Pneumatic artificial muscles (PAMs) and pneumatic cylinders are two types of main pneumatic actuators in robots. There are different shapes and specifications for different types of actuators to meet various application requirements.

Compared with other kinds of driving units, the PAMs have unique advantages in flexibility and bionics. The PAMs are contractile and linear motion engines operated by gas pressure. Their core element is a flexible reinforced closed membrane, and it attaches at both ends to fittings by which mechanical power is transferred to a load. As the membrane is inflated by gas, the membrane contracts axially and thereby exerts a pulling force on its load. The force and motion thus generated by this type of actuator are linear and unidirectional [7]. Therefore, when the pneumatic artificial muscle (PAM) is inflated, its shape will thicken radially and its length will be shortened, which results in axial displacement and tension. Otherwise, there will be no output force and bearing capacity for the PAM in the non-inflated state. This working principle is very similar to the human muscle. The characteristics of the PAM are mainly summarized as high power-weight ratio, high degree of energy transformation, good flexibility, good self cushioning ability, good bionics, no piston structure inside, no lubrication, no crawling phenomenon, easy miniaturization, and easy to realize low-speed movement. Therefore, the PAM receives much attention

L. Zhao et al., *Pneumatic Servo Systems Analysis*, Advances in Industrial Control,
https://doi.org/10.1007/978-981-16-9515-5_1

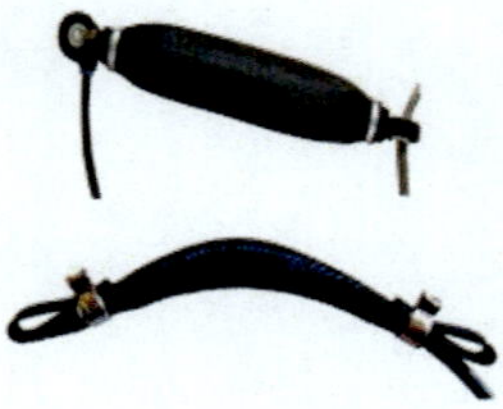

(a) Air muscle

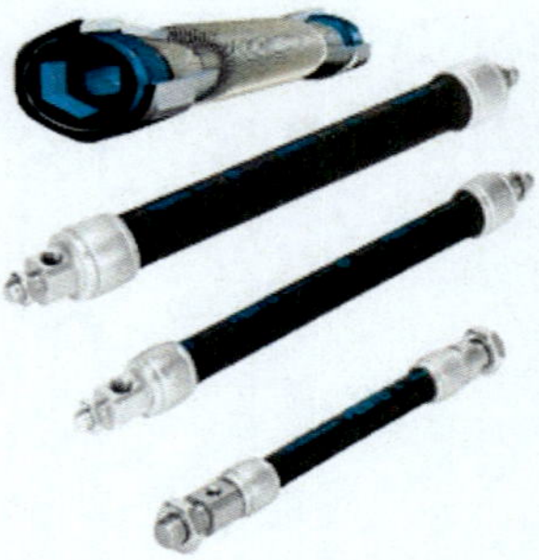

(b) Fluid muscle

Fig. 1.1 Different types of PAMs

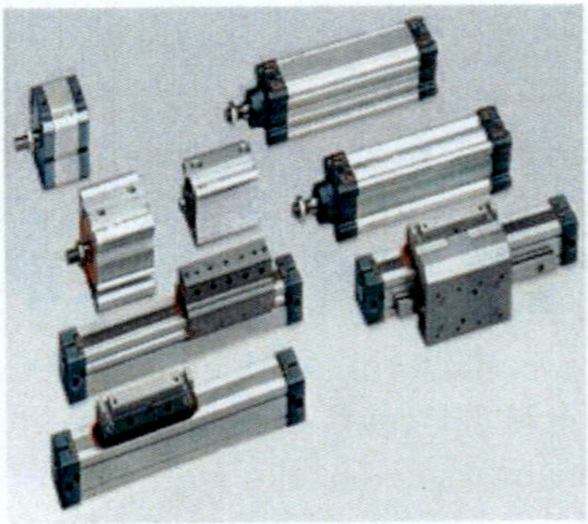

(a) Pneumatic cylinders

(b) Bellows cylinders

Fig. 1.2 Different types of pneumatic cylinders

from researchers in robotics and industrial automation. The two types of representative PAMs include air muscle produced by Shadow in London in Fig. 1.1a and fluid muscle produced by Festo in Germany in Fig. 1.1b.

In linear reciprocating motion, the pneumatic cylinder is a typical actuator in the industrial scene. It converts the pressure energy of compressed air into mechanical energy and makes the driven mechanism works as expected. With the expansion of application fields and the development of new technologies, the types of cylinders are increasing, such as piston cylinder, plunger cylinder, diaphragm cylinder, rod cylinder, rodless cylinder, bellows cylinder, and so on. Different types of pneumatic cylinders are shown in Fig. 1.2a. The bellows cylinder produced by Festo is shown in Fig. 1.2b, which is used as a spring element or for reducing oscillations. The pneumatic cylinder is widely used in semiconductor, printing, robot, food processing, and other fields.

Fig. 1.3 Pneumatic robot

1.1.1 Application of Pneumatic System

Since late twentieth century, more and more attention has been paid to the automation and labor-saving in various fields of production and manufacturing. Due to its unique advantages, pneumatic systems are more and more widely used in the field of industrial mechanization and automation. Typical applications of the PAMs and the pneumatic cylinders are introduced in the following.

In the area of biorobotic and Industrial applications, the PAMs can produce high torque at low and medium speeds and can be easily installed without transmission device. Due to its light weight, the PAMs can be used as the actuator of portable machinery. In addition, the PAMs have natural compliance and impact resistance. It is a suitable solution for driving industrial machinery and industrial robots, especially safe industrial robots with human–computer interaction. Different pneumatic robots have been designed by PAMs. Bipedal robot in [8] could achieve vertical jumping, soft landing from a one-meter drop and postural control during standing. For the RoboThespian in Fig. 1.3 at [9], much of its human-like agility relies on the pneumatic fluidic muscles developed by Festo. The movement of the arms and hands is taken on by a total of two or eight fluidic muscles respectively in different sizes.

In the automobile manufacturing industry, most modern automobile manufacturing production lines have adopted pneumatic technology, especially in the welding production line, such as each car body moving and positioning, the welding gun moving, and welding spot control. In automobile assembly line, various pneumatic actuators and pneumatic control systems are adopted for tire installation, glass installation, engine delivery and assembly, and so on. Figure 1.4 shows a pneumatic servo system for automobile spot welding of SMC corporation.

In the area of production automation, the use of pneumatic system can also improve labor productivity and reduce production costs. With the increasing popularity of production automation, pneumatic systems have been used in various fields of production and manufacturing. In some assembly production lines of industries such as bicycles, watches, and washing machines, many operation processes of workpieces on the production line are automatically completed by pneumatic systems. Pneumatic air-jet looms, printing machine, automatic cleaning machines also use pneumatic systems to effectively reduce the heavy and monotonous human labor. In short, pneumatic technology is an environmental protection and low-cost automation technology. The pneumatic systems are important means to realize production

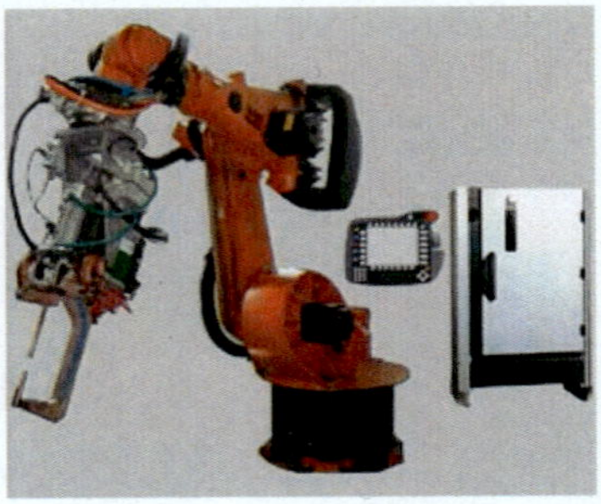

Fig. 1.4 Pneumatic welding

(a) Pneumatic assisted manipulator (b) Hot stamping machine

Fig. 1.5 Pneumatic systems in automobile manufacturing industry

automation, which makes them important in production and life. Figure 1.5a shows a pneumatic assisted manipulator driven by a pneumatic cylinder. The pneumatic assisted manipulator is a product of Suzhou Luokeweige Automation Co., Ltd. of China. Figure 1.5b shows a pneumatic hot stamping machine driven by a pneumatic cylinder.

1.1.2 Control Methods Overview of Typical Pneumatic Servo System

In the pneumatic systems, pneumatic control elements, pneumatic actuators, sensors, and other devices [10] are important components, which can compose a typical pneumatic closed-loop system shown in Fig. 1.6. The pneumatic control elements mainly include various proportional valves and on-off valves. The pneumatic executive elements mainly include pneumatic artificial muscles, pneumatic cylinders, pneumatic grippers, and vacuum cups. The sensors mainly include displacement sensors, pressure sensors, angle sensors, and speed sensors.

PAMs are pneumatic executive elements that are used to design various novel flexible mechanisms by domestic and foreign researchers, such as pneumatic dexter-

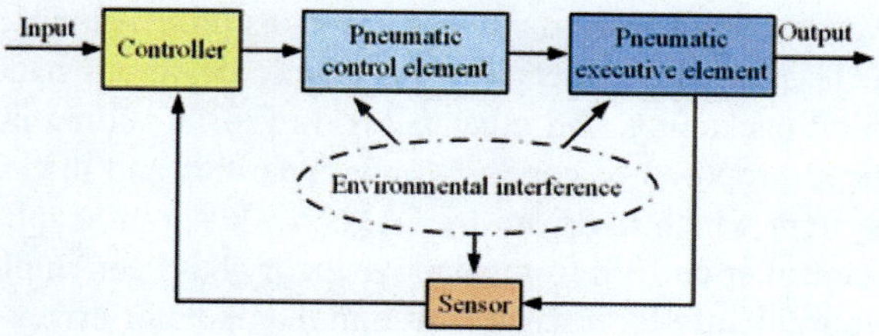

Fig. 1.6 A typical pneumatic closed-loop system

ous hands, pneumatic bionic mechanical arms, pneumatic rehabilitation apparatuses, robotic systems [11, 12]. In addition, a highly flexible manipulator [13], a hand rehabilitation gloves [14], and an ankle rehabilitation robot [15] are all driven by PAMs. To achieve desired movements, the above mechanisms have high requirements for modeling and control. Some accurate dynamic models of PAMs have been established by methods such as an optimizing parameter identification [16], an artificial neural network [17], an empirical model of mechanical properties [18], and so on. In order to achieve good tracking performances, many researchers have tried to control the PAMs by fuzzy control [19], robust control [20], sliding mode control [21], model predictive control [108], and other control methods. However, these control methods still have some limitations under conditions of coupling structures, 1 oad changes, and external disturbances. It is a challenge for us to improve the control accuracy of pneumatic flexible systems. In this book, three experimental platforms driven by PAMs are studied, which are a pneumatic manipulator platform for a pneumatic manipulator system, a pneumatic dexterous hand platform for a pneumatic dexterous hand system and a pneumatic motion simulation platformfor a pneumatic motion simulation system.

The pneumatic manipulator platform is a bionic device that can perform specific operations. There are many advantages for the pneumatic manipulator system driven by PAMs, such as high power–weight ratio, application safety, simple structure, and environmental protection. The pneumatic manipulator system is applied widely in robotics [22], industrial automation [23], and so on. The pneumatic manipulator system has many functions in practice, such as fitness equipment, accident rescue, and rehabilitation training. It is a novel pneumatic equipment which is also applied in scientific research teaching, algorithm verification, advancing nursing, and a lot of fields. In [24], a manipulator driven by PAMs has been designed to deal with uncertainties and disturbance in tracking control via a reduced-order ESO. In [25], a manipulator driven by PAMs has been presented to study positioning control performance by using active disturbance rejection controller. In [26], a manipulator driven by PAMs has devoted research as a primary work of the multi-DoF manipulator system.

The pneumatic dexterous hand is a bionic device which can grasp objects in a particular position under different working conditions. There are many satisfactory advantages for the pneumatic dexterous hand driven by PAMs, such as strong degree of flexibility, high safety performance, low cost of processing, and good safety performance. Originally, traditional pneumatic dexterous hand is widely applied in various fields, such as industrial robots [109], bionic robot [110], and space robots [27]. Nowadays, a class of pneumatic dexterous hand is also used in health care, food packaging, and other fields. In [28], a sliding mode controller with ESO has been proposed to cope with nonlinearities and disturbances for a dexterous hand system which is driven by PAMs. In [29], an adaptive recurrent neural networks controller derived from adaptive theory has been implemented to adjust parameters conveniently for a serial two-joint manipulator driven by PAMs. In [30], a two-axis planar articulated robot driven by four PAMs has been presented to study a system compensated by nonlinear effects.

The pneumatic motion simulation platform is a mechanism with PAMs which can achieve a desired motion trajectory with three degrees of freedom. Originally, the motion simulation system was used in military field, such as flight simulation training, chariot road test, and submarine swaying test. Nowadays, a class of similar motion simulation system is widely applied in rehabilitation treatment, theme park, dynamic cinema , and so on [31]. The motion simulation platform has many functions, such as uniform speed, acceleration, and turning. The motion simulation system mainly uses hydraulic, pneumatic, or electric motor servo system. Due to advantages of rapid response, low cost and high power–weight ratio, PAMs are widely used in the motion simulation platform [32]. A two-axis planar articulated motion robot driven by four PAMs has been introduced in [30]. An adaptive wearable ankle motion robot manufactured by PAMs has been presented in [33]. In [34], an adaptive robust posture controller has been designed for a parallel motion platform driven by PAMs with a redundant degree of freedom. In [35], an integrated direct/indirect adaptive robust controller has been proposed to improve control performance for a parallel motion platform driven by PAMs.

Pneumatic cylinders are another type of executive elements for pneumatic system. Pneumatic cylinders can be applied in flexible manufacturing systems, surgery assisted robots, and other industries because it has a certain sense of internal pressure and a certain flexibility. However, it is difficult by traditional classical control theories to achieve satisfactory performances for pneumatic cylinder servo systems due to some negative effects caused by complex modeling, various friction forces, air compressibility, and so on. Therefore, many researchers have put forward a lot of control strategies to improve control accuracy for pneumatic cylinder servo systems [36]. An adaptive robust control has been applied by Professor Guoliang Tao of Zhejiang University to track the synchronization trajectories of multiple cylinders [37]. A multifaceted sliding mode controller has been proposed by YiChang Tsai in Taiwan science and Technology University for a pneumatic cylinder servo system with variable load [2]. An adaptive backstepping sliding mode controller has been designed by R. A. Rahman in Canada to improve the position control accuracy for a low consumption cylinder [38]. In fact, the accuracy of modeling and the robustness of system are always difficult problems for research of complex pneumatic cylinder servo systems. How to realize an accuracy control of the pneumatic cylinder servo system is a subject with high value of application under conditions of incomplete models, external disturbances, and so on. In this book, two experimental platforms driven by pneumatic cylinders are studied, which are a pneumatic rod cylinder servo platform and a pneumatic right angle composite motion platform.

The pneumatic rod cylinder servo platform is a common product which achieves straight reciprocating motion. The pneumatic rod cylinder servo system is applied widely in medical equipments, military equipments, robots, and other fields due to their advantages of clean, safety, low cost, quick reaction, and so on [36, 39]. In addition, due to the flexibility of the pneumatic rod cylinder servo system platform, it is also used in flexible devices, such as automatic door operators and pneumatic surgical robot. The pneumatic rod cylinder servo system platform has many functions in practice, such as precise fixed-point and reciprocating handling [40]. A pneumatic rod cylinder servo system which is applied in the wearable robotics has been established in [41] to study precise position control. A human walking assist system, which is mainly made up of pneumatic rod cylinder servo system, has been presented in [6]. A pneumatic rod cylinder servo system has been used to simulate a handling device under loads [42]. Meanwhile, control methods of the pneumatic rod cylinder servo system platform have been proposed by more and more scholars [43–45].

The pneumatic right angle composite motion platform is an automatic production equipment which can perform certain operations in three-dimensional space. For various advantages, such as rapid response, reliable operation, and high power–weight ratio, the platform driven by rodless cylinders is used extensively in food fields, health care, industrial automation, and so on [1, 46]. The pneumatic right angle composite motion system has many functions in practice, such as location, periodically motion and complexly trajectory tracking [47]. Various improved PID controllers with simple structures and easy implementations have been designed in some similar pneumatic actuated system platforms [39]. A sliding mode controller has been designed to ensure good performances for a pneumatic cylinder motion system due to its strong robustness and high tolerance for uncertainties [48]. A backstepping control technique has also been introduced for the stability analysis and controller design problems of the nonlinear pneumatic system [40].

1.2 Problems Studied in This Book

Nonlinear control will be studied based on the following five experimental platforms on pneumatic servo systems in this book: Pneumatic manipulator platform, pneumatic dexterous hand platform, pneumatic motion simulation platform, pneumatic rod cylinder servo system platform and pneumatic right angle composite motion platform. The nonlinear control methods are designed for the five experimental platforms which are described by five pneumatic servo systems: Pneumatic manipulator system, pneumatic dexterous hand system, pneumatic motion simulation system, pneumatic rod cylinder servo system, pneumatic right angle composite motion system. Moreover, the effectiveness of the nonlinear control methods are also shown by experimental results on the five pneumatic servo systems.

1.2.1 Pneumatic Manipulator System

Pneumatic manipulator platform, which can achieve rapid response and precise control, is an experimental equipment with one degree of freedom driven by PAMs. Due to highly nonlinear and time-varying behavior caused by the elasticity and compressibility of air, it is difficult to achieve accurate control in PAM systems. The faultiness of PAM systems has promoted many researches into a number of control strategies, such as PID method [49], adaptive robust control strategy [50], and sliding mode control [51]. In order to deal with nonlinearities and uncertainties, the ADRC method is designed to estimation and compensation these nonlinearities and uncertainties. As an important part of the ADRC, ESO can be used to estimate the strongly nonlinear in the nonlinear or uncertain systems. Both unknown dynamics and disturbance are assumed to be total disturbance which will be estimated in real time with an observer in the ADRC [52]. Since the design of an ESO does not need an accurate mathematical model, it has been widely used in many complex systems, such as permanent magnet synchronous motor servo systems [53], hypersonic vehicles [54], and nonlinear multi-agent systems [55]. Some strategies on the ADRC are adopted with the linear ESO to deal with the strongly nonlinear for the pneumatic manipulator platform.

In recent years, various control methods are popular to solve the issues about uncertainties of the model and strong nonlinearities for PAMs, such as [49, 50, 56, 57]. Nevertheless, it is difficult to get an accurate control in PAMs systems for the existence of highly nonlinear and time-varying dynamics caused by elasticity and compressibility of air. It is an important problem that how to design efficient ESO to estimate nonlinear terms, external disturbances and unmodeled dynamics for the PAMs systems. Owing to good estimation performance of ESO, different kinds of ESOs are used to deal with uncertainties and nonlinearities. In [58], a kind of linear ESOs for nonlinear time-varying plants is proposed, and a complete stability analysis of the proposed linear ESO is also presented. A sort of adaptive ESOs for nonlinear disturbed systems are adopted in [51, 59], and implemented to air–fuel ratio control of gasoline engines. Owing to some states that can be acquired in real time, these states need not be obtained. Therefore, in order to reduce the estimated burden of ESO, a reduced-order ESO is designed to estimate and compensate uncertainties and nonlinearities for the pneumatic manipulator platform.

In pneumatic manipulator platform, there exist highly nonlinear and time-varying dynamics in PAMs systems, which is caused by elasticity and compressibility of air. Therefore, it is necessary to improve the performances of PAMs via an algorithm that is deal with uncertainties and nonlinearities. It is important to design efficient controllers for the pneumatic manipulator platform. ESO is used to estimate the nonlinearities and uncertainties, and a good tracking effect is achieved by designing various controllers to compensate nonlinear terms. Due to sliding mod control can guarantee a performance of trajectory tracking for PAM driven joint systems with good robustness property [60], it has been applied to pneumatic systems in recent years. In [61], a terminal sliding mode controller is designed to achieve the robust control of PAMs robot manipulators. In [51], a sliding mode control based on nonlinear disturbance observer has been designed to obtain good tracking performance. In [62], a high-order sliding mode controller is employed to assure convergence of electro-pneumatic systems. Therefore, it is necessary to improve the performances of the pneumatic manipulator platform by adopting a sliding mode controller that combines with the estimate of the disturbances via a super-twisting ESO.

Summarizing the above results, two challenges still remain despite all the reported literatures on exploring control methods for many mechanisms of PAMs. The one challenge is that compressibility of air and nonlinear elasticity of rubber tube bring highly nonlinearities and time-varying behavior. The other challenging is improving control precision and response rapidity for the pneumatic manipulator platform via ADRC control approach. The second challenge is improving control precise and response rapidity for the pneumatic manipulator system via ADRC algorithm. The two challenges are important in both theory and practice, which motivated us to carry on this research work. Moreover, as an important part of the ADRC, the TD is investigated to provide a desired transient and differential trajectory for a desired input signal. The TD is also considered as a filter used in the control method. In addition, convergence analysis of the nonlinear ESO, the reduced-ESO and the super-twisting ESO are performed, which is a challenging in theory. Finally, experimental results show that the control methods have advantages in dealing with nonlinearities and uncertainties. In this part, we will mainly investigate improving the performances for the pneumatic manipulator platform.

1.2.2 Pneumatic Dexterous Hand System

Pneumatic dexterous hand platform is an experimental equipment with two degrees of freedom driven by PAMs. There exists an obvious advantage for pneumatic actuators compared with over rigid actuators, when robots need to interact with human beings [63]. However, compressibility of air and nonlinear elasticity of rubber tube bring many properties for PAMs, such as highly nonlinearities and time-varying behavior [64]. Hence, more and more attention are attracted to deal with the nonlinear problem of PAMs. Some conventional PID controllers are introduced to obtain trajectory tracking control for certain mechanisms driven by PAMs [53, 57]. In recent years, ADRC technology is increasingly used to solve nonlinear problems. As the core part of ADRC, ESO is used to estimate the nonlinearity in kinds of systems. Note that nonlinear ESOs have been proposed to deal with uncertain terms in nonlinear systems. In [65], a nonlinear ESO is presented to estimate nonlinear terms of the two-link flexible manipulator system. In [66], a nonlinear ESO is introduced to estimate uncertainties and external disturbances for a missile guidance system. Some strategies on the ADRC are adopted with the nonlinear ESO to deal with nonlinearities for the dexterous hand driven by PAMs.

As good characteristics of bionic muscle, PAMs are widely applied in robotic hands. In recent years, various control methods are researched to solve the issues about joint coupling, uncertainties of the model and strong nonlinearities for PAMs [11], such as PID method [57], adaptive robust control method [67] and sliding mode control method [62], and so on. To the best of our knowledge, it is hard to achieve precise position control in the pneumatic dexterous hand system because of strong nonlinearity mainly caused by compressibility of air and coupling between joints. Sliding mode control algorithm can achieve fast response speed and good control accuracy. In addition, the sliding mode control algorithm is applied in many pneumatic fields, such as pneumatic artificial muscle manipulator system [57] and electro-pneumatic systems [62]. The sliding mode control method can guarantee a good performance in trajectory tracking for joint systems and it also has good robustness property [60]. Besides, ADRC algorithm possess good resistance to external interference. The sliding mode controller based on ADRC method is an effective algorithm to solve the issues. Therefore, it is necessary to improve the performances of the pneumatic dexterous hand system by adopting sliding mode combined with ADRC algorithm.

The research of control strategy is the core of precise control for PAMs, which are essential for precise control of a pneumatic dexterous hand system. However, it is difficult to achieve the good position control performance due to the asymmetrical hysteresis nonlinear property, compressibility of air and friction forces [2, 68]. Besides, coupling problems are caused by tendons which connect PAMs and manipulator [2, 69]. In order to solve these negative problems of the system, many control methods are proposed in existing literatures, such as PID control based on neural network method [57], switching model predictive control method [108], adaptive recurrent neural networks control method, and so on. However, most of the methods require highly complete mathematical model which are obtained hardly in the PAMs. ADRC proposed by Jingqing Han in 1990s is composed of TD, ESO and nonlinear state error feedback controller [70]. TD is designed to arrange a transition process to curb overshot situation [111]. ESO is introduced to estimate the disturbances in the system, which gives ADRC the properties of strong disturbance rejection [112]. ADRC has been currently used in many difference systems, such as 1-kw H-bridge DC-DC power converter [113], flywheel energy-saving systems [71] and autonomous land vehicle systems [72]. Therefore, ADRC is appropriate to deal with the problems on the two-joint manipulator system.

In summary, two challenges for the control of pneumatic dexterous hand system are achieving high precision control and solving joints coupling problems. The two challenges are important in both theory and practice, which motivated us to carry on this research work. An nonlinear controller is proposed based on a backstepping method, which is used to solve nonlinearities in the trajectory tracking process. In addition, a sliding mode controller based on a nonlinear ESO is used to deal with nonlinearities and coupling disturbances in the trajectory tracking procedure for improving the control performance. Furthermore, a nonlinear controller based on a nonlinear ESO is introduced to estimate strong nonlinear owing to coupling and disturbances. Finally, experimental results are shown that the control methods have strong advantages in dealing with various uncertainties for the pneumatic dexterous hand platform. In this part, we will mainly investigate improving performances for the pneumatic dexterous hand platform.

1.2.3 Pneumatic Motion Simulation System

Pneumatic motion simulation platform is a mechanism of PAMs with three degrees of freedom, which is used to simulate desired motions. However, there exists strongly nonlinear in the mechanism of PAMs because of inflatable deformation, air elasticity, and compressibility [32]. Moreover, varying loads always exist in control systems driven by PAMs [68]. It has drawn much attention of researchers for dealing with the varying loads in recent years. Therefore, tracking control for nonlinear or uncertain systems is an important and practical problem which has attracted a lot of attentions [73, 74]. As an important part of the ADRC, ESO can be used to estimate the strongly nonlinear in the nonlinear or uncertain systems. Linear ESOs have been proposed to deal with uncertain terms in nonlinear systems [46, 75]. The linear ESO has been applied to estimate nonlinear variable cutting load for a variant dynamics in a fast tool servo system [76]. In [46], the linear ESO has been designed to estimate both internal and external disturbances actively for a micro-electro-mechanical system. Some strategies on the ADRC are adopted with the linear ESO to deal with the strongly nonlinear for the mechanism of PAMs.

Studies of control methods are very popular for some simulation mechanisms driven by PAMs in recent years, please refer to [33, 77, 78] and the references therein. It is an important issue that how to design efficient and ordinary controllers for the nonlinear pneumatic muscle systems [78]. Owing to strong robustness and high tolerance, some sliding mode control approaches are used for the nonlinear pneumatic muscle systems [79]. In order to improve control performances and robustness on accurate trajectory tracking of the nonlinear pneumatic muscle systems, a sliding mode controller has been designed based on a nonlinear disturbance observer in [60]. Though many methods are developed to accomplish control task precisely by nonlinear observers, there is still much room for further investigation. The ESO is designed to estimate nonlinear terms, external disturbances and unmodeled dynamics of nonlinear systems [80]. Therefore, it is necessary to improve the performances of mechanisms driven by PAMs by adopting the ADRC with the nonlinear ESO.

The researches on PAMs are divided into three categories: model research [32], structure research, [30] and control strategy research [81]. The accurate mathematical models for the PAM mechanisms are obtained difficultly due to elastic forces of rubber, forces of friction, the end interface of PAMs, and other uncertain factors. Therefore, it is necessary to improve the performances of PAMs via an algorithm which is less dependent on mathematical models. As the kernel of ADRC, the ESO is designed to estimate nonlinear terms, external disturbances and unmodeled dynamics of nonlinear systems [80]. During its development, the ESO has been split into two typical forms: Nonlinear ESO and linear ESO. The nonlinear ESO is applied in various researches due to the good estimation efficiency of nonlinear structures

[82]. Considering the flexibility of design for complicated systems, linear structures are not an ideal choice. In addition, high gains are needed for the linear ESO to estimate total disturbances, which may appear a phenomenon of peaking [83]. In addition, a class of adaptive ESO has been proposed in [59]. The adaptive ESO combines both advantages of theoretical completeness in conventional linear ESO and good performances in conventional nonlinear ESO. Hence, the adaptive ESO is an appropriate method to estimate nonlinear parts of pneumatic systems. Moreover, backstepping control for nonlinear practical systems has been well studied in [84]. Complex industrial processes have also been considered for using the backstepping control techniques [85]. However, to the best of our knowledge, very few results are available on positioning control for mechanisms of PAM systems via adaptive ESO and backstepping control techniques.

Summarizing the above results, we arrive at the conclusion that several challenges still remain despite all the reported literatures on exploring control methods for pneumatic motion simulation systems. One of the challenges is to solve the problem of the strongly nonlinear for the pneumatic motion simulation system. The second challenge is to improve control performances for the pneumatic motion simulation system using the nonlinear ESO. The last challenge is to improve control precise and response rapidity for the pneumatic motion simulation system via adaptive ESO and backstepping approach. These challenges are important in both theory and practice, which motivate us to carry on this research work. Moreover, as an important part of the ADRC, the TD is investigated to provide a desired transient and differential trajectory for a desired input signal. The TD is also considered as a filter used in the control method. In addition, convergence analysis problems on the linear ESO, the nonlinear ESO, and the adaptive ESO are performed, which is a challenge in nonlinear control theory. Finally, experimental results are shown that the control methods have strong advantages in dealing with various uncertainties for the pneumatic motion simulation platform. In this part, we will mainly investigate to improve the control performances for the pneumatic motion simulation system.

1.2.4 Pneumatic Rod Cylinder Servo System

Pneumatic rod cylinder servo platform is a mechanism platform which is used to conduct experiments for position control and trajectory tracking of pneumatic rod cylinder, which is widely used in industrial automation production. Due to strong nonlinearity which is caused by air compressibility, large friction forces and nonlinear properties of servo valves [86], it is difficult to acquire a precise positioning control for pneumatic servo systems. Therefore, it is worth to study on positioning control of the pneumatic rod cylinder servo system. ADRC is a good method which is widely used in pneumatic servo systems. ESO is designed to estimate and compensate total disturbances of the pneumatic servo system. The nonlinear controller is applied to ensure satisfied performances of the closed-loop system. At present, a lot of control strategies have been put forward to improve these negative problems of pneumatic servo systems, please refer to [87]. It is important to reduce negative effects caused by nonlinearity of servo valves, various friction forces, air compressibility, and so on [10] in the control of the pneumatic rod cylinder servo system. The SMC is an effective control approach which has been widely applied to various complicated systems such as nonlinear active suspension vehicle systems [88] and trajectory tracking of nonholonomic wheeled mobile robots [89]. A nonsingular fast terminal SMC method has been presented to avoid the singularity problem, and it has a faster convergence rate than a nonsingular terminal SMC method [90]. However, very few results are available on the extended state observer-based finite-time tracking nonsingular fast terminal SMC for pneumatic servo systems. Therefore, it is necessary and worthwhile to study how to improve control effect of the pneumatic rod cylinder servo system.

The ESO is the core of the ADRC, which has an ability to reduce requirements of highly integrated mathematical models for uncertain systems [70] during designing of controllers. Moreover, both the strong nonlinearity and the external disturbance are also solved by the estimation ability of the ESO [91]. At present, a generalized linear ESO has been presented to estimate a complicated disturbance [92]. A robust tracking controller has been proposed for a wheeled mobile robot based on the generalized linear ESO [93]. What's more, excellent controllers are necessary to get good performances for complicated systems with suitable observers, such as a sliding mode controller [43]. The integral sliding mode (ISM) control method has also been widely used in various complicated systems, such as a chain of uncertain fractional-order integrator [92] and a second-order multi-agent system [94]. As we know, the ISM tracking control method is seldom applied in pneumatic cylinder servo systems with varying loads via the generalized nonlinear ESO. Therefore, the validity of the proposed method needs to be verified through our further research works.

Pneumatic rod cylinder servo systems have been investigated widely because of their various advantages, such as low cost, simple structure, and convenient maintenance [95]. Meanwhile, pneumatic rod cylinder servo systems are used extensively in mechanical operations, medical areas, automation systems, and other fields [36]. However, friction and varying loads are hindrances to get an accurate positioning control for a closed-loop system. To reduce the burden of the friction and varying loads, many researchers have focused on optimized ADRC methods. Power integrator techniques have been used in controller design to obtain higher robustness and finite-time stability. However, to the best of our knowledge, the effectiveness of the super-twisting ESO (STESO) associated with the power integrator-based finite-time controller has yet not been fully investigated for nonlinear systems. This problem is important and challenging in both theory and practice, which motivated us to carry on this research work.

Summarizing the above results, we arrive at the conclusion that several challenges still remain despite many reported literatures on exploring control methods for pneumatic rod cylinder servo systems. One of such challenges is to improve performances for the pneumatic rod cylinder servo system using the NFTSM. The second challenge is improving control precise and response rapidity for the pneumatic rod cylinder servo system via a generalized nonlinear ESO and an integral sliding mode tracking control method. The last challenge is to solve the problem on burden of the friction and varying loads and achieve finite-time control for the pneumatic rod cylinder servo system. These challenges have great effect on both theory and practice, which encourage us to do more researches. At last, many experiments are done via pneumatic rod cylinder servo system platform. The results prove that the control methods solve the control problem in the pneumatic rod cylinder servo system effectively. In this part, we will mainly investigate how to improve the control performances of the pneumatic rod cylinder system.

1.2.5 Pneumatic Right Angle Composite Motion System

Pneumatic right angle composite motion platform is used for magic cube positioning and homing. A key technology for the magic cube positioning and homing is position control of a pneumatic rodless cylinder system. However, there exists strong nonlinear in the pneumatic rodless cylinder system because of air elasticity and compressibility [32]. Moreover, varying loads always exist in position control of the pneumatic rodless cylinder system. Much attention of researchers has been drawn to deal with the varying loads in recent years. Position control for nonlinear or uncertain systems such as the pneumatic rodless cylinder system, which is an important and practical problem, has also attracted a lot of attentions [44, 73]. The ADRC method is less dependent on the accurate mathematical model and has strong anti-interference ability [70, 96, 97]. Active disturbance rejection controller consists of tracking differentiator, extended state observer, and nonlinear state error feedback controller. Extended state observer is not based on an accurate model, therefore it can be applied to a wide variety of plants [71, 80, 98, 99]. Extended state observer regards the nonlinear part as the extended state and estimates the nonlinear part, so it is an appropriate method for pneumatic rodless cylinder system. Owing to the great advances in nonlinear control theory, nonlinear controllers become commonly used in industrial applications [67]. Different from most existing nonlinear controllers, a non-smooth function is incorporated into the nonlinear state error feedback controller such that small errors produce high gains, and large errors produce small gains [80, 100]. However, to the best of our knowledge, very few results are available on active disturbance rejection position control of pneumatic systems without pressure states. These problems are important and challenging in both theory and practice, which motivated us to carry on this research work.

As there are 3 freedoms of right angle composite motion platform, it is widely applied as robotic actuators. Researchers have adopted many methods to improve the control performance of pneumatic rodless cylinder servo position systems. In recent years, various of improved proportional integral derivative controllers with the simple structure and easy implementation are designed in pneumatic actuated systems [87, 101]. Complex industrial processes are also considered by using the backstepping control techniques in [85, 102, 103]. There exists a lot of space to improve the speed of response for the pneumatic servo system with high positioning accuracy via the backstepping control techniques. Backstepping control technique is also introduced for the stability analysis and controller design of nonlinear pneumatic systems in [40, 104]. To the best of our knowledge, it is hard to achieve precise position control in pneumatic servo systems because of strong nonlinearity mainly caused by compressibility of air. Moveover, pressure sensors are usually not too expensive [43], and it is easy to install sensors in pneumatic servo systems. It is possible to get the chamber pressures of the cylinder which are used in the controller designed. Therefore, it is feasible to improve the performances of pneumatic right angle composite motion platform by adopting the backstepping controller with the nonlinear ESO.

There exists negative influence for actuator saturation on positioning accuracy of pneumatic servo systems. Actuator saturation is often overlooked in control design problems. However, the actuator saturation is inevitable in many practical control systems due to the limitation of actuators or inherent physical constraints of systems. Actuator saturation has a great negative effect on the control performance of a system, even cause undesirable inaccuracy and leading instability [105–107]. The accurate position control of pneumatic right angle composite motion system is really a challenge problem, which limits the application of pneumatic right angle composite motion system in industry. However, the pneumatic right angle composite motion system has a good advantage in linear servo control where the electromagnetic interference needs to be avoided in some practical occasions. Therefore, to further improve the control accuracy, actuator saturation is considered to design the nonlinear controller for the reason of that actuator saturation is inevitable in pneumatic right angle composite motion system. Furthermore, a linear ESO is employed to estimate and compensate unknown nonlinear disturbances. Experiment results illustrate that the proposed controller significantly improves the positioning accuracy compared with [51, 59, 61]. The actuator saturation is proposed to deal with the problems for the pneumatic right angle composite motion platform.

Summarizing the above results, challenges still remain despite all the reported literatures on exploring control methods for many mechanisms of pneumatic rodless cylinder systems. The first challenge is that compressibility of air and changing friction bring highly nonlinearities. The second challenge is improving control precision and response rapidity for the pneumatic rodless cylinder system via ADRC approach. The last challenge is that actuator saturation is considered to design the nonlinear controller for the pneumatic rodless cylinder system. These challenges are important in both theory and practice, which motivated us to carry on this research work. An linear controller is proposed based on a backstepping method, which is used to solve nonlinearities in the trajectory tracking process. In addition, a linear state error feedback controller based on a linear ESO is used to deal with nonlinearities disturbances in the trajectory tracking procedure for improving the control performance. Furthermore, a nonlinear ADRC controller based on a linear ESO is introduced to estimate strong nonlinear owing to disturbances. Finally, experimental results are shown that the control methods have strong advantages in dealing with various uncertainties for the pneumatic servo systems. In this chapter, we will mainly investigate to improve control performances for the pneumatic right angle composite motion platform.

1.3 Conclusion

Pneumatic servo systems play the important role in mechanical operations, medical areas, automation systems, and many other fields. Therefore, more and more attention has been paid to the position control of the pneumatic servo system. It is a key idea to estimate the uncertain factors effectively by using suitable control methods for pneumatic servo systems. The control methods are requested to automatically detect the real-time effect of the model, the external disturbances and then automatically compensate it. In this book, the five experimental platforms are discussed on pneumatic servo systems: Pneumatic manipulator system, pneumatic dexterous hand system, pneumatic motion simulation system, pneumatic rod cylinder servo system, pneumatic right angle composite motion system. The results in this book will doubtless promote the development of control technology for pneumatic servo systems.

References

1. Bone GM, Shu N (2007) Experimental comparison of position tracking control algorithms for pneumatic cylinder actuators. IEEE-ASME Trans Mechatron 12(5):557–561
2. Tsai Y-C, Huang A-C (2008) Multiple-surface sliding controller design for pneumatic servo systems. Mechatronics 18(9):506–512
3. Kirihara K, Saga N, Saito N (2008) Upper limb rehabilitation support device using a pneumatic cylinder. In: Annual conference of IEEE industrial electronics, pp 1287–1292
4. Wang J, Pu J, Moore P (1999) Accurate position control of servo pneumatic actuator systems: an application to food packaging. Control Eng Pract 7(6):699–706
5. Lee L-W, Li I-H (2012) Wavelet-based adaptive sliding-mode control with H_∞, tracking performance for pneumatic servo system position tracking control. IET Control Theory Appl 6(11):1699–1714
6. Morichika T, Kikkawa F, Oyama O, Yoshimitsu T (2007) Development of walking assist equipment with pneumatic cylinder. In: SICE annual conference, pp 1058–1063
7. Daerden F, Lefeber D (2002) Pneumatic artificial muscle: actuators for robotics and automation. Eur J Mech Environ Eng 47:11–21
8. Georgios A, Georgios N, Stamatis M (2011) A survey on applications of pneumatic artificial muscles. In: 19th mediterranean conference on control and automation. Aquis Corfu Holiday Palace, Corfu, Greece, pp 20–23
9. https://festo.com/group/zh/cms/11921.htm
10. Rahmat MF, Salim SNS, Sunar NH, Faudzi AM, Ismail ZH, Huda K (2012) Identification and non-linear control strategy for industrial pneumatic actuator. Int J Phys Sci 7(17):2565–2579
11. Situm Z, TrsliC P, TriviC D, Stahan V, Brezak H, SremiC D (2015) Pneumatic muscle actuators within robotic and mechatronic systems. In: Proceedings of international conference fluid power, pp 175–188
12. Jahanabadi H, Mailah M, Zain MZM, Hooi HM (2011) Active force with fuzzy logic control of a two-link arm driven by pneumatic artificial muscles. J Bionic Eng 8(4):474–484
13. Deimel R, Brock O (2013) A compliant hand based on a novel pneumatic actuator. In: IEEE international conference on robotics and automation, pp 2047–2053
14. Polygerinos P, Lyne S, Wang Z, Nicolini LF, Mosadegh B, Whitesides GM, Walsh CJ (2013) Towards a soft pneumatic glove for hand rehabilitation. In: IEEE/RSJ international conference on intelligent robots and systems, pp 1512–1517

15. Jamwal PK, Xie SQ, Tsoi YH, Aw KC (2010) Forward kinematics modelling of a parallel ankle rehabilitation robot using modified fuzzy inference. Mech Mach Theory 45(11):1537–1554
16. Sarosi J, Biro I, Nemeth J, Cveticanin L (2015) Dynamic modeling of a pneumatic muscle actuator with two-direction motion. Mech Mach Theory 85:25–34
17. Song C, Xie S, Zhou Z, Hu Y (2015) Modeling of pneumatic artificial muscle using a hybrid artificial neural network approach. Mechatronics 31:124–131
18. Wakimoto S, Misumi J, Suzumori K (2016) New concept and fundamental experiments of a smart pneumatic artificial muscle with a conductive fiber. Sens Actuators A: Phys 250:48–54
19. Chandrapal M, Chen X, Wang W, Hann C (2012) Nonparametric control algorithms for a pneumatic artificial muscle. Expert Syst Appl 39(10):8636–8644
20. Ba DX, Ahn KK (2015) Indirect sliding mode control based on gray-box identification method for pneumatic artificial muscle. Mechatronics 32:1–11
21. Aljodah A, Majeed LK (2018) Second order sliding mode controller design for pneumatic artificial muscle. J Eng 24(1):159–172
22. Yaegashi K, Saga N, Satoh T (2005) Control of robot arm using pneumatic artificial muscle with spherical joint. IEEE Int Conf Mechatron Autom 2:1093–1098
23. Zhou JY, Zhou RJ, Wang YY (2001) Robust nonlinear reduced-order dynamic controller design and its application to a single-link manipulator. IEEE Int Conf Robot Autom 2:1149–1154
24. Zhao L, Liu X, Wang T, Liu B (2018) A reduced-order extended state observerícbased trajectory tracking control for one-degree-of-freedom pneumatic manipulator. Adv Mech Eng 10(4):1–9
25. Zhao L, Li Q, Yang H, Li H (2015) Positioning control of a one-DOF manipulator driven by pneumatic artificial muscles based on active disturbance rejection control. In: Chinese control conference, pp 841–845
26. Zhao L, Li Q, Liu B, Cheng H (2019) Trajectory tracking control of a one degree of freedom manipulator based on a switched sliding mode controller with a novel extended state observer framework. IEEE Trans Syst Man Cybern: Syst 49(6):1110–1118
27. Yokokohji Y, Toyoshima T, Yoshikawa T (1993) Efficient computational algorithms for trajectory control of free-flying space robots with multiple arms. IEEE Trans Robot Autom 9(5):571–580
28. Zhao L, Cheng H, Wang T (2018) Sliding mode control for a two-joint coupling nonlinear system based on extended state observer. ISA Trans 73:130–140
29. Ahn KK, Anh HPH (2009) Design and implementation of an adaptive recurrent neural networks (ARNN) controller of the pneumatic artificial muscle (PAM) manipulator. Mechatronics 19(6):816–828
30. Hildebrandt A, Sawodny O, Neumann R, Hartmann A (2005) Cascaded control concept of a robot with two degrees of freedom driven by four artificial pneumatic muscle actuators. Am Control Conf 1:680–685
31. Yang H, Yu Y, Zhang J (2017) Angle tracking of a pneumatic muscle actuator mechanism under varying load conditions. Control Eng Pract 61:1–10
32. Chen M, Jiang B (2013) Robust attitude control of near space vehicles with time-varying disturbances. Int J Control Autom Sys 11(1):182–187
33. Jamwal PK, Xie SQ, Hussain S, Parsons JG (2014) An adaptive wearable parallel robot for the treatment of ankle injuries. IEEE-ASME Trans Mechatron 19(1):64–75
34. Zhu X, Tao G, Yao B, Cao J (2008) Adaptive robust posture control of parallel manipulator driven by pneumatic muscles with redundancy. IEEE-ASME Trans Mechatron 13(4):441–450
35. Zhu X, Tao G, Yao B, Cao J (2009) Integrated direct/indirect adaptive robust posture trajectory tracking control of a parallel manipulator driven by pneumatic muscles. IEEE Trans Control Syst Technol 17(3):576–588
36. Chen HI, Shih MC (2013) Visual control of an automatic manipulation system by microscope and pneumatic actuator. IEEE Trans Autom Sci Eng 10(1):215–218

37. Meng D, Tao G, Li A, Li W (2016) Precision synchronization motion trajectory tracking control of multiple pneumatic cylinders. Asian J Control 18(5):1749–1764
38. Rahman AR, He L, Sepehri N (2016) Design and experimental study of a dynamical adaptive backstepping-sliding mode control scheme for position tracking and regulating of a low-cost pneumatic cylinder. Int J Robust Nonlinear Control 26(4):853–875
39. Smaoui M, Brun X, Thomasset D (2006) A study on tracking position control of an electropneumatic system using backstepping design. Control Eng Pract 14(8):923–933
40. Rao Z, Bone GM (2008) Nonlinear modeling and control of servo pneumatic actuators. IEEE Trans Control Syst Technol 16(3):562–569
41. Taheri B, Case D, Richer E (2014) Force and stiffness backstepping-sliding mode controller for pneumatic cylinders. IEEE-ASME Trans Mechatron 19(6):1799–1809
42. Shen X, Zhang J, Barth EJ, Goldfarb M (2006) Nonlinear model-based control of pulse width modulated pneumatic servo systems. J Dyn Syst Meas Control 128(3):663–669
43. Liu Y, Kung T, Chang K, Chen S (2013) Observer-based adaptive sliding mode control for pneumatic servo system. Precis Eng 37(3):522–530
44. Shi S, Zhao S, Li J (2014) On tracking ability analysis of linear extended state observer for uncertain system. Comput Modell New Technol 18(2):175–179
45. Situm Z, Pavkovic D, Novakovic B (2004) Servo pneumatic position control using fuzzy PID gain scheduling. J Dyn Syst Measur Control 126(2):376–387
46. Qiu Z, Wang B, Zhang X, Han J (2013) Direct adaptive fuzzy control of a translating piezoelectric flexible manipulator driven by a pneumatic rodless cylinder. Mech Syst Signal Process 36(2):290–316
47. Hassan MY, Kothapalli G (2010) Comparison between neural network based PI and PID controllers. In: International multi-conference on systems, signals and devices, pp 1–6
48. Pandian SR, Takemura F, Hayakawa Y, Kawamura S (2002) Pressure observer-controller design for pneumatic cylinder actuators. IEEE-ASME Trans Mechatron 7(4):490–499
49. Ahn KK, Thanh TDC (2005) Nonlinear PID control to improve the control performance of the pneumatic artificial muscle manipulator using neural network. J Mech Sci Technol 19(1):106–115
50. Tang Z, Ge SS, Lee KP, He W (2016) Adaptive neural control for an uncertain robotic manipulator with joint space constraints. Int J Control 89(7):1428–1446
51. Xue W, Bai W, Yang S, Song K, Huang Y, Xie H (2015) ADRC with adaptive extended state observer and its application to air-fuel ratio control in gasoline engines. IEEE Trans Ind Electron 62(9):5847–5857
52. Przybyla M, Kordasz M, Madonski R, Herman P, Sauer P (2012) Active disturbance rejection control of a 2DOF manipulator with significant modeling uncertainty. Bull Pol Acad Sciences-Tech Sci 60(3):509–520
53. Nuchkrua T, Leephakpreeda T (2013) Fuzzy self-tuning PID control of hydrogen-driven pneumatic artificial muscle actuator. J Bionic Eng 10(3):329–340
54. Hu C, Liu Y (2014) Nonlinear attitude control based on extended state observer for hypersonic vehicles. In: IEEE Chinese guidance, navigation and control conference, pp 2474–2479
55. Qin W, Liu Z, Chen Z (2014) Formation control for nonlinear multi-agent systems with linear extended state observer. IEEE/CAA J Automatica Sinica 1(2):171–179
56. Ahn KK, Nguyen HTC (2007) Intelligent switching control of a pneumatic muscle robot arm using learning vector quantization neural network. Mechatronics 17(4–5):255–262
57. Tu DCT, Ahn KK (2006) Nonlinear PID control to improve the control performance of 2 axes pneumatic artificial muscle manipulator using neural network. Mechatronics 16(9):577–587
58. Zheng Q, Gao LQ, Gao Z (2007) On stability analysis of active disturbance rejection control for nonlinear time-varying plants with unknown dynamics. In: IEEE conference on decision and control, pp 3501–3506
59. Pu Z, Yuan R, Yi J, Tan X (2015) A class of adaptive extended state observers for nonlinear disturbed systems. IEEE Trans Ind Electron 62(9):5858–5869
60. Tsagarakis N, Iqbal J, Khan H, Caldwell D (2014) A novel exoskeleton robotic system for hand rehabilitation-conceptualization to prototyping. Biocybern Biomed Eng 34(2):79–89

61. Zhao L, Xia Y, Yang Y, Liu Z (2017) Multicontroller positioning strategy for a pneumatic servo system via pressure feedback. IEEE Trans Ind Electron 64(6):4800–4809
62. Girin A, Plestan F, Brun X, Glumineau A (2009) High order sliding-mode controllers of an electropneumatic actuator: application to an aeronautic benchmark. IEEE Trans Control Syst Technol 17(3):633–645
63. Shin D, Yeh X, Khatib O (2014) A new hybrid actuation scheme with artificial pneumatic muscles and a magnetic particle brake for safe human-robot collaboration. Int J Robot Res 33(4):507–518
64. Chang MK, Liou JJ, Chen ML (2011) T-S fuzzy model-based tracking control of a one-dimensional manipulator actuated by pneumatic artificial muscles. Control Eng Pract 19(12):1442–1449
65. Yang H, Yu Y, Yuan Y, Fan X (2015) Back-stepping control of two-link flexible manipulator based on an extended state observer. Adv Space Res 56(10):2312–2322
66. Zhu Z, Xu D, Liu J, Xia Y (2013) Missile guidance law based on extended state observer. IEEE Trans Ind Electron 60(12):5882–5891
67. Zhu X, Tao G, Yao B, Cao J (2008) Adaptive robust posture control of a parallel manipulator driven by pneumatic artificial muscles. Automatica 44(9):2248–2257
68. Vo-Minh T, Tjahjowidodo T, Ramon H, Brussel HV (2011) A new approach to modeling hysteresis in a pneumatic artificial muscle using the maxwell-slip model. IEEE-ASME Trans Mechatron 16(1):177–186
69. Linares-Flores J, Barahona-Avalos J, Sira-Ramirez H, Contreras-Ordaz MA (2012) Robust passivity-based control of a buckícboost-converter/dc-motor system: an active disturbance rejection approach. IEEE Trans Ind Appl 48(6):2362–2371
70. Han J (2009) From PID to active disturbance rejection control. IEEE Trans Ind Electron 56(3):900–906
71. Chang X, Li Y, Zhang W, Wang N, Xue W (2015) Active disturbance rejection control for a flywheel energy storage system. IEEE Trans Ind Electron 62(2):991–1001
72. Xia Y, Pu F, Li S, Gao Y (2016) Lateral path tracking control of autonomous land vehicle based on ADRC and differential flatness. IEEE Trans Ind Electron 63(5):3091–3099
73. Li H, Liao X, Chen G (2013) Leader-following finite-time consensus in second-order multi-agent networks with nonlinear dynamics. Int J Control Autom Syst 11(2):422–426
74. Sheng S, Sun C (2016) An adaptive attitude tracking control approach for an unmanned helicopter with parametric uncertainties and measurement noises. Int J Control Autom Syst 14(1):217–228
75. Huang C-E, Li D, Xue Y (2013) Active disturbance rejection control for the ALSTOM gasifier benchmark problem. Control Eng Pract 21(4):556–564
76. Wu D, Chen K (2009) Design and analysis of precision active disturbance rejection control for noncircular turning process. IEEE Trans Ind Electron 56(7):2746–2753
77. Aschemann H, Schindele D (2014) Comparison of model-based approaches to the compensation of hysteresis in the force characteristic of pneumatic muscles. IEEE Trans Ind Electron 61(7):3620–3629
78. Aschemann H, Schindele D (2008) Sliding-mode control of a high-speed linear axis driven by pneumatic muscle actuators. IEEE Trans Ind Electron 55(11):3855–3864
79. Lilly JH, Quesada PM (2004) A two-input sliding-mode controller for a planar arm actuated by four pneumatic muscle groups. IEEE Trans Neural Syst Rehabil Eng 12(3):349–359
80. Yang H, You X, Xia Y, Li H (2014) Adaptive control for attitude synchronisation of spacecraft formation via extended state observer. IET Control Theory Appl 8(18):2171–2185
81. Andrikopoulos G, Nikolakopoulos G, Manesis S (2014) Advanced nonlinear PID-based antagonistic control for pneumatic muscle actuators. IEEE Trans Ind Electron 61(12):6926–6937
82. Godbole AA, Kolhe JP, Talole SE (2013) Performance analysis of generalized extended state observer in tackling sinusoidal disturbances. IEEE Trans Control Syst Technol 21(6):2212–2223
83. Khalil HK (2002) Nonlinear systems. Prentice Hall, NJ

84. Wang T, Zhang Y, Qiu J, Gao H (2015) Adaptive fuzzy backstepping control for a class of nonlinear systems with sampled and delayed measurements. IEEE Trans Fuzzy Syst 23(2):302–312
85. Wang T, Gao H, Qiu J (2016) A combined fault-tolerant and predictive control for network-based industrial processes. IEEE Trans Ind Electron 63(4):2529–2536
86. Qiu Z, Shi M, Wang B, Xie Z (2012) Genetic algorithm based active vibration control for a moving flexible smart beam driven by a pneumatic rod cylinder. J Sound Vib 331(10):2233–2256
87. Gao X, Feng Z-J (2005) Design study of an adaptive Fuzzy-PD controller for pneumatic servo system. Control Eng Pract 13(1):55–65
88. Li H, Yu J, Hilton C, Liu H (2013) Adaptive sliding-mode control for nonlinear active suspension vehicle systems using T-S fuzzy approach. IEEE Trans Ind Electron 60(8):3328–3338
89. Yang J-M, Kim J-H (1999) Sliding mode control for trajectory tracking of nonholonomic wheeled mobile robots. IEEE Trans Robot Autom 15(3):578–587
90. Xu S, Chen C, Wu Z (2015) Study of nonsingular fast terminal sliding-mode fault-tolerant control. IEEE Trans Ind Electron 62(6):3906–3913
91. Li S, Yang X, Yang D (2009) Active disturbance rejection control for high pointing accuracy and rotation speed. Automatica 45(8):1854–1860
92. Li S, Yang J, Chen W-H, Chen X (2012) Generalized extended state observer based control for systems with mismatched uncertainties. IEEE Trans Ind Electron 59(12):4792–4802
93. Kang HS, Kim YT, Hyun CH, Park M (2013) Generalized extended state observer approach to robust tracking control for wheeled mobile robot with skidding and slipping. Int J Adv Robot Syst 10:1–10
94. Yu S, Long X (2015) Finite-time consensus for second-order multi-agent systems with disturbances by integral sliding mode. Automatica 54:158–165
95. Takosoglu JE, Dindorf RF, Laski PA (2009) Rapid prototyping of fuzzy controller pneumatic servo-system. Int J Adv Manuf Technol 40(3–4):349–361
96. Han J (1998) Auto-disturbances-rejection controller and its applications. Control Decis 13(1):19–23
97. Xia Y, Liu B, Fu M (2012) Active disturbance rejection control for power plant with a single loop. Asian J Control 14(1):239–250
98. Tang H, Li Y (2014) Development and active disturbance rejection control of a compliant micro-/nanopositioning piezostage with dual mode. IEEE Trans Ind Electron 61(3):1475–1492
99. Li S, Li J, Mo Y (2014) Piezoelectric multimode vibration control for stiffened plate using ADRC-based acceleration compensation. IEEE Trans Ind Electron 61(12):6892–6902
100. Erazo C, Angulo F, Olivar G (2012) Stability analysis of the extended state observers by Popov criterion. Theor Appl Mech Lett 2(4):1–4
101. Lee HK, Choi GS, Choi GH (2002) A study on tracking position control of pneumatic actuators. Mechatronics 12(6):813–831
102. Wang T, Qiu J, Gao H (2017) Adaptive neural control of stochastic nonlinear time-dealy systems with multiple constraints. IEEE Trans Syst Man Cybern: Syst 47(8):1875–1883
103. Wang T, Gao H, Qiu J (2016) A combined adaptive neural network and nonlinear model predictive control for multirate networked industrial process control. IEEE Trans Neural Netw Learn Syst 27(2):416–425
104. Ren H-P, Huang C (2013) Adaptive backstepping control of pneumatic servo system. In: IEEE international symposium on industrial electronics, pp 1–6
105. Hu T, Lin Z, Chen BM (2002) An analysis and design method for linear systems subject to actuator saturation and disturbance. Automatica 38(2):351–359
106. Hu T, Lin Z, Chen BM (2002) Analysis and design for discrete-time linear systems subject to actuator saturation. Syst Control Lett 45(2):97–112
107. Chen BM, Lee TH, Peng K, Venkataramanan V (2003) Composite nonlinear feedback control for linear systems with input saturation: theory and an application. IEEE Trans Autom Control 48(3):427–439

108. Andrikopoulos G, Nikolakopoulos G, Manesis S (2013) Pneumatic artificial muscles: a switching model predictive control approach. Control Eng Pract 21(12):1653–1664
109. Jacobsen SC, Wood JE, Knutti DF, Biggers KB (1984) The Utah-MIT dexterous hand: work in progress. Int J Robot Res 3(4):21–50
110. Li G, Jin J, Deschamps-Berger S, Sun Z, Zhang W, Chen Q (2014) Indirectly self-adaptive underactuated robot hand with block-linkage mechanisms. Int J Precis Eng Manuf 15(8):1553–1562
111. Bluethmann W, Ambrose R, Diftler M, Askew S, Huber E, Goza M, Rehnmark F, Lovchik C, Magruder D (2003) Robonaut: a robot designed to work with humans in space. Autonom Robot 14(2–3):179–197
112. Lovchik CS, Diftler MA (1999) The robonaut hand: a dexterous robot hand for space. In: IEEE international conference on robotics and automation, pp 907–912
113. Sun B, Gao Z (2005) A DSP-based active disturbance rejection control design for a 1-kW H-bridge DC-DC power converter. IEEE Trans Ind Electron 52(5):1271–1277

Chapter 2
Control Methods for Pneumatic Servo Systems

In engineering practice, there are many nonlinear problems in pneumatic servo systems, such as parameter uncertainty, dead zone, saturation, and so on. Therefore, the pneumatic servo system should be regarded as a nonlinear system rather than a simple linear system. Obviously, dealing with nonlinear problems based on a nonlinear system is a challenging problem, which is the key point to improve the accuracy, rapidity, and stability of position control in the pneumatic servo systems. At present, researchers have applied various nonlinear control methods to deal with the nonlinear problems in the pneumatic servo systems, such as active disturbance rejection control(ADRC), backstepping method, sliding mode control(SMC), and finite-time control method. Overview of the recent research on the above methods is given in this chapter.

2.1 Active Disturbance Rejection Control

At the end of 1980s, active disturbance rejection control is proposed by Han Jingqing. The ADRC is built on both a modern control theory and a classical regulation theory to analyze system characteristics. The ADRC not only inherits the essence of traditional PID control but also develops a new and practical control method based on special nonlinear effect, which does not depend on the precise mathematical model of a controlled object. The core idea of ADRC is to transform the controlled system with uncertainties, nonlinearities, and disturbances into a simple integral series standard form. Moreover, a total disturbance is a main concept to characterize comprehensive factors affecting the system stability. An extended state observer is used to estimate the total disturbance in real time, then an estimation of the total disturbance is com-

L. Zhao et al., *Pneumatic Servo Systems Analysis*, Advances in Industrial Control,
https://doi.org/10.1007/978-981-16-9515-5_2

pensated in a designed controller. Therefore, the ADRC greatly simplifies control design of the controlled system for dealing with multiple disturbances.

The ADRC is mainly composed of three parts: (1) Tracking Differentiator (TD); (2) Extended State Observer (ESO); (3) Nonlinear State Error Feedback (NLSEF). A basic control framework is briefly described in the following.

2.1.1 Tracking Differentiator

The TD is an important part of the ADRC to extract continuous and differential signals reasonably from discontinuous or random noise measurement signals in practical engineering. Main advantage of TD is that fast tracking is realized under the premise of basically no overshoot, which means that it solves the contradiction between overshoot and quickness.

A common form of the TD is shown as follows:

$$\begin{cases} f_h(t) = f_{han}(\eta(t), v_2(t), r_0, h_0) \\ \eta(t) = v_1(t) - v(t) \\ \dot{v}_1(t) = v_2(t) \\ \dot{v}_2(t) = v_3(t) \\ \quad\vdots \\ \dot{v}_n(t) = f_h(t), \end{cases} \tag{2.1}$$

where $v(t)$ is the input signal, h_0 is the integral step, $v_1(t)$ is the filtered and tracked signal of $v(t)$, $v_2(t)$ is the differential signal of $v(t)$. Besides, $f_{han}(\eta(t), v_2(t), r_0, h_0)$ is a nonlinear function defined as

$$\begin{cases} d = r_0 h_0 \\ d_0 = h_0 d \\ y(t) = \eta(t) + h_0 \mu(t) \\ a_0(t) = \sqrt{d^2 + 8r_0|y(t)|} \\ a(t) = \begin{cases} \mu(t) + \frac{(a_0(t) - d)}{2}\mathrm{sign}(y(t)), & |y(t)| > d_0 \\ \mu(t) + \frac{y(t)}{h_0}, & |y(t)| \le d_0 \end{cases} \\ f_{han}(\cdot) = -\begin{cases} r_0 \mathrm{sign}(a(t)), & |a(t)| > d \\ r_0 \frac{a(t)}{d}, & |a(t)| \le d \end{cases} \end{cases} \tag{2.2}$$

where $\eta(t)$ and $\mu(t)$ are variables, r_0 and h_0 are the velocity factor and the filter factor, respectively. Note that the larger r_0 is, the faster a tracking speed will be. The larger h_0 is, the better a filtering effect will be. The differential signal is independent on a signal generation source, which is not dependent on the mathematical model of the signal generation source. The function of differential tracker is to arrange the transition process and obtain smooth input signals according to constraints of a

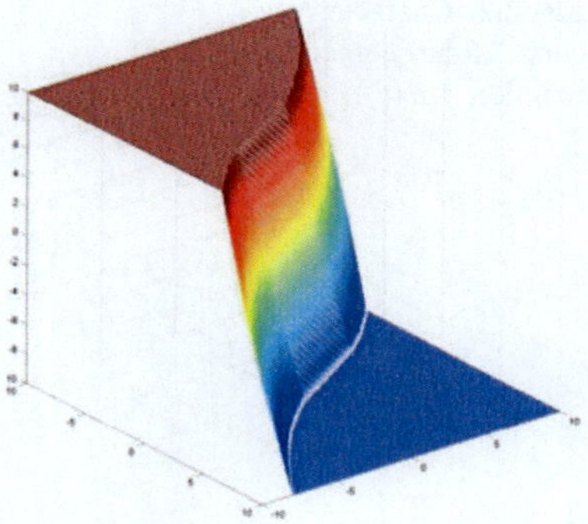

Fig. 2.1 Characteristic curve of the nonlinear function $f_{han}(\eta(t), v_2(t), r_0, h_0)$

reference input. The characteristic curve of the nonlinear function (2.2) is shown as in Fig. 2.1.

2.1.2 Extended State Observer

By extending the total disturbance affecting the output of the controlled object into a new extended state variable, the order of an original system will be extended. For the extended system, the ESO is designed to observe the total disturbance. A mathematical expression of n-order ESO is given as follows:

$$\begin{cases} e(t) = z_1(t) - y(t) \\ fe_1(t) = fal(e(t), \alpha_1, \delta) \\ fe_2(t) = fal(e(t), \alpha_2, \delta) \\ \quad\vdots \\ \dot{z}_1(t) = z_2(t) - \beta_{01} e(t) \\ \dot{z}_2(t) = z_3(t) - \beta_{02} fe_1(t) \\ \quad\vdots \\ \dot{z}_n(t) = z_{n+1}(t) - \beta_{0n} fe_{n-1}(t) + bu(t) \\ \dot{z}_{n+1}(t) = -\beta_{0(n+1)} fe_n(t) \end{cases} \tag{2.3}$$

in which a nonlinear function fal $(e(t), \alpha, \delta)$ is shown as

$$\text{fal}(e(t), \alpha, \delta) = \begin{cases} \frac{e(t)}{\delta^{1-\alpha}}, & |e(t)| \leq \delta \\ |e(t)|^{\alpha} \operatorname{sign}(\mathrm{e(t)}), & |e(t)| > \delta, \end{cases} \tag{2.4}$$

where $e(t)$ is the estimated error, α and δ are adjustable parameters and $z_1(t)$, $z_2(t), \ldots, z_n(t)$ are the estimated value of each state variable. The extended state $z_{n+1}(t)$ is an estimated value of the total disturbance. Note that fal$(e(t), \alpha, \delta)$ is a

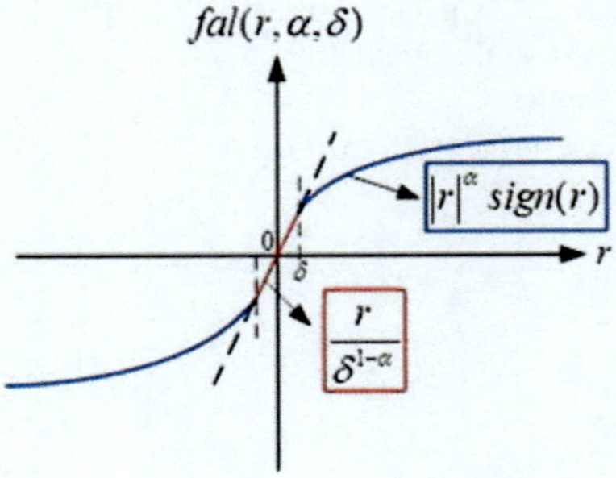

Fig. 2.2 Characteristic curve of the segment-wise function $fal(r, \alpha, \delta)$

power function with continuous linear segment near the origin, which can avoid the phenomenon of high frequency oscillation. Characteristic curve of the nonlinear function (2.4) is shown as in Fig. 2.2, in which $r = e(t)$ simplify. Note that there is a linear segment in neighborhood of an origin where error variable is small relatively. When error $|r| > \delta$, $fal(\cdot)$ is the concave function. That is, a small error corresponds to a large gain and a large error corresponds to a small gain. It has been discovered experimentally that the error is large relatively at the initial phase of control, a small gain makes the output smooth and avoids damaging equipments. A large gain speeds up the reaction rate when the error is small. Therefore, control signals are distributed reasonably to achieve energy-saving emission reduction effect.

2.1.3 Nonlinear State Error Feedback

Based on state errors generated by the TD (2.1) and total disturbance information by the ESO (2.3), a combined nonlinear control law is constructed to obtain good robustness and adaptability. Therefore, a NLSEF of the n-order system is written as follows:

$$\begin{cases} e_1(t) = v_1(t) - z_1(t) \\ e_2(t) = v_2(t) - z_2(t) \\ \vdots \\ e_n(t) = v_n(t) - z_n(t) \\ u_0(t) = f\left(e_1(t), e_2(t), \ldots, e_n(t), p\right), \end{cases} \tag{2.5}$$

where p is the control parameter and $f\left(e_1(t), e_2(t), \ldots, e_n(t), p\right)$ is an optional nonlinear function. While the nonlinear function and the parameters in the NLSEF are reasonably selected, good robustness will be achieved for the n-order system.

At present, the ADRC has been widely used in position control, attitude control, process control, and so on. The ADRC is not only used in precision control but also in disturbance rejection performance, which has completely surpassed PID control.

2.2 Backstepping

Backstepping is a recursive design method which decomposes a control design problem of a complex nonlinear system into one of a series of low-order subsystems. A good idea on dealing with a high-order system is to design a virtual control law to ensure some performance, such as stability and passivity. By gradually modifying the virtual control law and designing a real controller to realize the global adjustment or tracking, a desired performance index is achieved for the complex nonlinear system. The backstepping design method is suitable for uncertain nonlinear systems with state linearization or strict parameter feedback, which can be easily realized by symbolic algebra software.

Consider the following single input single output system as

$$\begin{cases} \dot{x}_1(t) = x_2(t) + f_1\left(x_1(t)\right) \\ \dot{x}_2(t) = x_3(t) + f_2\left(x_1(t), x_2(t)\right) \\ \quad \vdots \\ \dot{x}_i(t) = x_{i+1}(t) + f_i\left(x_1(t), \ldots, x_i(t)\right) \\ \quad \vdots \\ \dot{x}_n(t) = f_n\left(x_1(t), \ldots, x_n(t)\right) + u(t). \end{cases}$$

The main process is summarized in the following. Let $x_{i+1}(t)$ be virtual control for $\dot{x}_i(t)$ subsystem. Determining an appropriate virtual feedback $x_{i+1}(t) = \alpha_i(t)(i = 1, 2, \ldots, n-1)$, the $\dot{x}_i(t)$ subsystem is asymptotically stable. However, a solution of the system generally does not satisfy $x_{i+1}(t) = \alpha_i(t)$. Therefore, an error variable is introduced to hold that there will be some asymptotic characteristics between $x_{i+1}(t)$ and $\alpha_i(t)$ through the control action. Finally, asymptotic stabilization is realized for the system.

Example: Take a second-order system as an example

$$\begin{cases} \dot{x}_1(t) = x_2(t) + f_1\left(x_1(t)\right) \\ \dot{x}_2(t) = f_2\left(x_1(t), x_2(t)\right) + u(t). \end{cases} \tag{2.6}$$

Considering $x_2(t)$ as a virtual input and finding a suitable $\alpha_1(t)\left(x_1(t)\right)$ such that $x_2(t) = \alpha_1(t)\left(x_1(t)\right)$, $x_1(t)$ will be gradually stable at 0. Constructing an error as $e_1(t) = x_1(t)$, a corresponding energy function is $V_1(t) = \frac{1}{2}e_1^2(t)$ by a Lyapunov method. One has that

$$\dot{V}_1(t) = e_1(t)\dot{e}_1(t) = e_1(t)\left[x_2(t) + f_1\left(x_1(t)\right)\right].$$

Selecting $x_2(t) = -e_1(t) - f_1\left(x_1(t)\right)$ in the above formula, $\dot{V}_1(t)$ is negative definite. Furthermore, one is obtained that $\alpha_1(t)\left(x_1(t)\right) = -e_1(t) - f_1\left(x_1(t)\right)$. Next, it is proved that $x_1(t)$ and $e_2(t)$ is asymptotically stable at the origin point as well as $x_2(t)$ is asymptotically stable at $\alpha_1(t)$ by selecting $e_2(t) = x_2(t) - \alpha_1(t)$.

Constructing an energy function as $V_2(t) = V_1(t) + \frac{1}{2}e_2^2(t)$, it follows that $\dot{V}_2(t) = -e_1^2(t) + e_2(t)\left(e_1(t) + \dot{e}_2(t)\right)$. Note that $\dot{V}_2(t) = -e_1^2(t) - e_2^2(t)$ is guaranteed to be negative definite by selecting

$$e_1(t) + \dot{e}_2(t) = -e_2(t). \tag{2.7}$$

Combing the second-order system, one has that

$$\dot{e}_2(t) = \dot{x}_2(t) - \dot{\alpha}_1(t) = u(t) + f_2\left(x_1(t), x_2(t)\right) - \dot{\alpha}_1(t). \tag{2.8}$$

Considering Eqs. (2.7)–(2.8), it follows that

$$u(t) + f_2\left(x_1(t), x_2(t)\right) - \dot{\alpha}_1(t) = -e_1(t) - e_2(t).$$

Finally, a backstepping controller is deduced as follows:

$$u(t) = -f_2\left(x_1(t), x_2(t)\right) + \dot{\alpha}_1(t) - e_1(t) - e_2(t). \tag{2.9}$$

Based on the Lyapunov method, the second-order system (2.6) is stable by the proposed backstepping controller.

Backstepping is widely used in output regulation, robust control, adaptive control, and other control methods.

2.3 Sliding Mode Control

Sliding mode control has strong robustness for nonlinear systems because of its insensitivity to unknown parameters and disturbances. The sliding mode control has the advantages of fast response and simple physical implementation. However, a chattering problem occurs due to nonideal switching characteristics of sliding mode control. Therefore, various ways to solve the chattering problem, mainly including continuous function approximation method, reaching law method, disturbance observer method, and so on.

Suppose an one-order system as follows:

$$\dot{x}(t) = f(x(t), t) + b(x(t), t)u(t) + \beta(x(t), t), \tag{2.10}$$

where $x(t) \in R^n$, $b(x(t), t) \neq 0$, $u(t) \in R$ is the control input, $\beta(x(t), t)$ is external disturbances and internal uncertainty. Design the control input with a switching function $s(x(t))$ as follows:

$$u(t)=\begin{cases}u^{+}(x(t)), & s(x(t))>0\\ u^{-}(x(t)), & s(x(t))\le 0.\end{cases}\tag{2.11}$$

Note that there exist characteristics for the SMC, such that (1) Sliding mode exists; (2) The arrival condition is satisfied; (3) The stability of sliding mode movement is ensured. The designing of sliding mode control is summarized as follows. Firstly, a sliding mode switching function $s(x(t))$ is designed. Secondly, a discontinuous control law is found to ensure every point on a plane $s(x(t))=0$ called a sliding mode surface. Thirdly, a sliding mode control law is designed to attract the plant state to the sliding mode surface. When the system reaches the sliding mode surface from an initial state, the system movement is determined by $s(x(t))=0$. A Lyapunov method is used in designing of SMC. If the sliding mode arrival condition $\lim_{t\to+\infty} s(x(t))\dot{s}(x(t))\le 0$ is satisfied, there exists a Lyapunov function as

$$V(x_1(x(t)),x_2(x(t)),\ldots,x_n(x(t)))=[s(x_1,x_2,\ldots,x_n)]^2,\tag{2.12}$$

which $V(x_1(x(t)),x_2(x(t)),\ldots,x_n(x(t)))$ is positive definite and its derivative is negative semi-definite, the system is stable at $s(x(t))=0$.

Example: Consider a two order system as

$$\ddot{x}(t)=f(x(t))+u(t)+d(t),\tag{2.13}$$

where $f(x(t))$ is a known function, $u(t)$ is the control input, $d(t)$ is the unknown external disturbance which has an upper bound $|d(t)|\le D$. Design a sliding mode surface as

$$s(t)=\dot{e}(t)+\lambda e(t),\ \lambda>0,$$

where $e(t)$ is a tracking error held that $e(t)=x(t)-x_d(t)$ with $x_d(t)$ being a reference input signal. Moreover, it is obtained that

$$\dot{s}(t)=-\varepsilon\,\mathrm{sign}(s(t))-Ks(t),$$

where $\varepsilon>0$ and $K>0$. Furthermore, a SMC is designed as

$$u(t)=-\varepsilon\,\mathrm{sign}(s(t))-Ks(t)-f(x(t))+\ddot{x}_d(t)-\lambda\dot{e}(t)-D.\tag{2.14}$$

Select a Lyapunov function which is defined as

$$V(t)=\frac{1}{2}s^2(t).$$

Differentiating $V(t)$ with respect to time, one has that

$$
\begin{aligned}
\dot{V}(t) &= s(t)\dot{s}(t) \\
&= s(t)(\ddot{e}(t) + \lambda\dot{e}(t)) \\
&= s(\ddot{x}(t) - \ddot{x}_d(t) + \lambda\dot{e}(t)) = s(t)(f(x(t)) + u(t) + d(t) - \ddot{x}_d(t) + \lambda\dot{e}(t)) \\
&= s(t)(-\varepsilon\text{sign}(s(t)) - Ks(t) - D + d(t)) \leq -\varepsilon|s(t)| - Ks^2(t) \leq 0.
\end{aligned}
$$

Based on the Lypunov method, the two-order system is stable by the proposed SMC. Note that sliding mode movement includes reaching phase and sliding phase. The movement from any initial state to the sliding mode surface is called the reaching phase. The dynamic quality of the reaching phase can be improved by using the proposed reaching law. The general reaching law is given as follows:

$$
\dot{s}(t) = -\varepsilon\text{sign}(s(t)) - f(s(t)),\ \varepsilon > 0,
$$

where $f(0) = 0$ satisfying $s(t)f(s(t)) > 0, s(t) \neq 0$.

The SMC draws more attention to wide applications in practice, such as robot control, motor control, and spacecraft control. In practical application, chattering phenomenon of the sliding mode control is a serious problem due to system inertia and other reasons. Eliminating the chattering of the SMC is of great significance for practical engineering application. More attention is focused on solving the chattering by combining different control methods, for example, neural sliding mode control, adaptive sliding mode control, and fuzzy sliding mode control. In addition, the fast reaching law is another research point in order to shorten the arrival time and improve the dynamic performance of the system.

2.4 Finite Time Stability

With the rapid development of science and technology, requirements on control performance of advanced control industry are not limited to stability. As an important performance index of a control system, convergence performance is one of the most concerned parts for an efficient control scheme. Most of the current research results on complex control system only consider asymptotic stability, that is, system states converge to equilibrium points when time tends to infinity. However, there is an asymptotically stable system with very bad transient performance, such as excessive overshoot, which has a very bad impact in engineering, and even cannot be applied at all, such as the system with saturation nonlinearity. Therefore, finite-time stability with rapid convergence performance is studied for nonlinear systems to design control laws for realizing the finite-time convergence of nonlinear system states. The finite stability means that the system state variables can converge to the desired equilibrium state in finite time to achieve the finite-time stability of the system.

A definition of the finite-time stability is given in the following.

Definition 2.1 Given a scalar $T > 0$ and $E_t(t \in [0, T], x = 0 \in E_t)$, $E_0(x = 0 \in E_0, E_0 \subseteq E_t)$. The system is finite-time stability about (E_0, E_t, T) if the system satisfy

$$x(0) \in E_0 \Rightarrow x(t) \in E_t, \forall t \in [0, T].$$

It can be seen from the above definition that the definition of finite-time stability contains three elements: Given time interval, initial stability condition bound and system state bound. The difference between finite-time stability and traditional Lyapunov asymptotic stability is that: (a) Finite-time stability only considers the system performance in the time interval $[0, t]$. The asymptotic stability is the system performance in the infinite time interval. (b) Finite-time stability requires the state of the system to remain within a specific limit range in a specific time interval. Moreover, asymptotic stability requires that the state is attractive and there is no constraint on its motion range.

A criterion based on a Lyapunov method is given in the following.

Lemma 2.2 *(Finite-time Lyapunov stability criterion) Consider a system*

$$\dot{x}(t) = f(x(t)),\ f(0) = 0,\ t_0 = 0,\ x_0 \triangleq x(0). \tag{2.15}$$

If there exists a positive and continuous function $V(x(t))$ in the domain definition U and satisfies

$$\dot{V}(x(t)) \leq -\alpha V^p(x(t)),\ x(t) \in \mathrm{U}\backslash\{0\}, \tag{2.16}$$

where α and p are positive constants which satisfy $p \in (0, 1)$. The system is finite-time stable. That is, the system state can converge to the equilibrium point in finite time, and the convergence time is satisfied with

$$T \leq \frac{1}{\alpha(1-p)} V^{1-p}(x_0).$$

Example: Consider a linear time invariant system

$$\begin{cases} \dot{x}(t) = Ax(t) + Bu(t) + Gw(t),\ x(0) = x_0 \\ \dot{w}(t) = Fw(t) \\ y(t) = Cx(t) + Dw(t), \end{cases} \tag{2.17}$$

where $x(t)$ is the system state, $u(t)$ is the control input of the system, and $w(t)$ is the external system disturbance input of the system. For the linear time invariant system (2.17), the design of finite-time bounded controller is given by

$$u(t) = Kx(t),\ K = LQ_1^{-1}. \tag{2.18}$$

For the linear time invariant system (2.17), positive numbers c_1, c_2, δ, T and $c_1 < c_2$ are given. There is a nonnegative constant α, a matrix L and two positive definite symmetric matrices Q_1 and Q_2. The following condition is satisfied with

$$\begin{bmatrix} AQ_1 + BL + Q_1A^T + L^TB^T - \alpha Q_1 & G \\ G^T & F^TQ_2 + Q_2F - \alpha Q_2 \end{bmatrix} < 0$$
$$\frac{c_1}{\lambda_{\min}\left(\tilde{Q}_1\right)} + \lambda_{\max}(Q_2)\,\delta < \frac{c_2 e^{-\alpha T}}{\lambda_{\max}\left(\tilde{Q}_1\right)}. \tag{2.19}$$

It is obtained that the closed-loop system is finite-time bounded in (c_1, c_2, δ, T, R). The proof process is given in the following.

For the linear time invariant system (2.17), let $\hat{A} = A + BK$. To make it finite-time bounded on (c_1, c_2, δ, T, R), it is only necessary to have an appropriate matrix to satisfy the following Lemma.

Lemma 1 *For the system (2.17) with* $u(t) = 0$*, there exists*

$$\begin{cases} \dot{x}(t) = Ax(t) + Gw(t), \ x(0) = x_0 \\ \dot{w}(t) = Fw(t). \end{cases} \tag{2.20}$$

For the system (2.20), given a positive number c_1, c_2, δ, T *and* $c_1 < c_2$*, we have that the positive symmetric matrix* $R \in R^{n\times n}$*. If there exist a positive constant* α*, two positive symmetric matrices* P_1 *and* P_2 *satisfy*

$$\begin{bmatrix} A^TP_1 + P_1A - \alpha P_1 & P_1G \\ G^TP_1 & F^TP_2 + P_2F - \alpha P_2 \end{bmatrix} < 0, \tag{2.21}$$

$$\lambda_{\max}\left(\tilde{P}_1\right)c_1 + \lambda_{\max}(P_2)\,\delta < \lambda_{\min}\left(\tilde{P}_1\right)c_2 e^{-\alpha T}, \tag{2.22}$$

where $\widetilde{P}_1 = R^{-1/2}P_1R^{-1/2}$, $\lambda_{\max}$ *and* $\lambda_{\min}$ *denote the maximum and minimum eigenvalues of the corresponding matrix, respectively. The system (2.20) is finite-time bounded on* (c_1, c_2, δ, T, R).

Letting $Q_1 = P_1^{-1}$ and $Q_2 = P_2$, consider the following equation:

$$\lambda_{\max}(Q) = \frac{1}{\lambda_{\min}\left(Q^{-1}\right)}.$$

For the linear time invariant closed-loop systems, Eq. (2.21) in Lemma 1 is expressed as

$$\begin{bmatrix} \hat{A}^TQ_1^{-1} + Q_1^{-1}\hat{A} - \alpha Q_1^{-1} & Q_1^{-1}G \\ G^TQ_1^{-1} & F^TQ_2 + Q_2F - \alpha Q_2 \end{bmatrix} < 0.$$

Multiply the above equation left and right by a diagonal matrix (Q_1, I), where I is the diagonal matrix with appropriate dimension, which can be obtained by calculation

$$\begin{bmatrix} Q_1\hat{A}^T + \hat{A}Q_1 - \alpha Q_1 & G \\ G^T & F^TQ_2 + Q_2F - \alpha Q_2 \end{bmatrix} < 0.$$

Put $\hat{A} = A + BK$ with $K = LPQ_1^{-1}$ is substituted into the above formula to obtain condition (2.19). Therefore, condition (2.19) is equivalent to two conditions in Lemma 1. That is, the designed controller is a finite-time bounded controller.

Asymptotic stability meets the requirements of practical engineering, but it depicts the steady-state performance of a system and cannot reflect the transient performance of the system. Finite-time stability is a stability concept which is different from the general stability in order to study the transient performance of the system. Finite-time convergence control widely exists in many fields, such as aerospace control system, robot control system, motor control system, machining control system, and so on.

2.5 Nonlinear Functions

In the system modeling, some laws of motion and nonlinear functions are considered, which are given in the following.

Rotation law of rigid body are introduced as follows:

$$T(t) = J\ddot{\theta}(t) + b_v\dot{\theta}(t), \tag{2.23}$$

where $T(t)$ is the moment, J is the moment of inertia for the rigid body, b_v is the coefficient of damping, $\theta(t)$ is the rotation angle.

Motion law of rigid body are introduced as follows:

$$ma(t) = F(t), \tag{2.24}$$

where m is the mass of the rigid body, $a(t)$ is the acceleration of the rigid body, $F(t)$ is the force on the rigid body.

Friction has a considerable impact on the pneumatic cylinder servo positioning control, which can cause the pneumatic cylinder to crawl at low speed. The friction of pneumatic rod cylinder consists of a viscous friction, a coulomb friction, and a Stribeck effect of stationary force transition between positive and negative velocities. A smooth transition process is simulated by using the Stribeck effect described by the hyperbolic sigmoid function and a differential Gaussian function in the case of dry friction. Each part is modeled independently and their mutual influences are ignored.

Viscous friction, which is also known as a linear damping, is quite easy to handle in modeling. The damping term is given as follows:

$$\tilde{F}_v(\dot{y}(t)) = F_v(\dot{y}(t)) = C\dot{y}(t), \tag{2.25}$$

where $\tilde{F}_v(\dot{y}(t))$ is the viscous friction, $\dot{y}(t)$ is the speed, C is the linear damping factor which describes the proportional relationship between the viscous friction and a velocity.

Coulomb friction, which is also known as a dry friction, is one of friction effects observed by Coulomb in 1821. Coulomb friction does not depend on absolute values

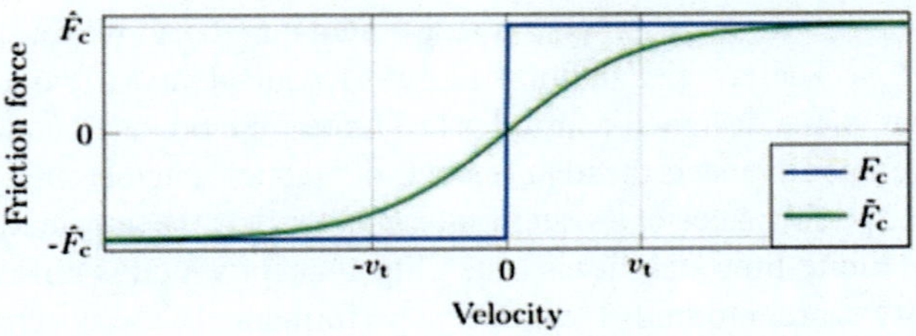

Fig. 2.3 The differences between the original and modified Coulomb friction

of velocities but changes its sign by the positive and negative of the velocity. The friction $F_c(\dot{y}(t))$ is usually described as follows:

$$F_c(\dot{y}(t)) = \mu F_n \text{sign}(\dot{y}(t)) = \hat{F}_c(\dot{y}(t)), \tag{2.26}$$

where F_n is the normal force, μ is the friction coefficient, $\hat{F}_c$ is the peak load of coulomb friction with $\hat{F}_c = \mu F_n$. Because property of the transformation function given by the velocity symbol is not expected to be seen, the symbolic function will be approximated as a hyperbolic tangent function with the transition velocity v_t [1]. Therefore, a modified Coulomb friction $\tilde{F}_c(\dot{y}(t))$ is written as follows:

$$\tilde{F}_c(\dot{y}(t)) = \hat{F}_c \tanh(\frac{\dot{y}(t)}{v_t}). \tag{2.27}$$

The differences between original and modified Coulomb friction are shown in Fig. 2.3. The steeper force transition makes the modified Coulomb function match with the original function well. However, it is realistic that a smoother transition can guarantee powerful numerical effects. The choice of transition velocity depends not only on combinations of sampling times and solvers, but also on combinations of state observers and models of actual application. In general, the transition velocity v_t has to be adapted to given system conditions.

Stribeck friction describes stiction between two surfaces caused by adhesive forces. It depends on many factors such as material, geometry, and temperature. The general form of the friction $F_\Delta(\dot{y}(t))$ is expressed as follows:

$$F_\Delta(\dot{y}(t)) = (\hat{F}_s - \hat{F}_c) g(\dot{y}(t)) \text{sign}(\dot{y}(t)) = \tilde{F}_\Delta g(\dot{y}(t)) \text{sign}(\dot{y}(t)), \tag{2.28}$$

where $g(\dot{y}(t))$ is the nonlinear function which describes attenuation characteristics of the Stribeck curve, $\hat{F}_s$ is the peak load of Stribeck and $\tilde{F}_\Delta = \hat{F}_s - \hat{F}_c$. According to [1], $g(\dot{y}(t))$ is introduced by a Gaussian Bell method as follows:

$$g(\dot{y}(t)) = e^{-(\frac{\dot{y}(t)}{v_S})^2}. \tag{2.29}$$

Then equation (2.29) is modified as follows:

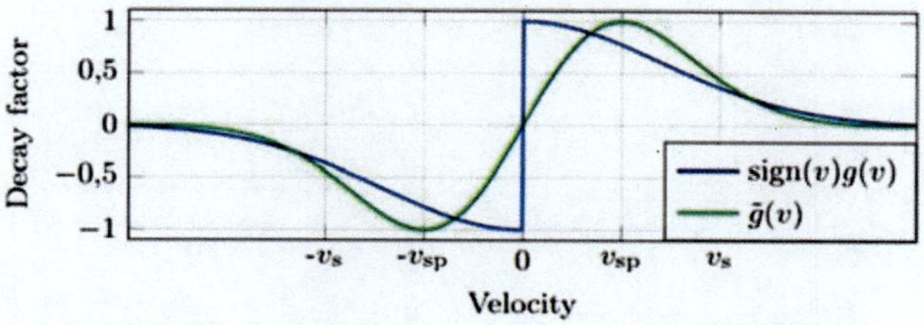

Fig. 2.4 The differences between the two functions $g(\dot{y}(t))$ and $\tilde{g}(\dot{y}(t))$

$$\tilde{g}(\dot{y}(t)) = \frac{\dot{y}(t)}{v_{sp}} e^{-(\frac{\dot{y}(t)}{\sqrt{2}v_{sp}})^2 + \frac{1}{2}}, \tag{2.30}$$

where $\tilde{g}(\dot{y}(t))$ is the transformed function with a symbolic function, v_s is the Stribeck velocity, v_{sp} is the peak velocity of Stribeck which is defined as frictional attenuation.

The differences between two functions $g(\dot{y}(t))$ and $\tilde{g}(\dot{y}(t))$ are shown in Fig. 2.4 for a peak velocity v_{sp} half size of the Stribeck velocity v_s. The flatter the slope at zero crossing is chosen, the more the peak of the approximated friction moves away from its usually position at zero velocity. The Stribeck effect is ignored well if a very flat slope is required. Otherwise, there is not only a wrong model behavior at standstill, but also a peak of friction at higher velocities which falsifies the model. If the Stribeck effect is not ignored, the additional force amplitude $\hat{F}_\Delta$ will change accordingly. Because the Coulomb friction $\tilde{F}_c(\dot{y}(t))$ is no longer a constant, the gap between Stribeck peak and Coulomb friction widens with the velocity v_{sp}. One has that

$$\hat{\hat{F}}_\Delta = \hat{F}_s - \hat{F}_c \tanh(\frac{v_{sp}}{v_t}) - C v_{sp}. \tag{2.31}$$

For completeness, the viscous friction at the peak velocity v_{sp} is subtracted even if its value is not significant for small velocities. According to the modified Gauss Bell (2.30) and the additional peak force (2.31), the following modified Stribeck friction is get as

$$\tilde{F}_\Delta(\dot{y}(t)) = \hat{\hat{F}}_\Delta \tilde{g}(\dot{y}(t)). \tag{2.32}$$

By adding up all three friction effects, the continuously differentiable friction model $F(t)$ is gotten as follows:

$$\begin{aligned} F(t) = \tilde{F}_\Sigma(\dot{y}(t)) &= \tilde{F}_c(\dot{y}(t)) + \tilde{F}_\Delta(\dot{y}(t)) + \tilde{F}_v(\dot{y}(t)) \\ &= \hat{F}_c \tanh(\frac{\dot{y}(t)}{v_t}) + (\hat{F}_s - \hat{F}_c \tanh(\frac{v_{sp}}{v_t}) \\ &\quad - C v_{sp}) \frac{\dot{y}(t)}{v_{sp}} e^{-(\frac{\dot{y}(t)}{\sqrt{2}v_{sp}})^2 + \frac{1}{2}} + C\dot{y}(t). \end{aligned} \tag{2.33}$$

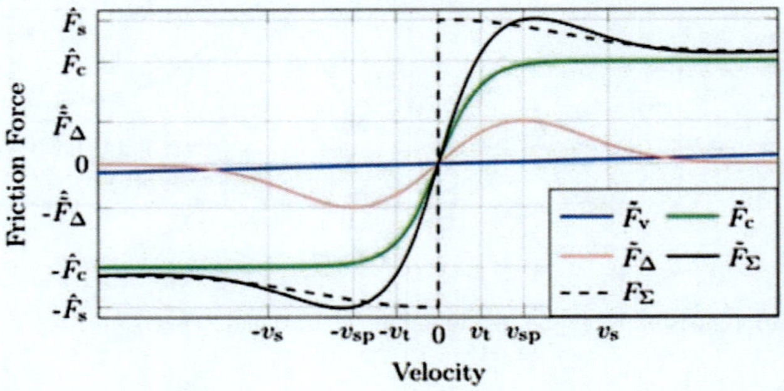

Fig. 2.5 The comparison of the continuously differentiable friction model with each part and the frictional force $\tilde{F}_{\sum}(\dot{y}(t))$

In Fig. 2.5, the comparison of the continuously differentiable friction model with each part and the frictional force $\tilde{F}_{\sum}(\dot{y}(t)) = \tilde{F}_c(\dot{y}(t)) + \tilde{F}_\Delta(\dot{y}(t)) + \tilde{F}_v(\dot{y}(t))$ are shown. The function approximation of the small sample time is achieved by using $v_s = 2v_{sp} = 4v_t$.

Reference

1. Specker T, Buchholz M, Dietmayer K (2014) A new approach of dynamic friction modelling for simulation and observation. In: International federation of automatic control, pp 4523–4528

Part I
Pneumatic Manipulator System

Chapter 3
Platform Introduction

3.1 Application Background

Pneumatic manipulator driven by PAMs are used in various fields, such as robot manipulators, rescue robots, and some rehabilitation robots. In Fig. 3.1a, a aneumatic dual-arm robot is developed by Beijing Institute of technology. Due to the special structure of its joints, the human body develops astoundingly large forces and scope of movement in almost every direction. The arm can grasp objects in a defined manner and position them with high precision. In Fig. 3.1b, a dual-arm robot driven by PAMs is developed by Tokyo Institute of technology [1]. In order to make rescue workers far away from scenes of disaster, a remote control of the dual-arm robot is used to ensure the personal safety of the rescue workers. As the advantages of the PAMs are rapid response, low cost, high power–weight ratio, and flexibility, the pneumatic manipulator driven by PAMs are also widely used in food processing, medical care, bionic robot, and so on. Therefore, designing and control technologies of highly bionic pneumatic robot have attracted wide attention in recent years.

3.2 Platform Structure

3.2.1 Platform Components

A pneumatic dual-arm robot driven by PAMs shown in Fig. 3.2 is designed by professor Tao Wang's team in Beijing Institute of Technology. Each arm of the pneumatic dual-arm robot has seven degrees of freedom, including three degrees of freedom for shoulder joint, two degrees of freedom for elbow joint, and two degrees of freedom for wrist joint. A pneumatic schematic diagram of pneumatic dual-arm robot is shown in Fig. 3.3.

L. Zhao et al., *Pneumatic Servo Systems Analysis*, Advances in Industrial Control,
https://doi.org/10.1007/978-981-16-9515-5_3

(a) Pneumatic dual-arm robot of BIT. (b) Dual arm robot

Fig. 3.1 Manipulator driven by PAMs

Fig. 3.2 Pneumatic dual-arm robot

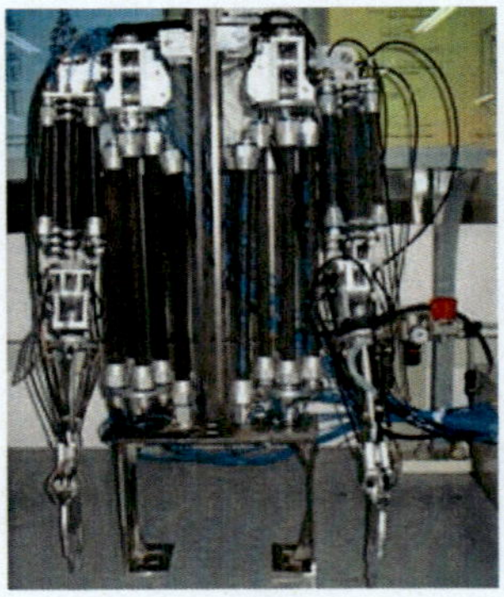

Fig. 3.3 Pneumatic schematic diagram of pneumatic dual-arm robot

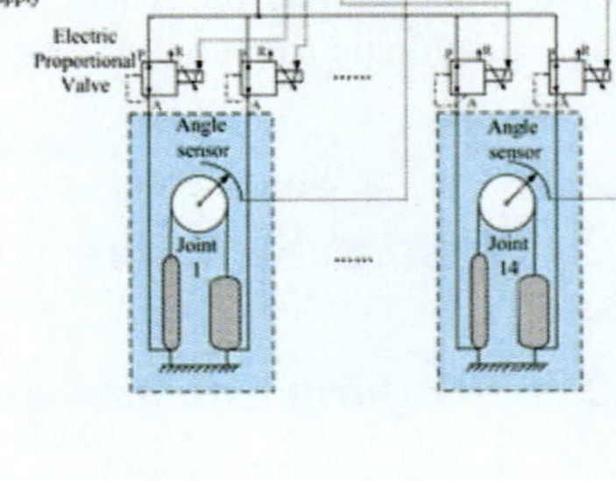

During the process of accomplishing various tasks, position and posture control of single joint is the key of motion control for the pneumatic dual-arm robot. Therefore, an experimental platform which is an experiment platform for the pneumatic manipulator system in this chapter is presented shown in Fig. 3.4. Moreover, a pneumatic manipulator driven by PAMs for one-DoF joint of the pneumatic dual-arm robot is shown in Fig. 3.5.

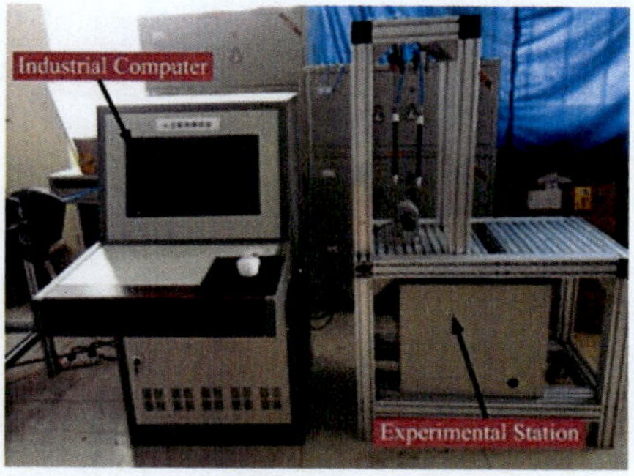

Fig. 3.4 Pneumatic manipulator platform

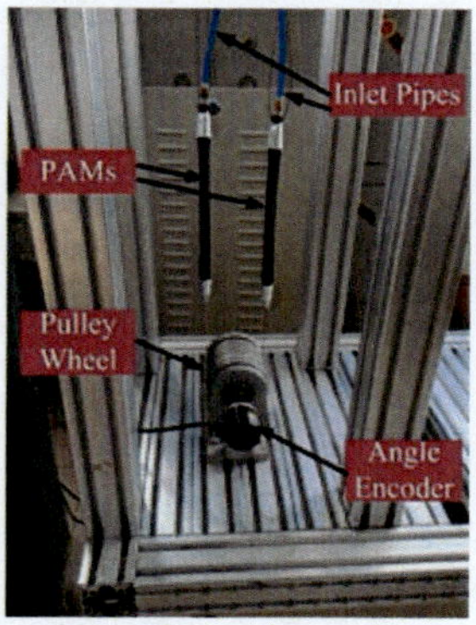

Fig. 3.5 One-DoF pneumatic manipulator

The pneumatic manipulator platform is designed as a pneumatic experiment platform with one-DoF joint of the pneumatic dual-arm robot. It consists of two parts: a control section and an operating section. The control section includes an industrial personal compute (IPC) (Advantech, 610H), a D/A board (Advantech, PCI-1727, 12 channels, and 14 bits) and a counting board (Advantech, PCI-1784, four 32-bit encoder counters). The D/A board and counting board are installed in the IPC. Moreover, the D/A board is used to transmit the expected voltage signals to two proportional valves to control pressures of PAMs, and the counting board is adopted to receive a feedback signal from an angle encoder. Besides, the operating section includes a pneumatic manipulator, two PAMs, an angle encoder (OMRON, E6B2-CWZ3E, resolution 2000 P/R), and two pressure proportional valves (SMC, ITV0050, output precision $\pm 6\%$, pressure range $0.001 \sim 0.9$MPa). The extension or contraction of each PAM is controlled via a proportional valve, and the relative motion of those two PAMs drives the pneumatic manipulator to produce rotation of a pulley wheel. The rotational angle could be measured by the angle encoder as feedback of the control system. In addition, the prototypes of PAMs are *Mckibben* type PAMs. Main components of the pneumatic manipulator platform are shown in Table 3.1.

Table 3.1 Main components of the pneumatic manipulator platform

Component name	Component model	Main performance indicators
PAM	Self-made	Initial diameter: 20 mm
		Initial length: 240 mm
		Working pressure: 0 ~ 0.35 MPa
Pressure proportional valve	ITV0050(SMC)	Output voltage: 0 ~ 5 VDC
		Working pressure: 0 ~ 1.0 MPa
D/A input	PCL-1727	12 path D/A output
		Output range: 0 ~ 10 V
Counter card	PCL-1784	Orthogonal frequency: 1.0 Mpa
		Pulse frequency: 2.4 Mpa
		32-bit counter

Table 3.2 Structure sizes of the pneumatic manipulator platform

PAM	Diameter	20 mm
	Windings number rayon	4.2
	Total length of rayon	300 mm
	Thickness of rubber tube	1 mm
	Rubber tube's modulus of elasticity	0.2 Mpa
One-DoF pneumatic manipulator	Diameter	80 mm
	Mass	1.30 kg
Frictional coefficient	0.015	
Modifying factor	13	

The experimental platform is driven by two PAMs to drive the rotation of the pneumatic manipulator. Furthermore, some structure sizes of the pneumatic manipulator platform are shown in Table 3.2.

3.2.2 Control Circuit

The pneumatic manipulator platform is driven by the two PAMs. Pressure proportional valves are used in the pneumatic circuit of the pneumatic manipulator platform. The pressure proportional valves are pressure relief valve whose pressures are proportional to the change of input voltage. The rated voltage and standard rated flow

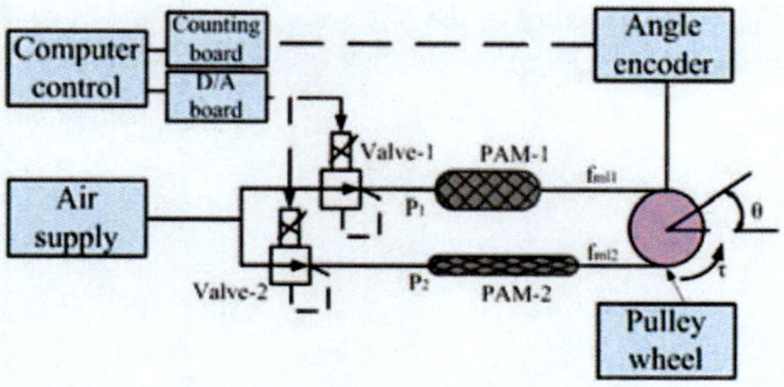

Fig. 3.6 A hardware control block diagram of the pneumatic manipulator system

of the pressure proportional valves are 24 V and 6 L/min, respectively. Linearity and sensitivity of the pressure proportional valves are less than 6 and 1%, respectively.

A control process for pneumatic manipulator platform is shown in Fig. 3.6. In Fig. 3.6, computer sends signals to the two pressure proportional valves via the D/A board. Internal pressure of the two PAMs are regulated in real time by the two pressure proportional valves. Then rotation of the pneumatic manipulator is driven by the two PAMs. The angle encoder sends angle signal to the computer via the counting board.

3.3 System Model

The PAM has a structure of rayon mesh and rubber tube to form a closed volume so that it can produce pulling force and displacement by inflating or deflating. Moreover, the schematic diagram of the PAM mesh grid fiber is given in Fig. 3.7. It is shown from Fig. 3.7 that a structure is given for the mesh grid fiber, and it is necessary to derive the model of PAM. In the above of Fig. 3.7, the rayon line wraps around the PAM. If expanding the rayon line along the angle α, we will get the bottom of Fig. 3.7. Meanwhile, according to Fig. 3.7, two equations are shown in the following:

$$\begin{cases} L = b\cos(\alpha) \\ N\pi D = b\sin(\alpha), \end{cases}$$

where L is the actual length of the PAM, D is the diameter, b is the total length of rayon, N is the number of rayon windings, α is the angle of the rayon line to its central axis.

$$V = \frac{1}{4}\pi D^2 L,$$

where V is the approximate operating volume of PAM.

According to [2, 3], the mathematical model of the *Mckibben* type PAM is derived as follows:

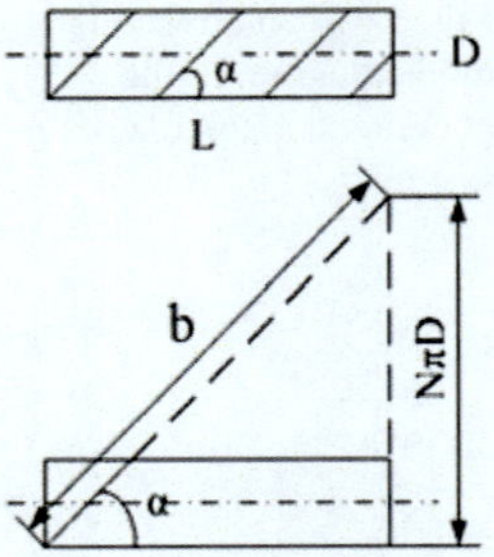

Fig. 3.7 Structure of the mesh grid fiber

$$f_m(L,p) = -p\frac{dV}{dL} = -p\frac{d(\frac{b^3\sin^2(\alpha)\cos(\alpha)}{4\pi N^2})}{d(b\cos(\alpha))} \tag{3.1}$$
$$= -p\frac{b^2}{4\pi N^2}(1-3\cos^2(\alpha))$$
$$= p\frac{3L^2-b^2}{4\pi N^2},$$

where $f_m(L,p)$ is the output force of PAM, p is the inflation pressure.

However, output force model (3.1) is an ideal expression. The model of PAM is further corrected by considering nonlinearities as the following:

$$f_{ml} = \begin{cases} \eta f_m + f_{r1} - f_{r2} - f_c, & p > p_d \\ f_{r1} - f_{r2}, & p \le p_d, \end{cases} \tag{3.2}$$

where η is a structure parameter, f_c is the friction [4], f_{r1} is the restoring force of rubber tube, f_{r2} is the binding force of rubber tube according to [3]. Dead-zone pressure is denoted by p_d, which is usually a very small value. The value of p_d is obtained by solving $\eta f_m(L,p_d) - f_{r2}(L) = 0$. Furthermore, f_c, f_{r1} and f_{r2} are calculated as

$$f_c = \left(\frac{1}{n_s}\right)\frac{\mu}{N}\left(\frac{b^2-L_0^2}{\sqrt{b^2-L^2}}\right)\frac{L_0^2}{L}ptanh(\Delta\dot{L})$$

$$f_{r1} = \frac{Et}{N}\sqrt{b^2-L^2}\left(\frac{L}{L_{\min}}-1\right)$$
$$f_{r2} = \frac{Et}{N}L^2\left(\frac{1}{\sqrt{b^2-L_{\max}^2}}-\frac{1}{\sqrt{b^2-L^2}}\right),$$

where t is thickness of the rubber tube, L_0 is initial length of the PMAs, $L_{\max}$ (or $L_{\min}$) is the maximum (or minimum) length that the PAM can be extended (or compressed), E is rubber tube's modulus of elasticity, μ is frictional coefficient, n_s represents the modifying factor, ΔL is defined as $\Delta L = 1/2(L_2 - L_1)$, L_1 and L_2 are the actual lengths of PAM-1 and PAM-2.

A dynamic model of the pneumatic manipulator is described as follows:

$$J\ddot{\theta}(t) + b_v\dot{\theta}(t) = \left[f_{ml1}(L_1, p_1) - f_{ml2}(L_2, p_2)\right]r, \tag{3.3}$$

where J is rotary inertia of pulley wheel with a mass of m and a radius of r, which is obtained as $J = 1/2mr^2$. b_v stands for the joint damping ratio, θ is the rotational angle of the pulley wheel, L_1 and L_2 are the length of PAM-1 and PAM-2, respectively. p_1 and p_2 are internal pressures of PAM-1 and PAM-2, respectively. f_{ml1} is output force of PAM-1, f_{ml2} is output force of PAM-2. f_{ml1} and f_{ml2} are calculated as in (3.2).

Then, multiple inputs system (3.3) with input variables p_1 and p_2 is reformulated to the following system with single input variable Δu.

$$\begin{cases} p_1 = p_0 + \Delta p = p_0 + \Delta u(t)K_0 \\ p_2 = p_0 - \Delta p = p_0 - \Delta u(t)K_0, \end{cases} \tag{3.4}$$

where p_0 is the pre-charged pressure of those two PAMs, Δp is the changing value of those two PAMs, Δu is the changing voltage value of the pressure proportional valves, and K_0 is the gain of the pressure proportional valve with the value of 0.1 MPa/V.

In addition, the relation between lengths L_0, L_1, L_2 and rotation angle θ is given as

$$\begin{cases} L_1 = L_0 - \theta r \\ L_2 = L_0 + \theta r. \end{cases} \tag{3.5}$$

Substituting (3.2), (3.4), and (3.5) into dynamic model (3.3), and ignoring the high-order terms of its Taylor series expended at $\theta = 0$ yield a state-space system as follows:

$$\begin{cases} \dot{x}_1(t) = x_2(t) \\ \dot{x}_2(t) = f_0(t) - b'_v x_2(t) + b_k(x_1(t), x_2(t))\Delta u(t) \\ y(t) = x_1(t), \end{cases}$$

where f_0 represents nonlinearity and uncertainty of the system model, and it is assumed continuously differentiable and bounded. b'_v is the system's hysteresis which is calculated by $b'_v = b_v/J$. System states are defined as $x_1(t) = \theta$ and $x_2(t) = \dot{\theta}$. The function $b_k(x_1, x_2)$ is the gain of $\Delta u(t)$. In the rest of this chapter, $b_k(x_1, x_2)$ is replaced by b_k for brevity.

Filter gain b_k is further divided into two parts, a constant term b_0 and a time-varying term $(b_k - b_0)$. Then, a dynamic model of the pneumatic manipulator system is rewritten as

Table 3.3 Parameters of the pneumatic manipulator system

Parameter	Value	Parameter	Value (mm)
N	4.2	b	300
μ	0.015	L_{max}	250
η	0.75	L_{min}	210
n_s	13	L_0	240
E	0.2 Mpa	t	1
m	1.3 kg	r	40

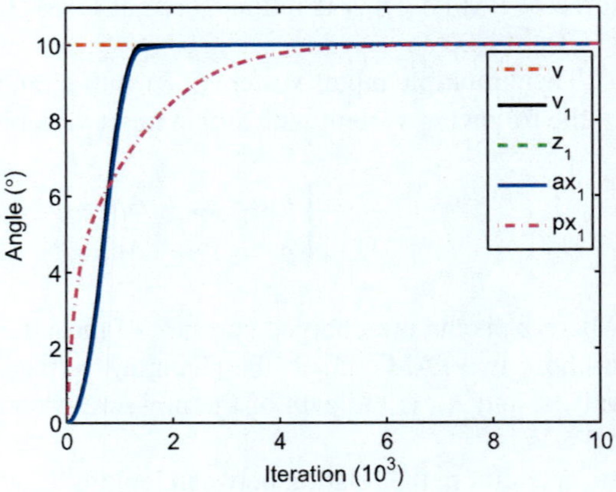

Fig. 3.8 Simulation results for tracking a step signal at 10°

$$\begin{cases} \dot{x}_1(t) = x_2(t) \\ \dot{x}_2(t) = f_t(t) - b_v' x_2(t) + b_0 \Delta u(t) \\ y(t) = x_1(t), \end{cases} \tag{3.6}$$

where $f_t(t) = f_0(t) + (b_k - b_0)\Delta u(t)$ represents the total disturbance of the pneumatic manipulator system.

3.4 Simulation and Results

In this section, a simulation is given to show the effectiveness of the dynamic system for the pneumatic manipulator system. Parameters of the pneumatic manipulator system (3.6) are given in Table 3.3.

According to the pneumatic manipulator system (3.6), a simulation is carried out. A PID control method and an active disturbance rejection control are used to control the pneumatic manipulator system (3.6). The simulation results of a step signal at 10° are shown in Fig. 3.8.

3.5 Conclusion

In this chapter, an pneumatic manipulator platform has been presented. The components of the pneumatic manipulator platform have been shown and listed in this chapter. Pneumatic circuit and control circuit for the pneumatic manipulator platform have been given to show the structure of the experimental platform. Moreover, a dynamic model of the pneumatic manipulator system has been established. Finally, effectiveness of the dynamic model has been proved by simulation.

References

1. http://www.kk.pi.titech.ac.jp/archives/research
2. Chou C-P, Hannaford B (1996) Measurement and modeling of McKibben pneumatic artificial muscles. IEEE Trans Robot Autom 12(1):90–102
3. Zhao L, Li Q, Yang H, Li H (2015) Positioning control of a one-DOF manipulator driven by pneumatic artificial muscles based on active disturbance rejection control. In: Chinese control conference, pp 841–845 (2015)
4. Makkar C, Hu G, Sawyer WG, Dixon WE (2007) Lyapunov-based tracking control in the presence of uncertain nonlinear parameterizable friction. IEEE Trans Autom Control 52(10):1988–1994

Chapter 4
Linear Feedback Control

4.1 Introduction

Owing to the good characteristics of rapidness response, safety operation, low cost, cleanliness, and high power–weight ratio, PAM is gradually popular in robot mechanism. However, due to highly nonlinear and time-varying behavior caused by the elasticity and compressibility of air, it is difficult to achieve accurate control in PAM systems. The faultiness of PAM systems has promoted many researches into a number of control strategies, such as PID method, adaptive control law, sliding mode approach, H_∞ technique, etc. A nonlinear PID controller united with neural networks is designed to improve the control performance of a PAM manipulator [1]. In [2, 3], a linearizing control scheme composed of fuzzy logic and linear PID is proposed for a pneumatic muscle system. In [4], a discontinuous projection-based adaptive robust control strategy is adopted to compensate for both parametric uncertainties and uncertain nonlinearities of a manipulator.Though controllers have been used to improve the pneumatic control performance, there are a lot of space for research on this issue, which motivates us to make an effort in this chapter.

In this chapter, we apply the ADRC approach to improve positioning performance of an pneumatic manipulator driven by PAMs. Firstly, a TD is used to deal with input signal to get a corresponding smooth signal and its differential signal. Secondly, the nonlinear behavior of PAM is estimated by ESO. Thirdly, a linear error feedback controller combining with compensation of disturbance is designed to ensure good response for the pneumatic manipulator system. Meanwhile, convergence analysis of ESO is proven by self-stable region approach, and the stability of the pneumatic manipulator system is also proved via Lyapunov's stability theory. Finally, experiment results are given to demonstrate the effectiveness of the developed techniques in this chapter.

The main contributions of this chapter are summarized as below:

i A TD is used to deal with input signal to get a corresponding smooth signal and its differential signal.

L. Zhao et al., *Pneumatic Servo Systems Analysis*, Advances in Industrial Control,
https://doi.org/10.1007/978-981-16-9515-5_4

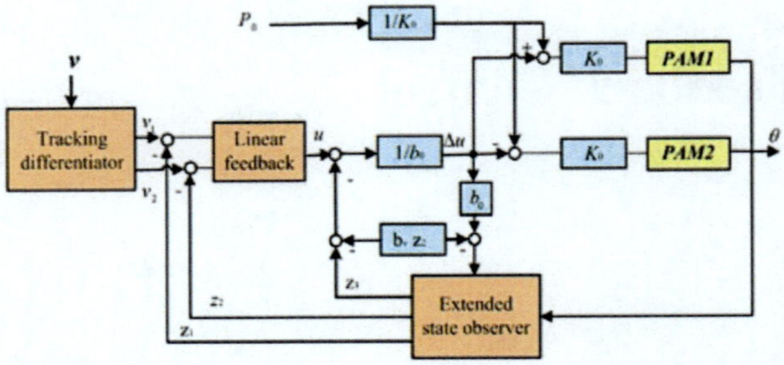

Fig. 4.1 The control structure of the pneumatic manipulator system based on ADRC

ii A linear error feedback controller combining with compensation of disturbance is designed to ensure good response.
iii Experimental results are given to demonstrate the effectiveness of the proposed method in this paper.

4.2 Main Results

4.2.1 Schematic Diagram of Control Method

In this chapter, ADRC method is used to study positioning problems for an pneumatic manipulator driven by PAMs. The control structure of the pneumatic manipulator system is shown in Fig. 4.1. According to Fig. 4.1, the ADRC method consists of three parts: a tracking differentiator, an ESO, and a linear feedback controller.

In the pneumatic manipulator system, the transient process is arranged by a TD which is designed as follows:

$$\begin{cases} f_h(t) = f_{han}(v_1(t) - v(t), v_2(t), r_0, h_0) \\ \dot{v}_1(t) = v_2(t) \\ \dot{v}_2(t) = f_h(t), \end{cases} \tag{4.1}$$

where h is step length, r_0 is a velocity factor, h_0 is a filter factor, $v(t)$ is input signal, $v_1(t)$ is the output which tracks $v(t)$, $v_2(t)$ is the differential signal of $v_1(t)$ with $\lambda(t) = v_1(t) - v(t)$.

4.2.2 Nonlinear Extended State Observer

In this chapter, the nonlinear and time-varying behavior of PAM is considered as a total disturbance which will be estimated by ESO. Considering equation (3.6), $f_T(t)$ is set as the total disturbance part and extended to another state variable. Letting

$f_T(t) = x_3(t)$, the pneumatic manipulator system (3.6) is rewritten as follows:

$$\begin{cases} \dot{x}_1(t) = x_2(t) \\ \dot{x}_2(t) = x_3(t) - b'_v x_2(t) + b_0 \Delta u(t) \\ \dot{x}_3(t) = w(t), \end{cases} \tag{4.2}$$

where $x_3(t)$ is continuous and differentiable, $w(t)$ is the derivative of $x_3(t)$. Therefore, a nonlinear ESO is designed as follows:

$$\begin{cases} e_1(t) = z_1(t) - x_1(t) \\ \dot{z}_1(t) = z_2(t) - \beta_{01} e_1(t) \\ \dot{z}_2(t) = z_3(t) - b'_v z_2(t) - \beta_{02} f_e(t) + b_0 \Delta u(t) \\ \dot{z}_3(t) = -\beta_{03} f_o(t), \end{cases} \tag{4.3}$$

where $z_1(t)$, $z_2(t)$, and $z_3(t)$ are the observations of $x_1(t)$, $x_2(t)$ and the total disturbance $x_3(t)$, respectively. β_{01}, β_{02}, and β_{03} are observer gains. In (4.3), $f_e(t) = fal(e_1(t), 0.5, 6h)$ and $f_o(t) = fal(e_1(t), 0.25, 6h)$. Both $f_e(t)$ and $f_o(t)$ are two different non-smooth functions. The function $fal(\cdot)$ is defined as

$$fal(e(t), c, \sigma) = \begin{cases} \frac{e(t)}{\sigma^{1-c}}, & |e(t)| \le \sigma \\ |e(t)|^c \text{sign}(e(t)), & |e(t)| > \sigma, \end{cases}$$

where σ is the length of linear region, c is the feedback exponent.

Based on the ESO (4.3) and the pneumatic manipulator system (3.6), making $e_1(t) = z_1(t) - x_1(t), e_2(t) = z_2(t) - x_2(t)$ and $e_3(t) = z_3(t) - x_3(t)$, the following error equations are given as

$$\begin{cases} \dot{e}_1(t) = e_2(t) - \beta_{01} e_1(t) \\ \dot{e}_2(t) = e_3(t) - b'_v e_2(t) - \beta_{02} f_e(t) \\ \dot{e}_3(t) = -w(t) - \beta_{03} f_o(t). \end{cases} \tag{4.4}$$

In the following, the self-stable region approach is adopted to analysis convergence of the error system (4.4).

Definition 4.1 Let a region Π with origin by its vertex. If the region Π has a property that all of the trajectories in Π will converge to the origin after a certain time, then the region Π is defined as SSR.

Considering the error system (4.4), the following two functions are given as

$$h_2(e_1(t), e_2(t)) = e_2(t) - \beta_{01} e_1(t) + k_1 g_1(e_1(t)) \text{sign}(e_1(t))$$

and

$$\begin{aligned} h_3(e_1(t), e_2(t), e_3(t)) = e_3(t) - b'_v e_2(t) - \beta_{02} f_e(t) - \beta_{01}(e_2(t) - \beta_{01} e_1(t)) \\ + k_2 g_2(e_1(t), e_2(t)) \text{sat}(h_2/g_1), \end{aligned}$$

where $g_1(e_1(t))$ and $g_2(e_1(t), e_2(t))$ are two arbitrary continuous positive functions with $g_1(0) = 0$ and $g_2(0, 0) = 0$, respectively. Moreover, k_1 and k_2 are two constants with $k_1 > 0$ and $k_2 > 0$.

Furthermore, two functions

$$\begin{cases} V_{h_2,g_1}(t) = \big(h_2^2(e_1(t), e_2(t)) - g_1^2(e_1(t))\big)/2 \\ V_{h_3,g_2}(t) = \big(h_3^2(e_1(t), e_2(t), e_3(t)) - g_2^2(e_1(t), e_2(t))\big)/2 \end{cases} \tag{4.5}$$

and two regions

$$\begin{cases} \Pi_3 = \big\{(e_1(t), e_2(t), e_3(t)) : |h_3(e_1(t), e_2(t), e_3(t))| \leq g_2(e_1(t), e_2(t))\big\} \\ \Pi_2 = \big\{(e_1(t), e_2(t), e_3(t)) : |h_2(e_1(t), e_2(t))| \leq g_1(e_1(t))\big\} \end{cases} \tag{4.6}$$

are given. There exists $V_{h_3,g_2}(t) \leq 0$ in region Π_3 and $V_{h_2,g_1}(t) \leq 0$ in region Π_2, respectively.

Theorem 4.2 *If the following inequality*

$$g_2(e_1(t), e_2(t)) > \frac{(k_1+1)^2}{(k_2-1)}\left|\frac{dg_1(e_1(t))}{de_1(t)}\right| |h_2(e_1(t), e_2(t))| \tag{4.7}$$

is satisfied, then all of the trajectories which do not converge to origin directly get into Π_2 and converge to origin in finite time in the region Π_2 in (4.6).

Proof Simplicity, $h_2(e_1(t), e_2(t))$ is denoted by $h_2(t)$, $h_3(e_1(t), e_2(t), e_3(t))$ is denoted by $h_3(t)$, $g_1(e_1(t))$ is denoted by $g_1(t)$, $g_2(e_1(t), e_2(t))$ is denoted by $g_2(t)$ and $g_1(e_1(t))\text{sign}(e_1(t))$ is denoted by $g_1(t)s$. Considering $V_{h_2,g_1}(t)$ in (4.5), there exists

$$\begin{aligned} \dot{V}_{h_2,g_1}(t) &= h_2(t)\dot{h}_2(t) - g_1(t)\dot{g}_1(t) \\ &= h_2(t)\left(\dot{e}_2(t) - \beta_{01}\dot{e}_1(t) + k_1\frac{dg_1(t)s}{de_1(t)}\dot{e}_1(t) - g_1\frac{dg_1(t)}{de_1(t)}\dot{e}_1(t)\right) \\ &\quad + k_1\frac{dg_1(t)s}{de_1(t)}(e_2(t) - \beta_{01}e_1(t)) - g_1(t)\frac{dg_1(t)}{de_1(t)}(e_2(t) - \beta_{01}e_1(t)) \\ &= h_2(t)\Bigg(h_2(t) - k_2g_2(t)\text{sat}(h_2(t)/g_1(t)) + k_1\frac{dg_1(t)s}{de_1(t)}(h_2(t) \\ &\quad - k_1g_1(t)s)\Bigg) - g_1(t)\frac{dg_1(t)}{de_1(t)}(h_2(t) - k_1g_1(t)s). \end{aligned}$$

Due to $V_{h_3,g_2}(t) < 0$ and $V_{h_2,g_1}(t) \geq 0$, we get $|h_3(e_1(t), e_2(t), e_3(t))| < g_2(e_1(t), e_2(t))$ and $h_2(t)(h_3(t) - k_2g_2(t)\text{sat}(h_2(t)/g_1(t))) \leq -|h_2(t)|(k_2-1)g_2(t)$. Then the following inequality

$$\begin{aligned}
\dot{V}_{h_2,g_1}(t) &\leq -|h_2(t)|(k_2-1)g_2(t) + h_2(t)k_1\frac{dg_1(t)s}{de_1(t)}(h_2(t)-k_1g_1(t)s) \\
&\quad - g_1\frac{dg_1(t)}{de_1(t)}(h_2(t)-k_1g_1(t)s) \\
&\leq -|h_2(t)|(k_2-1)g_2(t) + \left|h_2(t)k_1\frac{dg_1(t)s}{de_1(t)}(h_2(t)-k_1g_1(t)s)\right| \\
&\quad - g_1\frac{dg_1(t)}{de_1(t)}(h_2(t)-k_1g_1(t)s) \\
&\leq -|h_2(t)|(k_2-1)g_2(t) + \left|\frac{dg_1(t)}{de_1(t)}\right| |h_2(t)-k_1g_1(t)s|(k_1|h_2(t)| \\
&\quad + g_1(t)) \\
&\leq -|h_2(t)|(k_2-1)g_2(t) + h_2^2(t)(k_1+1)^2\left|\frac{dg_1(t)}{de_1(t)}\right| \qquad (4.8)
\end{aligned}$$

holds. If inequality (4.8) is established, then it will be known that $\dot{V}_{h_2,g_1}(t) < 0$ holds. In addition, by $V_{h_3,g_2}(t) < 0$ and $V_{h_2,g_1}(t) \geq 0$, there exist the following two cases.

Case 1:
If a trajectory $(e_1(t), e_2(t))$ arrives at a point $(\tilde{e}_1(t), \tilde{e}_2(t))$ that belongs to the boundary of Π_2 in an infinite time, then the point $(\tilde{e}_1(t), \tilde{e}_2(t))$ is the limit point of this trajectory. Moreover, it is easily known that $\dot{V}_{h_2,g_1}(t)(\tilde{e}_1(t), \tilde{e}_2(t)) = 0$ in the limit point. It is known that $\dot{V}_{h_2,g_1}(t) = 0$ holds in the origin point. That is, we have $(\tilde{e}_1(t), \tilde{e}_2(t)) = (0, 0)$. Therefore, all the trajectories will converge to origin directly.

Case 2:
By the definition of region Π_2, we get the following inequality

$$-k_1g_1(t)s - g_1(t) < e_2(t) - \beta_{01}e_1(t) < -k_1g_1(t)s + g_1(t).$$

Defining $V_1(t) = e_1^2(t)/2$, for $\forall e_1(t) \neq 0$, it is obtained that

$$\dot{V}_1(t) = e_1(t)\dot{e}_1(t) = e_1(t)(e_2(t) - \beta_{01}) < -(k_1-1)g_1(t)|e_1(t)| < 0.$$

All the analysis shows that $e_1(t) \to 0$ and $e_2(t) \to 0$. Hence, all the trajectories in Π_3 that do not converge to origin directly will enter into Π_2, and converge to the origin.

In the following, we take a consideration on those trajectories that out of Π_3.

Theorem 4.3 *If the following inequalities is satisfied with*

$$
\begin{aligned}
\beta_{01} > k_2k_3 - b_v' &+ \left| -b_v' \left(\frac{\beta_{01}}{k_3} - k_2 \right) - \frac{\beta_{02}}{k_3} f_e' + \beta_{01}k_2 - k_2^2 k_3 \right.\\
&\left. + \frac{k_1}{k_3} (g_1(t)s)' \right| + \left| b_v' \beta_{01} \frac{k_1}{k_3} + \beta_{02} f_e' \frac{k_1}{k_3} - k_1^2 k_2 (g_1(t)s)' \right| \\
&+ \frac{1}{|h_3(t)|}(M + \beta_{03}|f_o|) + \left| \frac{\partial g_2(t)}{\partial e_1(t)} + \frac{\partial g_2(t)}{\partial e_2(t)} \left(\frac{\beta_{01}}{k_3^2} - \frac{k_2}{k_3} \right) \right| \\
&+ \frac{k_1}{k_3^2} \left| \frac{\partial g_2(t)}{\partial e_1(t)} + \beta_{01} \frac{\partial g_2(t)}{\partial e_2(t)} \right| + \frac{1}{k_3} \left| \frac{\partial g_2(t)}{\partial e_2(t)} \right|
\end{aligned}
$$

then the trajectory that does not belong to the region Π_3 will be absorbed into Π_3, and bound in Π_3.

Proof Supposing $|w(t)| < M$, we make the following definition

$$
g_2(t) = \begin{cases} k_3|h_2(t)|, & |h_2(t)| \geq g_1(t) \\ k_3|g_1(t)|, & |h_2(t)| < g_1(t), \end{cases} \tag{4.9}
$$

where k_3 is a constant. By (4.7), it is known that

$$
k_3 > \frac{(k_1+1)^2}{k_2 - 1} \left| \frac{dg_1(t)}{de_1(t)} \right| .
$$

Moreover, we have

$$
\begin{aligned}
\dot{V}_{h_3,g_2}(t) &= h_3(t)\dot{h}_3(t) - g_2(t)\dot{g}_2(t) \\
&= h_3(t) \left(\dot{e}_3(t) - b_v' \dot{e}_2(t) - \beta_{02} f_e' \dot{e}_1(t) - \beta_{01}(\dot{e}_2(t) - \beta_{01}\dot{e}_1(t)) \right) \\
&\quad + h_3(t)k_2k_3 \left(\dot{e}_2(t) - \beta_{01}\dot{e}_1(t) + k_1(g_1(t)s)'\dot{e}_1(t) \right) \\
&\quad - g_2 \left(\frac{\partial g_2(t)}{\partial e_1(t)} \dot{e}_1(t) + \frac{\partial g_2(t)}{\partial e_2(t)} \dot{e}_2(t) \right) \\
&= h_3(t) \left(-w(t) - \beta_{03} f_o - b_v'(h_3(t) + (\beta_{01} - k_2k_3)h_2(t)) \right. \\
&\quad \left. -\beta_{01}k_1g_1(t)s \right) + h_3(t) \left(-\beta_{02} f_e'(h_2(t) - k_1g_1(t)s) - \beta_{01}h_3(t) \right. \\
&\quad \left. -k_2k_3h_2 \right) + h_3(t)k_2k_3 \left(h_3(t) + (\beta_{01} - k_2k_3)h_2(t) \right) \\
&\quad - \beta_{01}k_1g_1(t)s - h_3(t)k_2k_3(\beta_{01} - k_1(g_1(t)s)')(h_2(t) \\
&\quad - k_1g_1(t)s) - g_2(t)\frac{\partial g_2(t)}{\partial e_2(t)}(h_3(t) + \beta_{01}(h_2(t) - k_1g_1(t)s) \\
&\quad - k_2g_2(t)s) - g_2(t)\frac{\partial g_2(t)}{\partial e_1(t)}(h_2(t) - k_1g_1(t)s).
\end{aligned}
$$

Form (4.9), it is known that $g_2(t) < |h_3(t)|$, $g_2(t) = k_3h_2(t)$ and $k_3g_1(t) < g_2(t)$. The following inequality on $\dot{V}_{h_3,g_2}(t)$ is given as

$$\begin{aligned}\dot{V}_{h_3,g_2}(t) \le\; & h_3^2(t)(-b_v' - \beta_{01} + k_2k_3) + |h_3(t)h_2(t)|\,\big|-b_v'(\beta_{01} - k_2k_3) \\ & -\beta_{02}f_e' + \beta_{01}k_2k_3 - k_2^2k_3^2 + k_1(g_1(t)s)'\big| \\ & + |h_3(t)g_1(t)|\,\big|b_v'\beta_{01}k_1 + \beta_{02}f_e'k_1 - k_1^2k_2k_3(g_1(t)s)'\big| \\ & + |h_3(t)|(M + \beta_{03}|f_o|) + |h_3(t)g_2(t)|\left|\frac{\partial g_2(t)}{\partial e_2(t)}\right| \\ & + |g_2(t)h_2(t)|\left|\frac{\partial g_2(t)}{\partial e_1(t)}\right| + \frac{\partial g_2(t)}{\partial e_2(t)}(\beta_{01} - k_2k_3) \\ & + k_1|g_1(t)g_2(t)|\left|\frac{\partial g_2(t)}{\partial e_1(t)} + \beta_{01}k_1\frac{\partial g_2(t)}{\partial e_2(t)}\right|.\end{aligned}$$

Then, we have

$$\begin{aligned}\dot{V}_{h_3,g_2}(t) \le\; & h_3^2(t)(-b_v' - \beta_{01} + k_2k_3) + h_3^2(t)\Bigg|-b_v'\left(\frac{\beta_{01}}{k_3} - k_2\right) \\ & -\frac{\beta_{02}}{k_3}f_e' + \beta_{01}k_2 - k_2^2k_3 + \frac{k_1}{k_3}(g_1(t)s)'\Bigg| \\ & + h_3^2(t)\left|b_v'\beta_{01}\frac{k_1}{k_3} + \beta_{02}f_e'\frac{k_1}{k_3} - k_1^2k_2(g_1(t)s)'\right| + |h_3(t)|(M) \\ & + \beta_{03}|f_o| + h_3^2(t)\left|\frac{\partial g_2(t)}{\partial e_1(t)} + \frac{\partial g_2(t)}{\partial e_2(t)}\left(\frac{\beta_{01}}{k_3^2} - \frac{k_2}{k_3}\right)\right| \\ & + \frac{k_1}{k_3^2}h_3^2(t)\left|\frac{\partial g_2(t)}{\partial e_1(t)}\right| + \beta_{01}\frac{\partial g_2(t)}{\partial e_2(t)} + \frac{h_3^2}{k_3}\left|\frac{\partial g_2(t)}{\partial e_2(t)}\right|.\end{aligned}$$

Hence, β_{01}, β_{02} and β_{03} are satisfied with the inequality in Theorem 4.2, there exists $\dot{V}_{h_3,g_2}(t) < 0$.

If a trajectory $(e_1(t), e_2(t), e_3(t))$ arrives at a point $(\tilde{e}_1(t), \tilde{e}_2(t), \tilde{e}_3(t))$ which belongs to the boundary of Π_3 in an infinite time, then the point $(\tilde{e}_1(t), \tilde{e}_2(t), \tilde{e}_3(t))$ is the limit point of this trajectory. It is shown that $\dot{V}_{h_3,g_2}(\tilde{e}_1(t), \tilde{e}_2(t), \tilde{e}_3(t)) = 0$ in the limit point. It is known that $\dot{V}_{h_3,g_2}(t) = 0$ holds in the origin point. That is, we have $(\tilde{e}_1(t), \tilde{e}_2(t), \tilde{e}_3(t)) = (0, 0, 0)$. Therefore, all the trajectories out of Π_3 that not converge to origin directly would arrive at the boundary of the region Π_3, and enter the region Π_3.

4.2.3 Linear Error Feedback Controller

The combination of linear error feedback and compensation of extended state guarantee the desired system response. The errors between the TD (4.1) and the ESO (4.3) are written as follows:

$$\begin{cases} \gamma_1(t) = v_1(t) - z_1(t) \\ \gamma_2(t) = v_2(t) - z_2(t). \end{cases} \tag{4.10}$$

The errors between the given signal and output signal are written as

$$\begin{cases} s_1(t) = v_1(t) - x_1(t) \\ s_2(t) = v_2(t) - x_2(t). \end{cases} \tag{4.11}$$

Furthermore, we get the following expression:

$$\begin{cases} \gamma_1(t) = s_1(t) - e_1(t) \\ \gamma_2(t) = s_2(t) - e_2(t). \end{cases} \tag{4.12}$$

The linear error feedback controller is designed in the following

$$\begin{cases} u(t) = \beta_1\gamma_1(t) + \beta_2\gamma_2(t) \\ \Delta u(t) = (u(t) + b'_v z_2(t) - z_3(t))/b_0, \end{cases} \tag{4.13}$$

where β_1 and β_2 are two gains parameters.

Considering (4.11), (4.12), and (4.13), the error system is obtained as

$$\begin{cases} \dot{s}_1(t) = s_2(t) \\ \dot{s}_2(t) = v_3(t) - b'_v e_2(t) + e_3(t) - \beta_1(s_1(t) - e_1(t)) \\ \qquad\qquad -\beta_2(s_2(t) - e_2(t)), \end{cases} \tag{4.14}$$

where $v_3(t)$ is the derivative of $v_2(t)$, and it is bounded.

Theorem 4.4 *Consider the error system (4.14), the ESO (4.3) and the error feedback controller (4.13). The error system (4.14) is globally asymptotically uniformly bounded by choosing appropriate gains β_1 and β_2 in the linear error feedback controller (4.13).*

Proof The Lyapunov function is given as follows:

$$V(t) = \frac{1}{2}\big[2(s_1(t) - e_1(t)) + (s_2(t) - e_2(t))\big]^2 + 2(s_1(t) - e_1(t))^2.$$

Then the differential of $V(t)$ is written as

$$\begin{aligned} \dot{V}(t) &= \big[2(s_1(t) - e_1(t)) + (s_2(t) - e_2(t))\big]\big[2(s_2(t) - e_2(t) + \beta_{01}e_1(t)) \\ &\quad + \big(v_3(t) + \beta_{02} f_e(t) - \beta_1(s_1(t) - e_1(t)) - \beta_2(s_2(t) - e_2(t))\big)\big] \\ &\quad + 4(s_1(t) - e_1(t))(s_2(t) - e_2(t) + \beta_{01}e_1(t)) \\ &= -\big[2(s_1(t) - e_1(t)) + (s_2(t) - e_2(t))\big]\big[\beta_1(s_1(t) - e_1(t)) + \beta_2(s_2(t) \end{aligned}$$

$$
\begin{aligned}
&- e_2(t))\big] + 2\big[2(s_1(t) - e_1(t)) + (s_2(t) - e_2(t))\big](s_2(t) - e_2(t)) \\
&+ \big[2(s_1(t) - e_1(t)) + (s_2(t) - e_2(t))\big](2\beta_{01}e_1(t) + v_3(t) + \beta_{02}f_e(t)) \\
&+ 4(s_1(t) - e_1(t))(s_2(t) - e_2(t)) + 4\beta_{01}e_1(t)(s_1(t) - e_1(t)).
\end{aligned}
$$

Letting $\beta_1 = 2\beta$ and $\beta_2 = \beta$, and the following equation is gotten:

$$
\begin{aligned}
\dot{V}(t) = &-\beta\big[2(s_1(t) - e_1(t)) + (s_2(t) - e_2(t))\big]^2 + 8(s_1(t) - e_1(t))(s_2(t) \\
&- e_2(t)) + 2(s_2(t) - e_2(t))^2 + (s_1(t) - e_1(t))(8\beta_{01}e_1(t) + 2v_3(t) \\
&+ 2\beta_{02}f_e(t)) + (s_2(t) - e_2(t))(2\beta_{01}e_1(t) + v_3(t) + \beta_{02}f_e) \\
= &\,(2-\beta)\big[2(s_1(t) - e_1(t)) + (s_2(t) - e_2(t))\big]^2 - 8(s_1(t) - e_1(t))^2 \\
&+ (s_1(t) - e_1(t))(8\beta_{01}e_1(t) + 2v_3(t) + 2\beta_{02}f_e(t)) \\
&+ (s_2(t) - e_2(t))(2\beta_{01}e_1(t) + v_3(t) + \beta_{02}f_e(t)).
\end{aligned}
$$

Let

$$
\begin{aligned}
H = &\,|(s_1(t) - e_1(t))(8\beta_{01}e_1(t) + 2v_3(t) + 2\beta_{02}f_e(t)) \\
&+ (s_2(t) - e_2(t))(2\beta_{01}e_1(t) + v_3(t) + \beta_{02}f_e(t))|.
\end{aligned}
$$

Based on the convergence analysis of the ESO (4.3) and actual physical meaning, it is easy to know that the H is bound. Then an inequality about the differential of V is given as follows:

$$
\dot{V}(t) < (2-\beta)\big[2(s_1(t) - e_1(t)) + (s_2(t) - e_2(t))\big]^2 - 8(s_1(t) - e_1(t))^2 + H.
$$

That is, there exists $\dot{V}(t) < 0$ by adjusting the value of gains β. Hence, the error system (4.14) is globally asymptotically uniformly bounded.

Remark 4.5 Not only convergence analysis of ESO (4.3) but also stability analysis of the error system have been investigated in this chapter. Thereby, the linear error feedback controller (4.13) makes parameters adjustment much easier than regular ADRC. Furthermore, both simulation and experiment results are given to demonstrate the effectiveness of the developed ADRC approach for pneumatic actuators in the following.

4.3 Experiments and Results

In this chapter, some compared experiment results of the pneumatic manipulator system controlled by ADRC and PID approach are given. The experiment results of a step signal at 10° are shown in Fig. 4.2.

In Fig. 4.2a, $v(t)$ is the given step signal at 10°. $ax_1(t)$ and $px_1(t)$ are two system outputs via using the ADRC and the PID approach, respectively. $z_1(t)$ is the observation of $ax_1(t)$. From Fig. 4.2a, we can see that $px_1(t)$ has a faster response in the

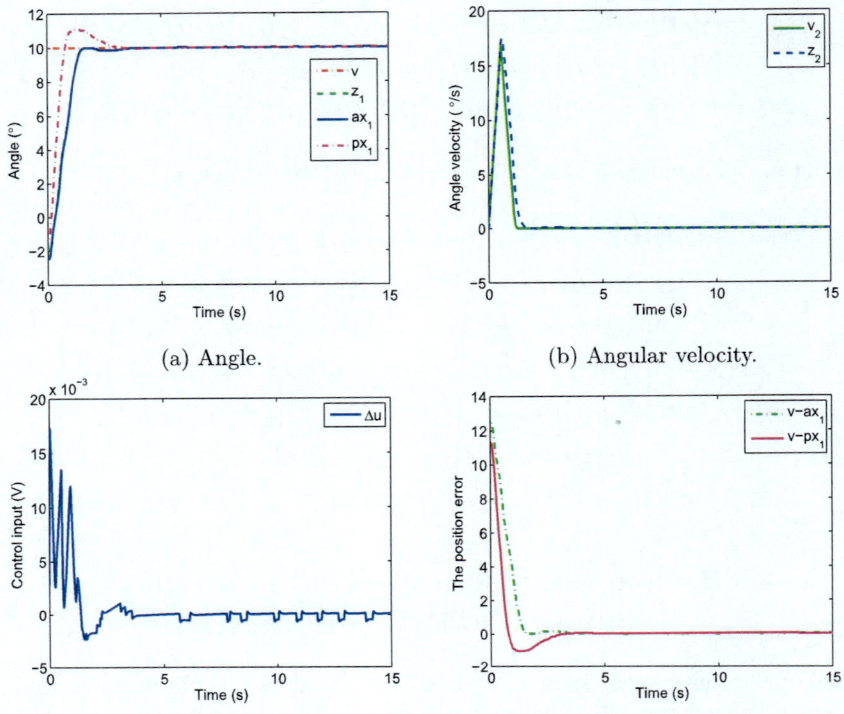

(a) Angle.

(b) Angular velocity.

(c) Control input.

(d) The position error.

Fig. 4.2 Compared experiment results for tracking a step signal at 10° based on the active disturbance rejection control and PID approach

beginning, but a overshoot phenomenon compared to $ax_1(t)$. Moreover, $ax_1(t)$ gets a smoother and faster convergence performance than $px_1(t)$ to track the given signal $v(t)$ in positioning control.

In Fig. 4.2b, $v_2(t)$ tracks the differential signal of the given signal $v(t)$. $z_2(t)$ is the observation of the output $ax_2(t)$ by using the ADRC approach. The coincidence of $v_2(t)$ and $z_2(t)$ proves the good performance in positioning and observing.

Control input $\Delta u(t)$ is the voltage variation in the pneumatic system, and it is shown in Fig. 4.2c. In positioning control, the value of $\Delta u(t)$ should reach a steady state at $0V$.

In Fig. 4.2d, there shows two errors between the given signal $v(t)$ and the two different outputs ($ax_1(t)$ and $px_1(t)$), respectively. From this figure, it can be seen that the ADRC has a smoother and faster convergence performance than PID approach too.

Remark 4.6 In the ADRC, the TD (4.1) is used to arrange the transition process, it can make the transition process more smooth and no overshoot. So the ADRC can overcome the contradiction between the rapidity and overshoot of tracking response, which is the unavoidable faultiness of PID approach. From simulation and experimental results, it is easy to draw the conclusion.

4.4 Conclusion

In this chapter, we have presented the ADRC method for an pneumatic manipulator system. This method is effective to solve the highly nonlinear behavior of the pneumatic manipulator system driven by PAMs. In addition, TD has been designed to get corresponding smooth signal and differential signal of reference input to avoid overshoot. ESO has been introduced to estimate the total disturbance in real time. Moreover, a linear error feedback controller combined with compensation of the total disturbance has been proposed to guarantee the performance of the system. The convergence analysis of ESO has been proven via self-stable region approach. The stability of the close-loop system has also been proved. According to this approach, there has been a better performance for tracking a step signal. Finally, experiment results have verified the effectiveness of the proposed controller than PID approach.

References

1. Ahn KK, Thanh TDC (2005) Nonlinear PID control to improve the control performance of the pneumatic artificial muscle manipulator using neural network. J Mech Sci Technol 19(1):106–115
2. Jahanabadi H, Mailah M, Zain MZM, Hooi HM (2011) Active force with fuzzy logic control of a two-link arm driven by pneumatic artificial muscles. J Bionic Eng 8(4):474–484
3. Nuchkrua T, Leephakpreeda T (2013) Fuzzy self-tuning PID control of hydrogen-driven pneumatic artificial muscle actuator. J Bionic Eng 10(3):329–340
4. Tang Z, Ge SS, Lee KP, He W (2016) Adaptive neural control for an uncertain robotic manipulator with joint space constraints. Int J Control 89(7):1428–1446

Chapter 5
Nonlinear Feedback Control

5.1 Introduction

Since pneumatic systems, commonly powered by compressed air or compressed inert gases, provide a lower cost, more flexible, or safer alternative to electric motors and actuators, this field has drawn lots of attentions [1]. Especially in pneumatic robots, flexibility of dexterous robot hands plays an important role in space exploration, industrial manufacture, and service robot industry. Due to gripping characteristics of low cost, safety operation, cleanliness, and high power–weight ratio, PAMs are popular in robot applications. Nevertheless, it is difficult to get an accurate control in PAMs systems for the existence of highly nonlinear and time-varying dynamics caused by elasticity and compressibility of air. In recent years, many control strategies are developed to deal with the problems on highly nonlinear, time-varying dynamics and compressibility of air. There exist a lot of control strategies to improve the performance of the PAMs system, such as PID methods [2], sliding mode approaches [3], switching algorithm [1], and adaptive robust control strategy [4]. However, most of the aforementioned studies do not deal with the dynamic uncertainties, transition processes and stability analysis directly, which motivates us to study a much more effective scheme.

In this chapter, a reduced-order ESO based on ADRC is proposed to deal with uncertainties for an pneumatic manipulator system. A TD has been designed to get a corresponding smooth signal of a given input signal to avoid overshoot and a differential signal as well. Moreover, a nonlinear error feedback controller has been adopted to guarantee a satisfying performance. Stability analysis of the reduced-order ESO and the closed-loop system has been proven by Lyapunov theory. Finally, experimental results exhibit the efficiency and advantages of the reduced-order ESO with an ADRC method for the one-DoF pneumatic manipulator driven by PAMs. The main contributions of this paper are summarized as follows:

i An application of a TD avoids the overshoot and smoothens transition process for an pneumatic manipulator system.

L. Zhao et al., *Pneumatic Servo Systems Analysis*, Advances in Industrial Control,
https://doi.org/10.1007/978-981-16-9515-5_5

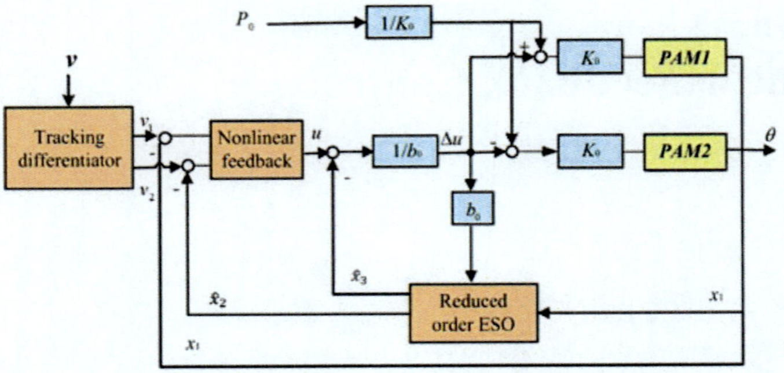

Fig. 5.1 An ADRC system diagram of the pneumatic manipulator system

ii A reduced-order ESO is implemented to estimate and compensate for the total disturbances. The lower the model order is and the fewer the parameters are, the easier stability analysis and parameter tuning are.

iii A nonlinear feedback controller combining with compensation of disturbances is designed to guarantee a satisfying performance of the closed-loop manipulator pneumatic system.

5.2 Main Results

5.2.1 *Schematic Diagram of Control Method*

An ADRC with a reduced-order ESO is designed to deal with the tracking problem for the pneumatic manipulator system introduced in the previous section. The diagram of the designed control system is given in Fig. 5.1.

TD is designed to generate a smooth track of any input signal and its differential signal of reference input to arrange transient process. The TD is usually implemented to pre-process the reference signal of control systems, aiming to avoid overshoot and optimize the system response. The TD is constructed as follows:

$$\begin{cases} f_h(t) = f_{han}(v_1(t) - v(t), v_2(t), r_0, h_0), \\ \dot{v}_1(t) = v_2(t), \\ \dot{v}_2(t) = f_h(t), \end{cases} \tag{5.1}$$

where $v(t)$ is the input reference signal, $v_1(t)$ is the output which tracks $v(t)$, $v_2(t)$ is the differential signal of $v_1(t)$, h is the step length, r_0 is a velocity factor, and h_0 is a filter factor.

5.2.2 Reduced-Order Extended State Observer

The nonlinear dynamics and uncertainties of the PAMs are regarded as a total disturbance. This disturbance is estimated by a reduced-order ESO. Considering equation (3.6), term $f_t(t)$ is regarded as an unknown and extended variable. Defining $f_T(t) = x_3(t)$, the pneumatic manipulator system (3.6) is rewritten as follows:

$$\begin{cases} \dot{x}_1(t) = x_2(t) \\ \dot{x}_2(t) = x_3(t) + b_0 \Delta u(t) \\ \dot{x}_3(t) = w(t) \\ y(t) = x_1(t), \end{cases} \tag{5.2}$$

where $x_3(t)$ is continuous and differentiable. Its derivative $w(t)$ is assumed to be bounded.

In this section, a reduced-order ESO is designed to estimate the total disturbance which is produced by the nonlinear and uncertain dynamics of the system. The reduced-order ESO for the pneumatic manipulator system (5.2) is given as follows:

$$\begin{cases} \dot{z}_2(t) = -\beta_1 z_2(t) + z_3(t) + (\beta_2 - \beta_1^2)x_1(t) + b_0 \Delta u(t), \\ \dot{z}_3(t) = -\beta_2 z_2(t) - \beta_1 \beta_2 x_1(t), \\ \hat{x}_2(t) = z_2(t) + \beta_1 x_1(t), \\ \hat{x}_3(t) = z_3(t) + \beta_2 x_1(t), \end{cases} \tag{5.3}$$

where $\hat{x}_2(t)$ and $\hat{x}_3(t)$ are estimates of $x_2(t)$ and $x_3(t)$, respectively, $z_2(t)$ and $z_3(t)$ are two intermediate variables of the reduced-order ESO (5.3), both β_1 and β_2 are two positive gains of observer (5.3). Since output $x_1(t)$ is directly measured by an angle encoder, there is no need to estimate $x_1(t)$. Compared to the conventional ESO, the structure of the reduced-order ESO (5.3) only contains the estimations of $x_2(t)$ and $x_3(t)$.

Based on the ESO (5.3) and the system description (5.2), define $e_2(t) = \hat{x}_2(t) - x_2(t)$ and $e_3(t) = \hat{x}_3(t) - x_3(t)$, the error system is obtained as follows:

$$\begin{cases} \dot{e}_2(t) = e_3(t) - \beta_1 e_2(t), \\ \dot{e}_3(t) = -w(t) - \beta_2 e_2(t). \end{cases} \tag{5.4}$$

In the following part, the convergence analysis of system (5.4) is given.

Theorem 5.1 *Considering the error system (5.4) and choosing suitable parameters for β_1 and β_2 which are satisfied with*

$$(\beta_1 - \beta_2 + 1)^2 - \beta_1 + 3 < 0 \tag{5.5}$$

and β_1 is a large enough number, then system (5.4) is asymptotically uniformly stable. Meanwhile, the estimation errors $e_2(t)$ and $e_3(t)$ converge to zero, respectively.

Proof Consider the following Lyapunov function as

$$V_e(t) = (e_2(t) - e_3(t))^2 + e_2^2(t). \tag{5.6}$$

It is obvious that (5.6) is positive definite. The derivation of $V_e(t)$ is as follows:

$$\begin{aligned}\dot{V}_e(t) = {} & 2e_2(t)e_3(t) - 2(\beta_1 - \beta_2)e_2^2(t) + 2e_2(t)w(t) - 2e_3^2(t) \\ & + 2(\beta_1 - \beta_2)e_2(t)e_3(t) - 2e_3(t)w(t) + 2e_2(t)e_3(t) - 2\beta_1 e_2^2(t).\end{aligned}$$

Letting $\beta_1 - \beta_2 = \lambda$, and λ is an arbitrary real number. The equation $\dot{V}_e(t)$ is rewritten as

$$\begin{aligned}\dot{V}_e(t) &= -(2\beta_1 + 2\lambda)e_2^2(t) + 2e_2(t)w(t) \\ &\quad - 2e_3^2(t) - 2e_3(t)w(t) + (2\lambda + 4)e_2(t)e_3(t) \\ &= -(\beta_1 + 2\lambda)e_2^2(t) - e_3^2(t) + (2\lambda + 4)e_2(t)e_3(t) \\ &\quad - (\beta_1 - 1)e_2^2(t) + 2w^2(t) - e_3^2(t) - 2e_3(t)w(t) - w^2(t) \\ &\quad - e_2^2(t) + 2e_2(t)w(t) - w^2(t).\end{aligned}$$

Let $\Phi(t) = -(\beta_1 + 2\lambda)e_2^2(t) - e_3^2(t) + (2\lambda + 4)e_2(t)e_3(t)$. When the following inequality holds

$$(2\lambda + 4)^2 < 4(\beta_1 + 2\lambda), \tag{5.7}$$

then $\Phi(t)$ is negative. Rewrite (5.7) as

$$(\lambda + 1)^2 < \beta_1 - 3, \tag{5.8}$$

there always exists a suitable β_1 to satisfy with the inequality (5.8).

Moreover, it is obvious that $-e_3^2(t) - 2e_3(t)w(t) - w^2(t) \leq 0$ and $-e_2^2(t) + 2e_2(t)w(t) - w^2(t) \leq 0$ hold. Let $-e_3^2(t) - e_3(t)w(t) - w^2(t) = -\Gamma_1^2(t)$ and $-e_2^2(t) + e_2(t)w(t) - w^2(t) = -\Gamma_2^2(t)$, then the $\dot{V}_e(t)$ is rewritten as follows:

$$\dot{V}_e(t) = -(\beta_1 - 1)e_2^2(t) + 2w^2(t) + \Phi(t) - \Gamma_1^2(t) - \Gamma_2^2(t).$$

Wrapping up the aforementioned analysis yields a conclusion that $\dot{V}_e(t)$ is negative definite if there exists a large enough β_1, or in other words, system (5.4) is asymptotically uniformly stable if there exists a large enough β_1.

5.2.3 Nonlinear Error Feedback Controller

In order to stabilize the closed-loop pneumatic system, a nonlinear feedback controller with an ability of cancelling the total disturbances is introduced in this section. The nonlinear error feedback controller is designed as follows:

$$\Delta u(t) = \left(k_1 fal(s_1(t), c_1, \delta) + k_2 fal(s_2(t), c_2, \delta) - \hat{x}_3(t)\right)/b_0, \tag{5.9}$$

where k_1 and k_2 are two gains, c_1 and c_2 are tuning parameters satisfying $0 < c_1 < 1 < c_2$, $s_1(t)$ and $s_2(t)$ are defined as the tracking errors of the TD (5.1):

$$\begin{cases} s_1(t) = v_1(t) - x_1(t), \\ s_2(t) = v_2(t) - \hat{x}_2(t). \end{cases} \tag{5.10}$$

Meanwhile, $fal(\cdot)$ is a non-smooth function. It is defined as follows:

$$fal(e(t), c, \delta) = \begin{cases} \frac{e(t)}{\delta^{1-c}}, & |e(t)| \le \delta, \\ |e(t)|^c \text{sign}(e(t)), & |e(t)| > \delta, \end{cases} \tag{5.11}$$

where $\delta \in (0, 1)$ is the length of linear region, c is the feedback exponent.

For this closed-loop system, errors between the given signal and output signal are written as

$$\begin{cases} s_1(t) = v_1(t) - x_1(t), \\ r_2(t) = v_2(t) - x_2(t). \end{cases} \tag{5.12}$$

Thereby, $fal(s_1(t), c_1, \delta)$ and $fal(s_2(t), c_2, \delta)$ are denoted by fal_1 and fal_2 for brevity, respectively. Combining equations (5.10), (5.12), and (5.9) yields an error differential system as follows:

$$\begin{cases} \dot{s}_1(t) = r_2(t), \\ \dot{r}_2(t) = v_3(t) + e_3(t) - (k_1 fal_1 + k_2 fal_2), \end{cases} \tag{5.13}$$

where $v_3(t)$ is the derivative of $v_2(t)$, and it is bounded.

Theorem 5.2 *The closed-loop system (5.13) is asymptotically uniformly stabilized by the reduced-order ESO (5.3) and the nonlinear error feedback controller (5.9) with properly tuned gains k_1 and k_2.*

Proof A Lyapunov function for the closed-loop system (5.13) is given as follows:

$$V(t) = (s_1(t) + r_2(t) - e_2(t))^2 + (s_1(t) - e_2(t))^2. \tag{5.14}$$

Then the derivative of $V(t)$ is as follows:

$$\begin{aligned}\dot{V}(t) &= 2(s_1(t)+r_2(t)-e_2(t))(r_2(t)+v_3(t)-k_1 fal_1 - k_2 fal_2\\ &\quad + \beta_1 e_2(t)) + 2(s_1(t)-e_2(t))(r_2(t)-e_3(t)+\beta_1 e_2(t))\\ &= -2(s_1(t)+s_2(t))(k_1 fal_1 + k_2 fal_2) + 2r_2^2(t) + 2s_1(t)r_2(t)\\ &\quad - 2r_2(t)e_2(t) + 2(s_1(t)+r_2(t)-e_2(t))(v_3(t)+\beta_1 e_2(t))\\ &\quad + 2(s_1(t)-e_2(t))(r_2(t)-e_3(t)+\beta_1 e_2(t)).\end{aligned}$$

Let $k_1 = k_2 = k$ for the sake of simplicity, it is obvious that $(s_1(t) + s_2(t))(k_1 fal_1 + k_2 fal_2) = k(s_1(t) + s_2(t))(fal_1 + fal_2)$. Since function $fal(\cdot)$ is both monotonous and odd, the relation between fal_1 with $s_1(t)$ and fal_2 with $s_2(t)$ is described as

1. If $|s_1(t)| \leq 1$, there exists $|fal_1| \geq |s_1(t)|$; If $|s_1(t)| > 1$, then $|fal_1| < |s_1(t)|$ holds.
2. If $|s_2(t)| \leq 1$, there exists $|fal_2| \leq |s_2(t)|$; If $|s_2(t)| > 1$, then $|fal_2| > |s_2(t)|$ holds.

Moreover, considering the expression $(s_1(t) + s_2(t))(fal_1 + fal_2)$, four solutions are discussed in the following:

1. If $fal_1 \geq s_1(t)$ and $fal_2 \geq s_2(t)$, then $(s_1(t) + s_2(t))(fal_1 + fal_2) \geq (s_1(t) + s_2(t))^2$ holds.
2. If $fal_1 \geq s_1(t)$ and $fal_2 < s_2(t)$, then $(s_1(t) + s_2(t))(fal_1 + fal_2) > (s_1(t) + fal_2)^2$ holds.
3. If $fal_1 < s_1(t)$ and $fal_2 \geq s_2(t)$, then $(s_1(t) + s_2(t))(fal_1 + fal_2) > (fal_1 + s_2(t))^2$ holds.
4. If $fal_1 < s_1(t)$ and $fal_2 < s_2(t)$, then $(s_1(t) + s_2(t))(fal_1 + fal_2) > (fal_1 + fal_2)^2$ holds.

Define $\Upsilon(t)$ as the smallest term among $(s_1(t) + s_2(t))^2$, $(s_1(t) + fal_2)^2$, $(fal_1 + s_2(t))^2$ and $(fal_1 + fal_2)^2$, thus

$$k(s_1(t) + s_2(t))(fal_1 + fal_2) > k\Upsilon(t) \geq 0.$$

One has that

$$\begin{aligned}\dot{V}(t) &< -2k\Upsilon(t) + 2r_2^2(t) + 2s_1(t)r_2 - (t)2r_2(t)e_2(t)\\ &\quad + 2(s_1(t)+r_2(t)-e_2(t))(v_3(t)+\beta_1 e_2(t))\\ &\quad + 2(s_1(t)-e_2(t))(r_2(t)-e_3(t)+\beta_1 e_2(t)).\end{aligned}$$

In order to the guaranteed convergence of the reduced-order ESO (5.3), both $e_2(t)$ and $e_3(t)$ converge to zero. Meanwhile, $v_3(t)$, s_1(t) and $r_2(t)$ are bounded practically. Let

$$\begin{aligned}M(t) &= |2s_1(t)r_2(t) - 2r_2(t)e_2(t) + 2(s_1(t)+r_2(t)-e_2(t))(v_3(t)\\ &\quad + \beta_1 e_2(t)) + 2(s_1(t)-e_2(t))(r_2(t)-e_3(t)+\beta_1 e_2(t))|.\end{aligned}$$

Then $M(t)$ is also bounded. Thereby, the following expression holds with a large enough k:

$$\dot{V}(t) < -2k\Upsilon(t) + 2r_2^2(t) + M(t) < 0.$$

Because (5.14) is positive definite, and its differential is negative definite, closed-loop system (5.13) is asymptotically uniformly stabilized.

5.3 Experiments and Results

An ADRC system combining aforementioned TD (5.1), reduced-order ESO (5.3), and nonlinear error feedback controller (5.9) is written as follows:

$$\begin{cases} v_1(k+1) = v_1(k) + 0.01v_2(k), \\ v_2(k+1) = v_2(k) + 0.01 f_{han}\left(\lambda(k), v_2(k), 20, 0.02\right), \\ z_2(k+1) = z_2(k) + 0.01(-\beta_1 z_2(k) + z_3(k) + (\beta_2 - \beta_1^2)x_1(k) + b_0 \Delta u(k)), \\ z_3(k+1) = z_3(k) + 0.01(-\beta_2 z_2(k) - \beta_1\beta_2 x_1(k)), \\ \hat{x}_2(k+1) = z_2(k+1) + \beta_1 x_1(k), \\ \hat{x}_3(k+1) = z_3(k+1) + \beta_2 x_1(k), \\ \Delta u(k) = (k_1 fal(s_1(k), 0.75, 0.1) + k_2 fal(s_2(k), 1.5, 0.1) - \hat{x}_3(k))/b_0. \end{cases}$$

Based on the nonlinear error feedback controller (5.9), some comparison experimental results between the reduced-order ESO (5.3) and a traditional ESO are shown in Figs. 5.2 and 5.3. Note that more information on the traditional nonlinear ESO have been described in [5].

Figure 5.2 shows comparison results between the reduced-order ESO (5.3) and a traditional ESO for tracking a step signal at 10°. In Fig. 5.2a, v is the given step signal at 10°, v_1 tracks v, x_{1r} and x_{1t} are two system outputs via using the reduced-order ESO (5.3) and the traditional ESO, respectively. As we can see, x_{1r} has a faster response and a higher accuracy than that of x_{1t} to track the smooth signal v_1. In Fig. 5.2b, v_2 is the differential signal of v_1, z_{2r} and z_{2t} are the observations of x_2 by using the reduced-order ESO (5.3) and the traditional ESO, respectively. The control inputs Δu_r and Δu_t of two different ADRC methods are plotted in Fig. 5.2c. When the system output levels off, Δu_r has a smaller stable solution error than Δu_t. In Fig. 5.2d, e_r and e_t denote two position errors of the given signal with the system output by adopting two different ADRC methods. Figure 5.2d also indicates that the reduced-order ESO (5.3) has a faster response and a smaller position error than the traditional ESO.

Figure 5.3 shows comparison results between the reduced-order ESO (5.3) and the traditional ESO for tracking a sinusoidal signal at 5° amplitude, 0.4Hz. In Fig. 5.3a, v is the given sinusoidal signal. x_{1r} and x_{1t} are two system outputs by adopting

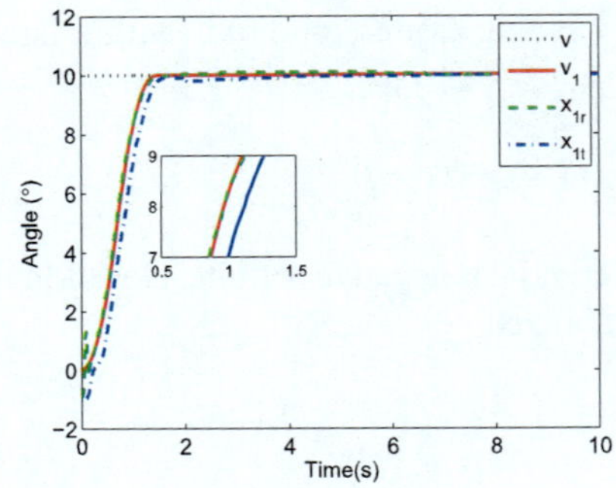

(a) Angle.

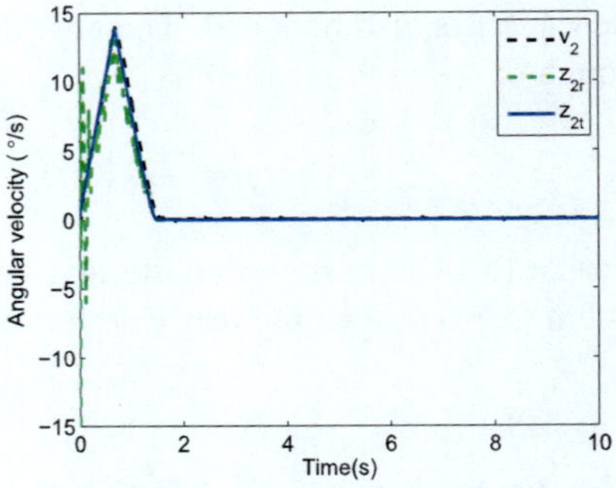

(b) Angle velocity.

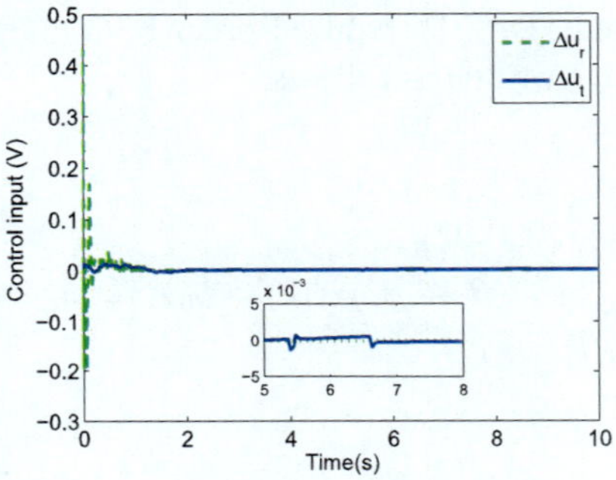

(c) Control input.

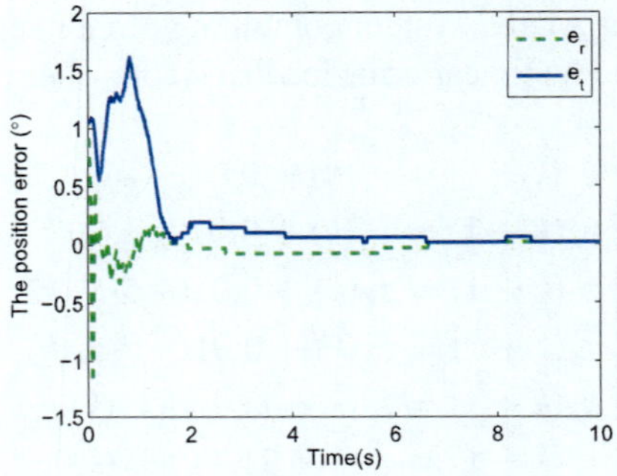

(d) The position error.

Fig. 5.2 Experimental comparison results for tracking a step signal at 10° between reduced-order ESO (5.3) and traditional ESO

the reduced-order ESO (5.3) and the traditional ESO, respectively. For tracking a dynamic response, it is clear to see that x_{1r} gets a faster response and a higher accuracy than x_{1t}. Figure 5.3b shows differential signals v_2, and z_{2r} and z_{2t} are the observations of x_2 by using the reduced-order ESO (5.3) and the traditional ESO, respectively. Figure 5.3c shows two control inputs Δu_r and Δu_t of two different ADRC methods. It is also seen that Δu_r has a smaller vibration than Δu_t. In Fig. 5.3d, it is much clear to observe that the reduced-order ESO (5.3) has a faster response and a smaller tracking error than the traditional ESO.

In conclusion, based on the TD (5.1) and the nonlinear error feedback controller (5.9), it is easily known that the reduced-order ESO (5.3) has an advantage on response speed and tracking steady error compared with the traditional ESO. That is, the reduced-order ESO (5.3) has a better performance than the traditional ESO based on ADRC for the pneumatic manipulator system (5.2).

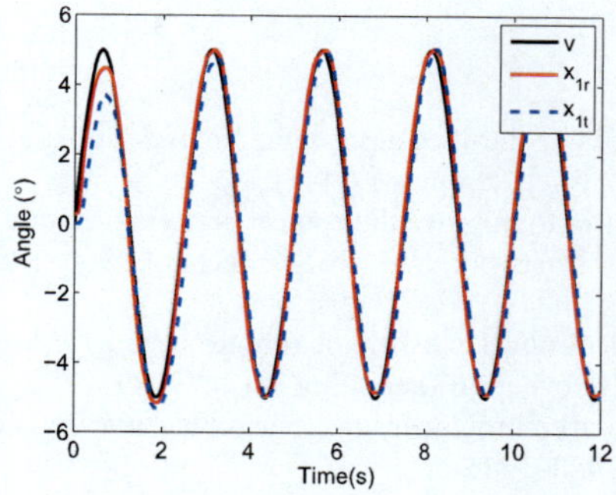

(a) Angle.

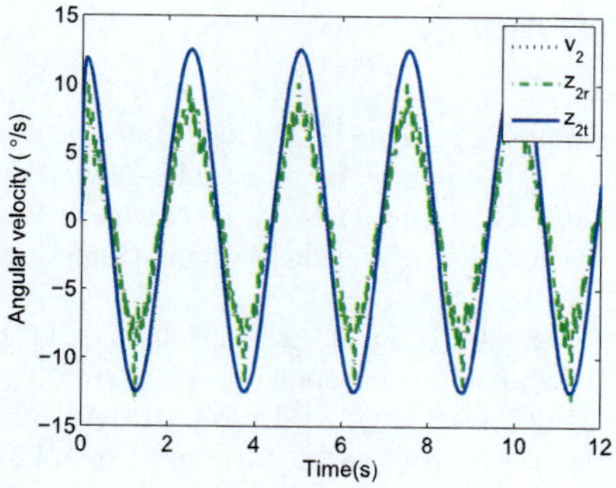

(b) Angle velocity.

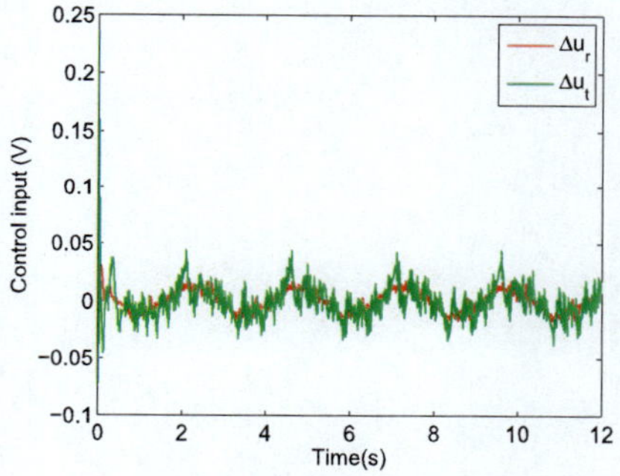

(c) Control input.

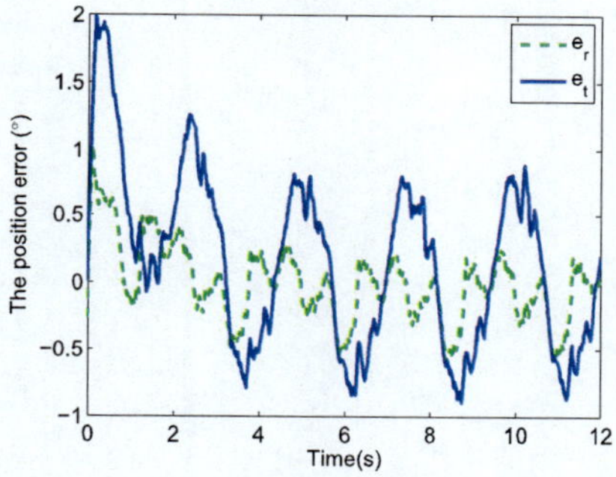

(d) The position error.

Fig. 5.3 Experimental comparison results for tracking a sinusoidal signal at 5° amplitude, 0.4Hz between reduced-order ESO (5.3) and traditional ESO

5.4 Conclusion

In this chapter, a reduced-order ESO based on ADRC has been proposed to deal with uncertainties in tracking control for an one-DoF manipulator pneumatic system. A TD has been designed to get a corresponding smooth signal of a given input signal to avoid overshoot and a differential signal as well. Moreover, a nonlinear error feedback controller has been adopted to guarantee a satisfying performance. Stability analysis of the reduced-order ESO and the pneumatic manipulator system have been proven by Lyapunov theory. Finally, experimental results have been exhibited the efficiency and advantages of the reduced-order ESO with an ADRC method for the one-DoF pneumatic manipulator system driven by PAMs.

References

1. Ahn KK, Nguyen HTC (2007) Intelligent switching control of a pneumatic muscle robot arm using learning vector quantization neural network. Mechatronics 17(4–5):255–262
2. Ahn KK, Thanh TDC (2005) Nonlinear PID control to improve the control performance of the pneumatic artificial muscle manipulator using neural network. J. Mech. Sci. Technol. 19(1):106–115
3. Tsagarakis N, Iqbal J, Khan H, Caldwell D (2014) A novel exoskeleton robotic system for hand rehabilitation-conceptualization to prototyping. Biocybern Biomed Eng 34(2):79–89
4. Tang Z, Ge SS, Lee KP, He W (2016) Adaptive neural control for an uncertain robotic manipulator with joint space constraints. Int J Control 89(7):1428–1446
5. Han J (2009) From PID to active disturbance rejection control. IEEE Trans Ind Electron 56(3):900–906

Chapter 6
Sliding Mode Control

6.1 Introduction

PAMs are widely used in manipulator systems due to their good characteristics of low cost, safety operation, cleanliness, and high power–weight ratio [1]. However, there exist highly nonlinear and time-varying dynamics in PAMs systems, which is caused by elasticity and compressibility of air. This feature deteriorates the accuracy of tracking control for PAMs systems. During the recent decades, a lot of approaches have been proposed to improve the tracking performance, such as PID methods, adaptive control laws, sliding mode techniques, neural networks control, etc. In [2], a fuzzy self-tuning PID control is adopted to get an accurate performance for PAM actuator. A nonlinear PID controller is designed by using neural networks for PAMs manipulators considering nonlinearity uncertainties and disturbances in [3]. In [4], an adaptive robust posture control is adopted to deal with the uncertainties in robotic manipulator systems. In [5], an adaptive robust control strategy is developed for the trajectory tracking of robotic manipulators. Beside that, in [6], a terminal sliding mode controller is designed to achieve the robust control of PAMs robot manipulators. In [7], a radial basis function neural network control algorithm is adopted to deal with the large uncertainty of robotic manipulators system.

In this chapter, trajectory tracking control is considered for a pneumatic manipulator system driven by PAMs. A sliding mode feedback controller is accordingly designed. In the proposed scheme, a novel extended state observer based on a generalized super-twisting algorithm is first designed to deal with internal uncertainties and external disturbances in the pneumatic manipulator system. Secondly, finite-time convergence of the closed-loop system with the proposed sliding mode technique is analyzed. Lastly, some experimental results are presented to show the validity and advantages of the proposed approach. The main contributions of this work are summarized as follows:

L. Zhao et al., *Pneumatic Servo Systems Analysis*, Advances in Industrial Control,
https://doi.org/10.1007/978-981-16-9515-5_6

i The traditional ESO and generalized super-twisting algorithm are combined to observe and estimate the total disturbances. The finite-time convergence and robustness of the ESO is proven via the super-twisting algorithm.
ii A novel switching sliding surface with the observation of the total disturbances is proposed to stabilize the closed-loop system. High approaching speed and robustness of the closed-loop system are guaranteed.
iii Experimental results of tracking exhibit the validity of the proposed approach. Meanwhile, experiments with same setup are implemented by applying some other existing methods, and their results are compared with that of the proposed method. Those results display advantages of the proposed approach.

6.2 Main Results

6.2.1 *Schematic Diagram of Control Method*

A schematic diagram of the finite-time trajectory tracking control for the pneumatic manipulator system (3.6) is shown in Fig. 6.1.

In order to tracking abruptly changed signals, such as the step signals and square wave signals, usually results in large overshoots. To avoid these undesirable dynamics in control, a nonlinear TD is introduced to arrange transient process by tracking smoothed input signal and its derivative. In this chapter, the TD is constructed as follows:

$$\begin{cases} f_h(t) = f_{han}(v_1(t) - v(t), v_2(t), r_0, h_0) \\ \dot{v}_1(t) = v_2(t) \\ \dot{v}_2(t) = f_h(t), \end{cases} \tag{6.1}$$

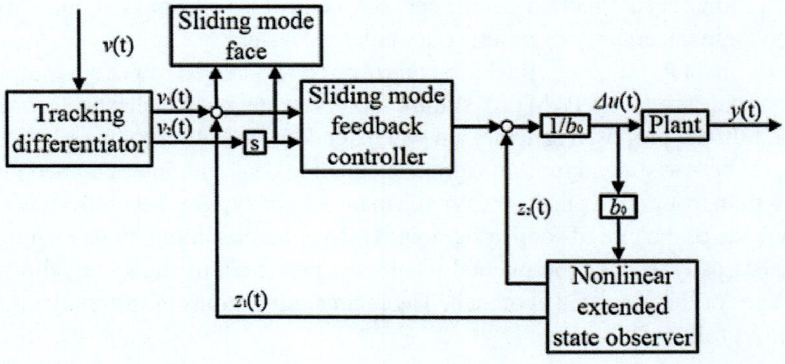

Fig. 6.1 Schematic diagram of the finite-time trajectory tracking control

where $v(t)$ is the input signal, $v_1(t)$ is output which tracks $v(t)$, $v_2(t)$ is the differential signal of $v_1(t)$, h is the step length, r_0 is a velocity factor, h_0 is a filter factor.

6.2.2 Nonlinear Extended State Observer

A stable switching sliding surface is designed in this section. First of all, the tracking error is defined as follows:

$$e_x(t) = x_1(t) - v_1(t). \tag{6.2}$$

Then, in regard to the second-order dynamic (3.6), the switching sliding surface is designed as

$$s(t) = \dot{e}_x(t) + \sigma(e_x(t)), \tag{6.3}$$

where $\sigma(e_x(t))$ is designed as a switching nonlinear function as follows:

$$\sigma(e_x(t)) = \begin{cases} q_1|e_x(t)|^{\frac{3}{2}}\text{sign}(e_x(t)) + q_2\text{sign}(e_x(t)), & |e_x(t)| > 1 \\ k_1|e_x(t)|\text{sign}(e_x(t)) + k_2\text{sign}(e_x(t)), & |e_x(t)| \leq 1, \end{cases} \tag{6.4}$$

where q_1, q_2, k_1, and k_2 are positive constants.

Moreover, $\text{sign}(e_x(t))$ is written as follows:

$$\text{sign}(e_x(t)) = \begin{cases} 1, & e_x(t) \to 0^+ \\ c, & e_x(t) = 0 \\ -1, & e_x \to 0^-, \end{cases}$$

where c is a constant, and $c \in [-1, 1]$. So the derivative of $\sigma(e_x(t))$, denoted by $g(e_x(t))$, is given as

$$g(e_x(t)) = \begin{cases} \frac{3}{2}q_1|e_x(t)|^{\frac{1}{2}}, & |e_x(t)| > 1 \\ k_1, & |e_x(t)| \leq 1. \end{cases} \tag{6.5}$$

Remark 6.1 The intrinsic feature of the proposed switching sliding surface does not affect the balance between convergence speed and stability. When $e_x(t) > 1$, the trajectory of $e_x(t)$ would have a high convergence speed which guarantees a short convergence time. Moreover, when $e_x(t) \leq 1$, the convergence speed of $e_x(t)$ would slow down and the chattering phenomenon would reduce. In addition, singular problem does not exist.

In this paper, the internal uncertainties and external disturbances of the manipulator driven by PAMs system are considered as a total disturbance to be estimated by another ESO. Then, amalgamating (3.6), (6.2), and (6.3) yields the following system:

$$\dot{s}(t) = g(e_x(t)) - v_3(t) + f_t(t) + b_0\Delta u(t), \tag{6.6}$$

where $v_3(t)$ is the derivative of $v_2(t)$ in TD (6.1), and $f_t(t)$ is the total disturbances that is estimated by the ESO.

Moreover, define system states $x_{s1}(t) = s(t)$ and $x_{s2}(t) = f_t(t)$, the system (6.6) is rewritten as

$$\begin{cases} \dot{x}_{s1}(t) = g(e_x(t)) - v_3(t) + x_{s2(t)} + b_0\Delta u(t) \\ \dot{x}_{s2}(t) = w(t). \end{cases} \tag{6.7}$$

It is assumed that $x_{s2}(t)$ is continuous and differentiable, and $w(t)$ is bounded with $|w(t)| \leq w_M$.

Let $e_1(t) = z_1(t) - x_{s1}(t)$ and $e_2(t) = z_2(t) - x_{s2}(t)$. So an ESO for (6.7) is designed as follows:

$$\begin{cases} \dot{z}_1(t) = g(e_x(t)) - v_3(t) + z_2(t) + b_0\Delta u(t) - \beta_1\phi_1(e_1(t)) \\ \dot{z}_2(t) = -\beta_2\phi_2(e_1(t)), \end{cases} \tag{6.8}$$

where $\phi_1(e_1(t))$ and $\phi_2(e_1(t))$ are nonlinear functions which are designed according to the generalized super-twisting algorithm [8] as

$$\phi_1(e_1(t)) = |e_1(t)|^{\frac{1}{2}}\text{sign}(e_1(t)) + e_1(t)$$
$$\phi_2(e_1(t)) = \frac{1}{2}\text{sign}(e_1(t)) + \frac{3}{2}|e_1(t)|^{\frac{1}{2}}\text{sign}(e_1(t)) + e_1(t).$$

Substituting (6.8) into (6.7) results in a error system:

$$\begin{cases} \dot{e}_1(t) = e_2(t) - \beta_1\phi_1(e_1(t)) \\ \dot{e}_2(t) = -\beta_2\phi_2(e_1(t)) - w(t). \end{cases} \tag{6.9}$$

Then we present a proof of stability and finite-time convergence of the proposed ESO (6.8) based on the error system (6.9).

Lemma 6.2 *[9] Consider a dynamics model $\dot{x}(t) = f(x(t))$, $f(0) = 0$ and $x \in R^n$. If there exist a positive definite function $V(x(t))$, and its derivative satisfies*

$$\dot{V}(x(t)) \leq -\gamma V^{\delta}(x(t)) - \nu V(x(t)), \tag{6.10}$$

where $\gamma > 0$, $\nu > 0$, $0 < \delta < 1$, the dynamics model is finite-time stable. Moreover, the convergence time is given by

$$T \leq \frac{1}{\nu(1-\delta)} ln \frac{\nu V^{1-\delta}(x_0(t)) + \gamma}{\gamma}. \tag{6.11}$$

In particular, when $\gamma > 0$, $\nu = 0$ *and* $0 < \delta < 1$, *then the dynamics model is also finite-time stable, and its convergence time is*

$$T \leq \frac{1}{\delta\gamma} V^{\delta}(x_0(t)). \tag{6.12}$$

According to Lemma 1, it is evident to obtain the dynamics model stable and its convergence time, which would be used in the following.

In the next stability analysis of error system (6.9), a Lyapunov function is given as

$$V_0(e(t)) = \zeta^T(t) P \zeta(t), \tag{6.13}$$

where $\zeta^T(t) = [\phi_1(e_1(t)), e_2(t)]$, and the derivative of $\zeta(t)$ is written as

$$\begin{aligned} \dot{\zeta}(t) &= \phi_1'(e_1(t)) \begin{bmatrix} -\beta_1\phi_1(e_1(t)) + e_2(t) \\ -\beta_2\phi_1(e_1(t)) + \frac{w(t)}{\phi_1'(e_1(t))} \end{bmatrix} \\ &= \phi_1'(e_1(t))(A\zeta(t) + B\rho(t)), \end{aligned}$$

where $\phi_1'(e_1(t)) = (\frac{1}{2|e_1(t)|^{\frac{1}{2}}} + 1)$, $\rho(t) = \frac{w(t)}{\phi_1'(e_1(t))}$. In addition, matrices A and B are given as

$$A = \begin{bmatrix} -\beta_1 & 1 \\ -\beta_2 & 0 \end{bmatrix}, \quad B = \begin{bmatrix} 0 \\ 1 \end{bmatrix}$$

Remark 6.3 Note that $V_0(e(t))$ is continuously differentiable everywhere except at the line $\{e_1(t) = 0\}$. For a positive definite P, $V_0(e(t))$ is a positive definite and radially unbounded function in R^2. In addition, the trajectories of error system (6.9) cannot stay on the line $\{e_1(t) = 0\}$ before reaching to the origin, so $\dot{V}_0(e(t))$ can be calculated in usual ways. Before the origin is reached, when the trajectory intersects the line $\{e_1(t) = 0\}$, the set of time instants is a zero measure. If the origin is reached at sometime T_t, then the trajectory would stay there.

An inequality of $\rho(t)$ and $\zeta(t)$ is given in the following, and it is helpful for the convergence analysis of the system (6.9) as

$$\rho^2(t) \leq \frac{w^2(t)\phi_1^2(e_1(t))}{\phi_1'^2(e_1(t))\phi_1^2(e_1(t))} \leq 4w_M^2\phi_1^2(e_1(t)) \tag{6.14}$$

That is

$$\varpi(\rho, \zeta) = \begin{bmatrix} \zeta(t) \\ \rho(t) \end{bmatrix}^T \begin{bmatrix} R & 0 \\ 0 & -1 \end{bmatrix} \begin{bmatrix} \zeta(t) \\ \rho(t) \end{bmatrix} \geq 0, \tag{6.15}$$

where $R = \begin{bmatrix} 4w_M^2 & 0 \\ 0 & 0 \end{bmatrix}$.

Theorem 6.4 *For a symmetric and positive definite matrix* $P = P^T > 0$ *and positive constant* $\epsilon > 0$, *if the following matrix inequality satisfies*

$$\begin{bmatrix} A^T P + PA + \epsilon P + R & PB \\ B^T P & -1 \end{bmatrix} \leq 0 \tag{6.16}$$

then all trajectories of error system (6.9) would converge to the origin in finite time. Let the quadratic Lyapunov function $V_0(e(t)) = \zeta^T(t)P\zeta(t) = V_0(t)$, *and a trajectory reaches the origin in a time shorter than*

$$T_0 = \frac{2}{\epsilon} ln\left(\frac{2}{\lambda_{\min}^{\frac{1}{2}}\{P\}} V^{\frac{1}{2}}(\zeta_0) + 1\right), \tag{6.17}$$

where ζ_0 *is the initial state.*

Proof For the Lyapunov function (6.13), the derivative of $V_0(t)$ is

$$\begin{aligned} \dot{V}_0(t) &= \phi_1'(e_1(t))\left\{\zeta^T(t)(A^T P + PA)\zeta(t) + \rho(t)B^T P\zeta(t) + \zeta^T(t)PB\rho(t)\right\} \\ &= \phi_1'(e_1(t)) \begin{bmatrix} \zeta(t) \\ \rho(t) \end{bmatrix}^T \begin{bmatrix} A^T P + PA & PB \\ B^T P & 0 \end{bmatrix} \begin{bmatrix} \zeta(t) \\ \rho(t) \end{bmatrix} \end{aligned}$$

It is easy to know that the following inequality is shown as

$$\lambda_{\min}\{P\}\|\zeta(t)\|_2^2 \leq \zeta(t)^T P\zeta(t) \leq \lambda_{\max}\{P\}\|\zeta(t)\|_2^2, \tag{6.18}$$

where $\|\zeta(t)\|_2^2$ is the Euclidean norm of $\zeta(t)$, and its expression is

$$\|\zeta(t)\|_2^2 = \zeta(t)_1^2 + \zeta(t)_2^2 = |e_1(t)| + 2|e_1(t)|^{\frac{3}{2}} + e_1^2(t) + e_2(t)^2. \tag{6.19}$$

Moreover, the following inequality is also satisfied with

$$|e_1(t)|^{\frac{1}{2}} \leq \|\zeta(t)\|_2 \leq \frac{V_0^{\frac{1}{2}}(t)}{\lambda_{\min}^{\frac{1}{2}}\{P\}}. \tag{6.20}$$

So considering (6.15), (6.16) and (6.20), it is easy to get

$$\begin{aligned} \dot{V}_0(t) &\leq \phi_1'(e_1(t)) \left\{ \begin{bmatrix} \zeta(t) \\ \rho(t) \end{bmatrix}^T \begin{bmatrix} A^T P + PA & PB \\ B^T P & 0 \end{bmatrix} \begin{bmatrix} \zeta(t) \\ \rho(t) \end{bmatrix} + \varpi(\rho, \zeta) \right\} \\ &= \phi_1'(e_1(t)) \begin{bmatrix} \zeta(t) \\ \rho(t) \end{bmatrix}^T \begin{bmatrix} A^T P + PA + R & PB \\ B^T P & -1 \end{bmatrix} \begin{bmatrix} \zeta(t) \\ \rho(t) \end{bmatrix} \\ &\leq \phi_1'(e_1(t)) \begin{bmatrix} \zeta(t) \\ \rho(t) \end{bmatrix}^T \begin{bmatrix} -\epsilon P & 0 \\ 0 & 0 \end{bmatrix} \begin{bmatrix} \zeta(t) \\ \rho(t) \end{bmatrix} \end{aligned}$$

$$\leq -\phi_1'(e_1(t))\epsilon V_0(t) = -\frac{1}{2|e_1(t)|^{\frac{1}{2}}}\epsilon V_0(t) - \epsilon V_0(t)$$
$$\leq -\frac{\epsilon\lambda_{\min}^{\frac{1}{2}}\{P\}}{2} V_0^{\frac{1}{2}}(t) - \epsilon V_0(t)$$

It is shown that the errors converge to zero in finite time, and the upper bound of convergence time is (6.17).

6.2.3 Sliding Mode Controller

In this section, we propose a sliding mode controller by taking advantage of the estimates from the ESO (6.8). Amalgamating the sliding surface (6.3) and the ESO (6.8), the sliding mode controller is designed as

$$\Delta u(t) = \frac{v_3(t) - g(e_x(t)) - z_2(t) - m_1\lfloor s(t)\rceil^{\alpha_1} - m_2\lfloor s(t)\rceil^{\alpha_2}}{b_0}, \tag{6.21}$$

where m_1 and m_2 are two positive gains of the sliding mode controller (6.21), $0 < \alpha_1 < 1$ and $\alpha_2 > 1$. Note that $\lfloor s(t)\rceil^{\alpha_1} = |s(t)|^{\alpha_1}\text{sign}(s(t))$ and $\lfloor s(t)\rceil^{\alpha_2} = |s(t)|^{\alpha_2}\text{sign}(s(t))$.

Theorem 6.5 *Considering the manipulator system (6.7), the sliding mode controller (6.21) with two appropriate gains m_1 and m_2, the closed-loop manipulator system can be stabilized in finite time.*

Proof First, the error $e_x(t)$ in (6.2), between the system output and the desired signal converges to the sliding surface in finite time, then the error $e_x(t)$ would converge to zero along the sliding surface in finite time too. Thereby, a Lyapunov function of $s(t)$ is given as

$$V_1(t) = \frac{1}{2}s^2(t). \tag{6.22}$$

Then, considering model (6.6) and sliding mode controller (6.21), the derivative of $V_1(t)$ is given as

$$\begin{aligned}\dot{V}_1(t) &= s(t)\dot{s}(t)\\ &= s(t)\left(g(e_x(t)) - v_3(t) + x_{s2}(t) + b_0\Delta u(t)\right)\\ &= s(t)\left(-e_2(t) - m_1\lfloor s(t)\rceil^{\alpha_1} - m_2\lfloor s(t)\rceil^{\alpha_2}\right)\\ &= -m_1|s(t)|^{\alpha_1+1} - m_2|s(t)|^{\alpha_2+1} - se_2(t).\end{aligned}$$

Let $-m_2|s(t)|^{\alpha_2+1} - se_2(t) = \Lambda(t)$, and it could achieve that $\Lambda(t) < 0$ by adjusting value of m_2, so the following inequality is satisfied with:

$$\begin{aligned}\dot{V}_1(t) &= -m_1|s(t)|^{(\alpha_1+1)} + \Lambda(t)\\ &= -2^{\frac{(\alpha_1+1)}{2}} m_1 V_1^{\frac{(\alpha_1+1)}{2}}(t) + \Lambda(t)\\ &< -2^{\frac{(\alpha_1+1)}{2}} m_1 V_1^{\frac{(\alpha_1+1)}{2}}(s_0).\end{aligned}$$

Thus all the trajectories out the sliding surface would converge to the sliding surface in finite time with an upper bound as follows:

$$T_1 \leq \frac{2^{\frac{(1-\alpha_1)}{2}}}{(\alpha_1+1)m_1} V_1^{\frac{(\alpha_1+1)}{2}}(s_0). \tag{6.23}$$

The proof on that the error $e_x(t)$ converges to zero along the sliding surface is presented in the following.

A Lyapunov function of $e_x(t)$ is shown as

$$V_2(t) = \frac{1}{2}e_x^2(t). \tag{6.24}$$

According to the analysis above, it is concluded that $s = \dot{e}_x + \sigma(e_x) = 0$ holds in finite time. Combining with (6.4), an upper bound of the derivative of V_2 is obtained as

$$\begin{aligned}\dot{V}_2(t) &= e_x(t)\dot{e}_x(t)\\ &= -e_x(t)\sigma(e_x(t))\\ &= -\begin{cases} q_1|e_x(t)|^{\frac{5}{2}} + q_2|e_x(t)|, & |e_x(t)| > 1\\ k_1|e_x(t)|^2 + k_2|e_x(t)|, & |e_x(t)| \leq 1\end{cases}\\ &< -\chi V_2^{\frac{1}{2}}(e_{x_0}),\end{aligned}$$

where $\chi = \min\{q_2, k_2\}$. The convergence time of $e_x(t)$ is obtained as

$$T_2 \leq \frac{2}{\chi} V_2^{\frac{1}{2}}(e_{x_0}) \tag{6.25}$$

Considering (6.17), (6.23) and (6.25), the total convergence time of $e_x(t)$ is obtained as follows:

$$\begin{aligned}T_T &= T_0 + T_1 + T_2\\ &\leq \frac{2}{\epsilon} ln\left(\frac{2}{\lambda_{\min}^{\frac{1}{2}}\{P\}} V^{\frac{1}{2}}(\zeta_0) + 1\right) + \frac{2^{\frac{(1-\alpha_1)}{2}}}{(\alpha_1+1)m_1} V_1^{\frac{(\alpha_1+1)}{2}}(s_0)\\ &+ \frac{2}{\chi} V_2^{\frac{1}{2}}(e_{x_0}).\end{aligned}$$

That is, the error $e_x(t)$ would converge to zero along the sliding surface in finite time.

Remark 6.6 In (6.21), m_1 and m_2 are two positive gains. If values of m_1 and m_2 are large, the convergence speed of the sliding mode feedback controller (6.21) would be fast. However, there would exist a stronger chattering phenomenon. On the contrary, if values of m_1 and m_2 are small, the tracking process would be smooth. But the convergence speed of the sliding mode feedback controller (6.21) would reduce. So we should choose appropriate values of m_1 and m_2 to make a trade-off between the convergence speed and the chattering phenomenon.

6.3 Experiments and Results

In this chapter, the proposed control scheme is implemented on the pneumatic manipulator system. Its tracking performance is tested under two scenarios, step input signal, and sinusoidal input signal. Furthermore, the proposed scheme is compared with two existing methods to verify its superiority.

In this experiment, the proposed scheme is tested with a step input signal at 10°, and the sampling period h is set as 0.01s. The setup of the parameters is listed in Table 6.1. Moreover, from Theorem 6.4 and the parameters listed in Table 6.1, the symmetric and positive definite matrix P is obtained as follows:

$$P = \begin{bmatrix} 20.72 & -50 \\ -50 & 255.22. \end{bmatrix}$$

The experimental results are plotted in Fig. 6.2. In Fig. 6.2a, $v(t)$ is the input step signal at 10°. $v_1(t)$ is a tracking signal of $v(t)$ generated by the TD (6.1), and v_1 is also the smoothed desired signal that the pneumatic manipulator system ought to track. $x_1(t)$ is output of the system by using the proposed approach. From Fig. 6.2a, it is easy to see that $x_1(t)$ has a fast response to track $v_1(t)$, and the convergence time is less than $0.7s$. Besides, $x_1(t)$ closely tracks with the goal signal $v_1(t)$. In Fig. 6.2b, $s(t)$ is the sliding surface. $z_1(t)$ is the observation of $s(t)$ via using the proposed ESO (6.8). From Fig. 6.2b, it can be seen that $z_1(t)$ has a good tracking performance with $s(t)$. Figure 6.2c shows the change of control input for the step signal. The control input is a voltage signal driving the manipulator system. Lastly, Fig. 6.2d shows the error of the desired signal $v_1(t)$ and the system output $x_1(t)$. From Fig. 6.2d, it is easy to know that the change trend of the error increases first and then decreases quickly. The error eventually levels off, and the residual error amplitude is less than 0.035°.

In this experiment, the proposed scheme is also tested with a sinusoidal input signal with a frequency of 0.5Hz, an amplitude of 5°, and the sampling period h is set as 0.01 s. The setup of the parameters is listed in Table 6.2. Moreover, from Theorem 6.5 and the parameters listed in Table 6.2, the symmetric and positive definite matrix P is obtained as follows:

Table 6.1 The sliding mode feedback controller (6.21) parameters for a step signal

Parameter	Value	Parameter	Value	Parameter	Value
r_0	100	h_0	0.05	q_1	30
q_2	10.5	k_1	55	k_2	0.5
β_1	50	β_2	10	m_1	29
m_2	5	α_1	0.5	α_2	1.5
b_0	3000	h	0.01	p_1	20.72
p_2	−50	p_4	255.022		

Table 6.2 The sliding mode controller (6.21) parameters for a sinusoidal signal

Parameter	Value	Parameter	Value	Parameter	Value
r_0	1000	h_0	0.05	q_1	0.4
q_2	1	k_1	0.6	k_2	0.8
β_1	60	β_2	30	m_1	22
m_2	5	α_1	0.7	α_2	1.5
b_0	4000	h	0.01	p_1	50.6
p_2	-50	p_4	102.67		

$$P = \begin{bmatrix} 50.6 & -50 \\ -50 & 102.67 \end{bmatrix}$$

The experimental results are plotted in Fig. 6.3. In Fig. 6.3a, $v(t)$ is the input sinusoidal signal with a frequency of 0.5Hz and an amplitude of 5°. Similarly, the system output x_1 tracks the desired signal v_1 rapidly and steadily. Figure 6.3b shows the sliding surface s and its observation z_1. It is obvious that the observed value z_1 closely tracks s. In Fig. 6.3c, the control input Δu is relatively smooth and it has a periodical fluctuation. Figure 6.3d shows the error between system output x_1 and desired signal v_1. As one can see from Fig. 6.3d, the error levels off quickly and the residual error amplitude is less than 0.3°. In this paper, the convergence times are less than 0.7 s, and the steady errors are small. While, the convergence times in [2, 5, 7] are almost 5s, which verifies the finite-time convergence of the proposed method.

Moreover, two existing sliding surface methods, denoted by Method 1 and Method 2, respectively, are implemented along with the proposed scheme under same scenario for comparison. Those two methods are briefly described as follows:

Method 1: the proposed sliding surface with a linear ESO structure that used in [10];

Method 2: a kind of linear sliding surface that used in [11] with the proposed ESO (6.8).

Results of a step input signal tracking experiment by implementing Method 1 and Method 2 are plotted in Fig. 6.4. In Fig. 6.4a, v is the input signal. x_{1_o} is the system output by using Method 1 and x_{1_m} is the system output by using Method 2. In

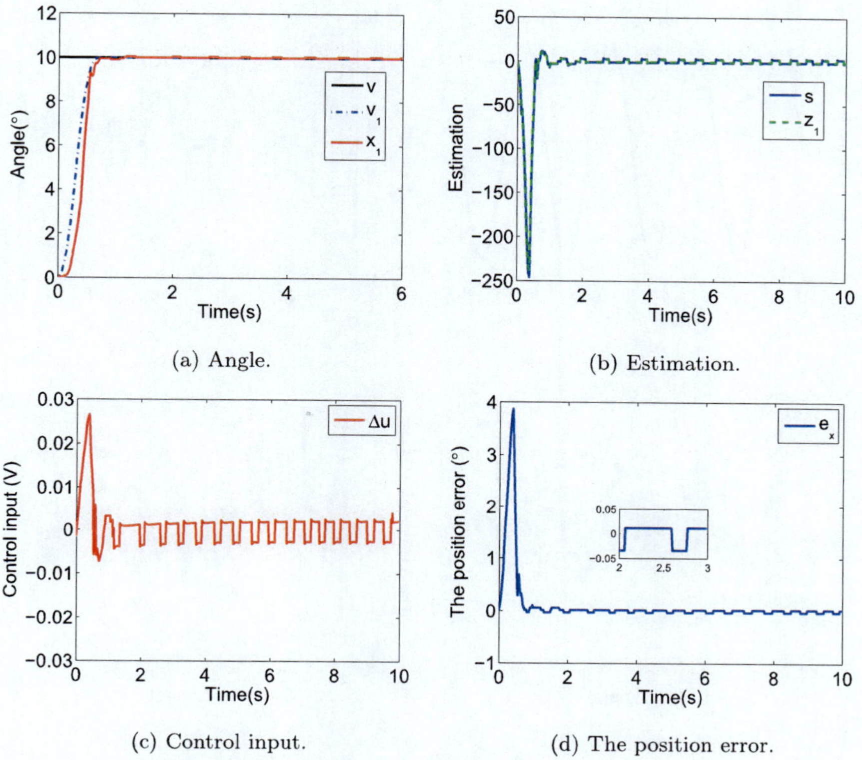

(a) Angle.

(b) Estimation.

(c) Control input.

(d) The position error.

Fig. 6.2 Experimental results for tracking a step signal at 10°

Fig. 6.4b, it shows the errors between the two methods with the desired signal. From Fig. 6.4, it is easy to see that the two methods' responses are slow for tracking the given signal, and the convergence time is almost at $3s$. Results of tracking a sinusoidal signal (0.4Hz, 5°) by implementing Method 1 and Method 2 are plotted in Fig. 6.5. In Fig. 6.5a, v is the input signal. x_{1_o} is the system output by using Method 1 and x_{1m} is the system output by using Method 2. Figure 6.5b shows two errors between the two methods with the desired signal. One can see from Fig. 6.5 that those two methods' performances are not satisfying in tracking a sinusoidal signal. Moreover, the residual error amplitudes between the desired signal and the two methods are bigger than 0.5°.

Remark 6.7 In general, from Figs. 6.2, 6.3, 6.4 and 6.5, those experimental results show the validity and advantages by using the proposed ESO (6.8) and sliding mode controller (6.21). First, the tracking response of the proposed method is much faster than that of those two comparative methods, and the convergence time is less than 0.7 s. Secondly, the steady error of the proposed method is smaller than that of the those two comparative methods.

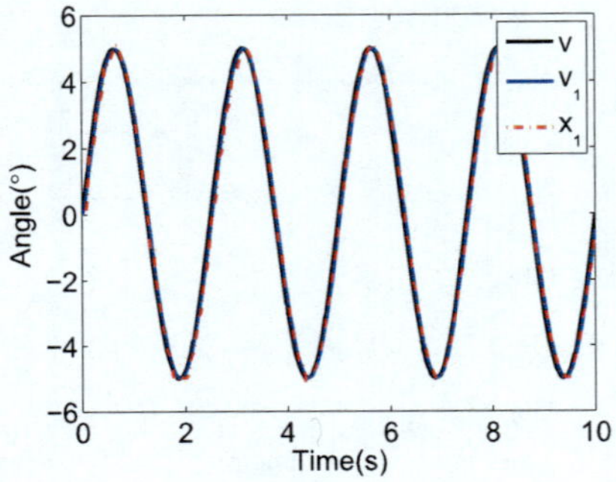

(a) Angle.

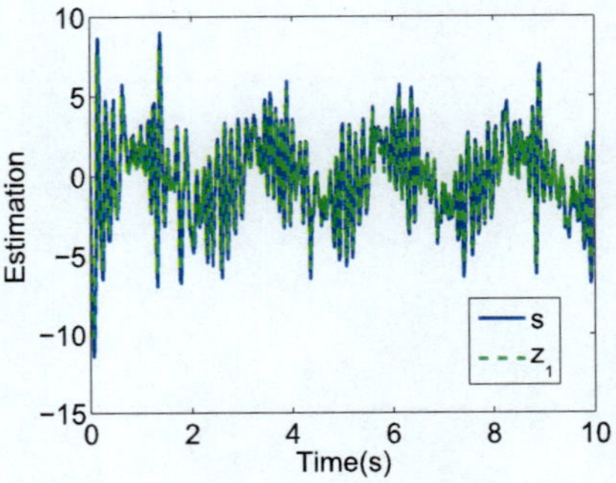

(b) Estimation.

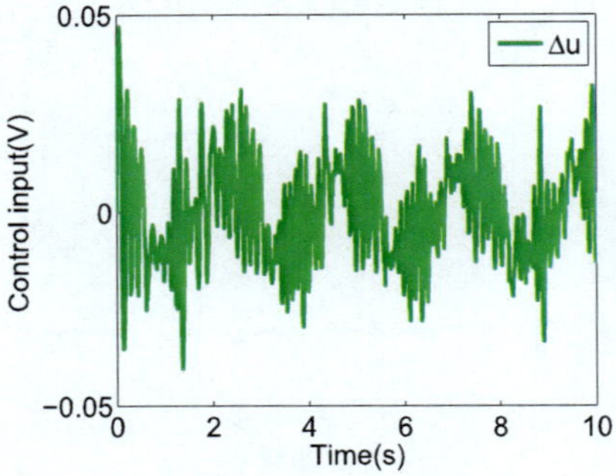

(c) Control input.

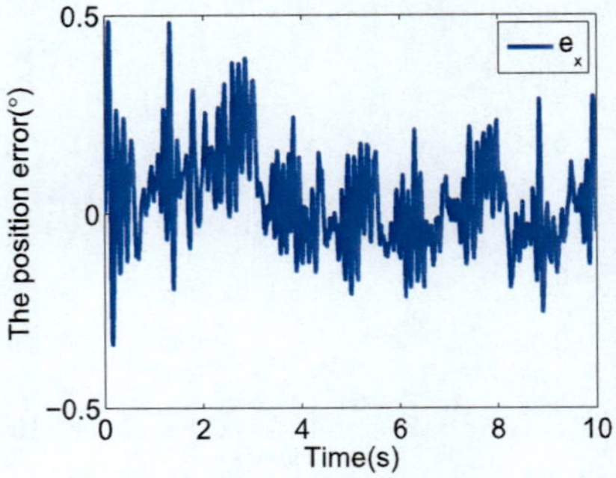

(d) The position error.

Fig. 6.3 Experimental results for tracking a sinusoidal signal at 0.4Hz, 5°

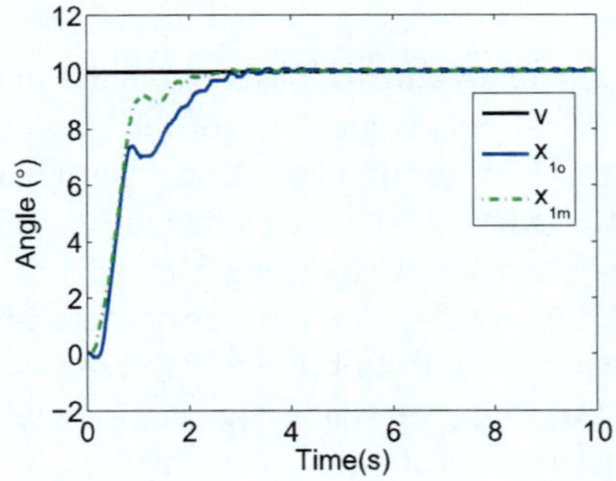

(a) Angle.

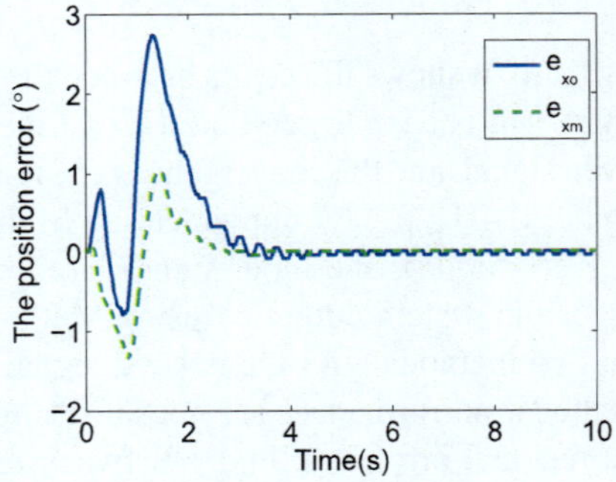

(b) The position error.

Fig. 6.4 Experimental comparison results for tracking a step signal at 10°

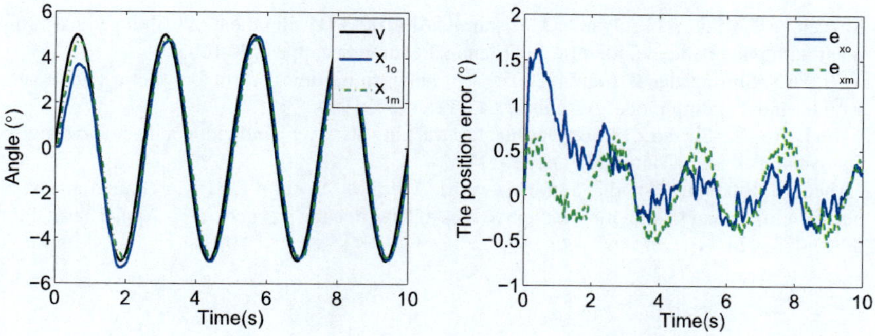

(a) Angle. (b) The position error.

Fig. 6.5 Experimental comparison results for tracking a sinusoidal signal at 0.4Hz, 5°

6.4 Conclusion

In this chapter, an ESO structure has been adopted to deal with internal uncertainties and external disturbances problems for the pneumatic manipulator system driven by PAMs. Meanwhile, a sliding mode controller based on an effective sliding surface and the observation of the total disturbances have been designed to guarantee stability of the closed-loop system. In addition, a TD is designed to arrange transient process and avoid overshoot phenomenon. Finite time's stability analysis of the proposed ESO and the closed-loop system has been proven by applying Lyapunov theory. Finally, the convergence times are less than 0.7s and the steady errors are small in experimental results, which have verified the validity and finite-time convergence of the proposed ADRC method for the pneumatic manipulator system driven by PAMs.

References

1. Ahn KK, Nguyen HTC (2007) Intelligent switching control of a pneumatic muscle robot arm using learning vector quantization neural network. Mechatronics 17(4–5):255–262
2. Nuchkrua T, Leephakpreeda T (2013) Fuzzy self-tuning PID control of hydrogen-driven pneumatic artificial muscle actuator. J Bionic Eng 10(3):329–340
3. Ahn KK, Thanh TDC (2005) Nonlinear PID control to improve the control performance of the pneumatic artificial muscle manipulator using neural network. J Mech Sci Technol 19(1):106–115
4. Zhu X, Tao G, Yao B, Cao J (2008) Adaptive robust posture control of a parallel manipulator driven by pneumatic artificial muscles. Automatica 44(9):2248–2257
5. Tang Z, Ge SS, Lee KP, He W (2016) Adaptive neural control for an uncertain robotic manipulator with joint space constraints. Int J Control 89(7):1428–1446
6. Zhao L, Xia Y, Yang Y, Liu Z (2017) Multicontroller positioning strategy for a pneumatic servo system via pressure feedback. IEEE Trans Ind Electron 64(6):4800–4809
7. Tu DCT, Ahn KK (2006) Nonlinear PID control to improve the control performance of 2 axes pneumatic artificial muscle manipulator using neural network. Mechatronics 16(9):577–587

8. Morichika T, Kikkawa F, Oyama O, Yoshimitsu T (2007) Development of walking assist equipment with pneumatic cylinder. In: SICE annual conference, pp 1058–1063
9. Yu S, Yu X, Shirinzadeh B, Man Z (2005) Continuous finite-time control for robotic manipulators with terminal sliding mode. Automatica 41(11):1957–1964
10. Yoo D, Yau SS-T, Gao Z (2007) Optimal fast tracking observer bandwidth of the linear extended state observer. Int J Control 80(1):102–111
11. Abedi M, Bakhtiari-Nejad F, Saffar-Avval M, Abedi M, Alasty A (2010) A comparative study between linear and sliding mode adaptive controllers for a hot gas generator. Appl Thermal Eng 30(5):413–424

Part II
Pneumatic Dexterous Hand System

Chapter 7
Platform Introduction

7.1 Application Background

With the maturity of robot arm technology and application, whether the robot can replace human and finish work as flexibly as human is a bottleneck for robot replacement of human. In the development of humanoid robot arm, how to make the end of manipulator execute as flexible as human hand has become a hot research direction. The humanoid dexterous hand with a variety of perception functions will assist the robot to further explore the market. In a dexterous hand, more degrees of freedom means that more drives are needed, which makes its structural design more complex. Human hand has experienced natural evolution, and its structure has been well optimized. Therefore, the structural parameters of human hand are important reference for designing dexterous hands. In Fig. 7.1, a software dexterous hand is studied by Beijing Institute of Technology. In the control of a pneumatic dexterous hand, attitude control and tracking control of single joint and multiple joints are keys, which are bases of flexible movement for the pneumatic dexterous hand. In order to make a better control performance of the pneumatic dexterous hand, a finger with two joints is designed, which is driven by four pneumatic artificial muscles (PAMs). During designing of the finger, statistical data, aesthetic proportions, and design convenience have been all taken into account, which makes the finger much more like a human finger. Moreover, the control of the finger has been deeply researched in this chapter.

7.2 Platform Structure

7.2.1 Platform Components

As shown in Fig. 7.2, a pneumatic dexterous hand platform is introduced, which is an experimental platform for pneumatic dexterous hand system driven by PAMs.

L. Zhao et al., *Pneumatic Servo Systems Analysis*, Advances in Industrial Control,
https://doi.org/10.1007/978-981-16-9515-5_7

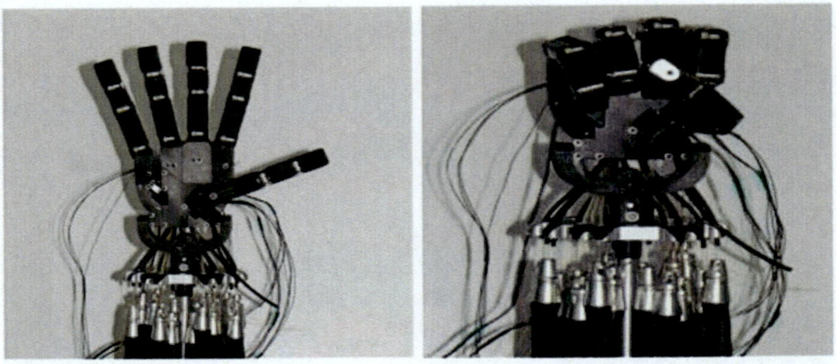

Fig. 7.1 Pneumatic dexterous hand

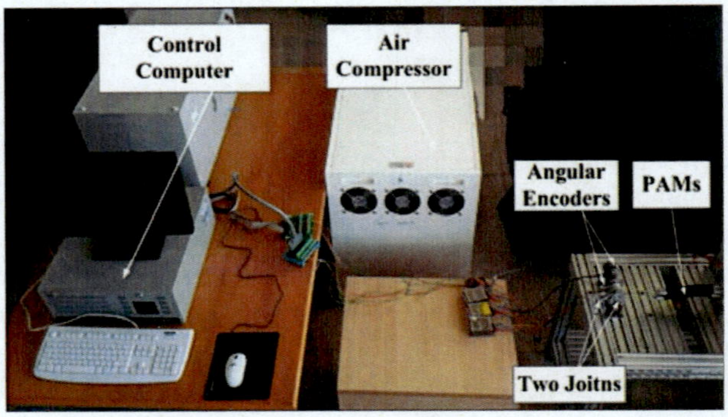

Fig. 7.2 Pneumatic dexterous hand platform

In the pneumatic dexterous hand platform, each joint is driven by a pair of PAMs with cables. When the PAMs diastole, the cables do not diastole in fact. To emulate the synergetic principle of agonist-antagonist muscles in human being, an artificial muscle will inflate while the other artificial muscle deflates in the same pair of actuators. If the joint needs to rotate counterclockwise, the PAMs need to deflate/inflate reversely as in the situation of clockwise. No matter the PAMs are inflated or deflated, the cables always are tensed. Thereby, the cables are fixed on each joint by screws so that the cables cannot skid on the joints. Figure 7.3 shows the diagrammatic drawing of the detail of a two-DoF two-joint tendon-driven finger actuated by four PAMs.

A 3D model is established via SolidWorks software according to the principle of Fig. 7.3, and an experiment setup is manufactured following the 3D model. The 3D model and manufactured setup are depicted in Figs. 7.4 and 7.5.

The considered pneumatic dexterous hand platform consists of two parts: the control terminal and the operating section of one-DoF pneumatic system. The control terminal includes an IPC (Advantech, 610H), a D/A board (Advantech, PCI-1727, 12 channels, and 14 bits), and a counting board (Advantech, PCI-1784, four 32-bit encoder counters). The D/A board and counting board are installed in the industrial

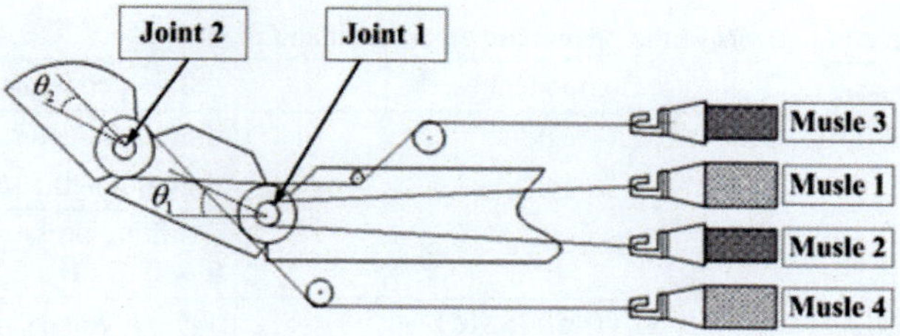

Fig. 7.3 Detail diagram of a two-DoF two-joint tendon-driven finger

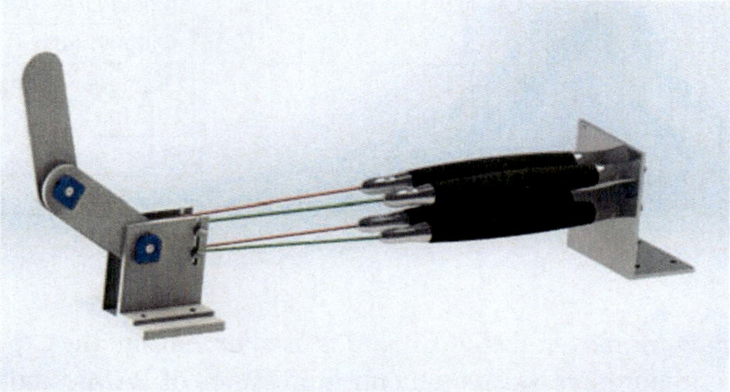

Fig. 7.4 3D design model

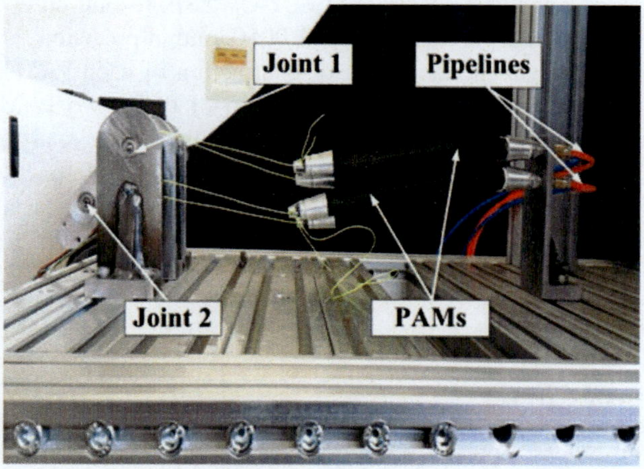

Fig. 7.5 3D design model

Table 7.1 Main components of the pneumatic dexterous hand platform

Component name	Component model	Main performance indicators
PAM	Self-made	Initial diameter: 10 mm
		Initial length: 150 mm
		Working pressure: $0 \sim 0.35$ MPa
Pressure proportional valve	ITV0050 (SMC)	Output voltage: $0 \sim 5$VDC
		Working pressure: $0 \sim 1.0$ MPa
D/A input	PCL-726	6 path D/A output
		Output range: $0 \sim 10$ V
Counter card	PCL-1784	Orthogonal frequency: 1.0 Mpa
		Pulse frequency: 2.4 Mpa
		24-bit counter

control computer. Moreover, the D/A board is used to transmit the expected voltage signals to two proportional valves to control pressures of PAMs, and the counting board is adopted to receive a feedback signal from an angle encoder. Besides, the operating section of one-DoF pneumatic system includes a one-DoF manipulator, two PAMs, an angle encoder (OMRON, E6B2-CWZ3E, resolution 2000 P/R), and two pressure proportional valves (SMC, ITV0050, output precision $\pm 6\%$, pressure range $0.001 \sim 0.9$MPa). The extension or contraction of each PAM is controlled via a proportional valve, and the relative motion of those two PAMs drives the manipulator to produce rotation of a pulley wheel, and the rotational angle could be measured by the angle encoder as feedback of the control system. In addition, the PAMs are prototypes of *Mckibben*-type PAMs. Furthermore, some structure sizes of the pneumatic dexterous hand platform are shown in Table 7.1. Furthermore, some structure sizes of the pneumatic dexterous hand platform are shown in Table 7.2.

7.2.2 Control Circuit

The pneumatic dexterous hand platform is driven by the four PAMs. Note that the pressure proportional valves are used in the pneumatic dexterous hand platform. A control process for the pneumatic dexterous hand driven by PAMs is shown in Fig. 7.6. In the pneumatic dexterous hand platform, the four pressure regulators receive control signals which come from the IPC by the D/A card. Internal pressures of the four PAMs are regulated in real time by the four pressure regulators. Then rotation of the double-joint manipulator is driven by the four PAMs via the two artificial tendons to obtain desired tracking trajectory. The two angular encoders gather angle signals which are sent to the IPC by the counting card.

Table 7.2 Structure sizes of the pneumatic dexterous hand platform

PAM	Diameter	20 mm
	Windings number rayon	2.7
	Total length of rayon	164 mm
	Thickness of rubber tube	1 mm
	Rubber tube's modulus of elasticity	0.2 Mpa
Joint 1	Diameter	14 mm
	Mass	0.158 kg
Joint 2	Diameter	11.5 mm
	Mass	0.027 kg
Length of joint	Joint 1	50 mm
	Joint 2	41 mm

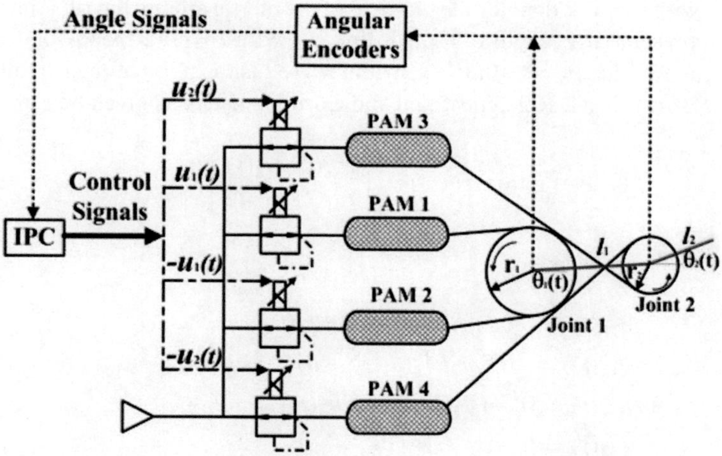

Fig. 7.6 Pneumatic circuit of the pneumatic dexterous hand experimental platform

7.3 System Model

Based on [1], the dynamic expression of the two-DoF two-joint finger system is expressed as

$$H(\theta(t))\ddot{\theta}(t) + C(\theta(t), \dot{\theta}(t))\dot{\theta}(t) + g(\theta(t)) = \tau(t), \tag{7.1}$$

where $\theta(t)$ is the joint angle vector with

$$\theta(t) = \left[\, \theta_1(t)\ \theta_2(t)\, \right]^T$$

in which $\theta_i(t)$ with $i = 1, 2$ is the angle of the ith joint as depicted in Fig. 7.2. Then $H(\theta(t))$ denotes a 2×2 inertia matrix as follows:

$$H(\theta(t)) = \begin{bmatrix} h_{11}(t) & h_{12}(t) \\ h_{21}(t) & h_{22}(t) \end{bmatrix} \tag{7.2}$$

with

$$\begin{aligned} h_{11}(t) &= I_{zz1} + I_{zz2} + m_1 r_{c1}^2 + m_2(l_1^2 + r_{c2}^2) \\ &\quad + m_2 l_1 l_2 \cos\theta_2(t) \\ h_{12}(t) &= I_{zz2} + m_2 r_{c2}^2 + 0.5 m_2 l_1 l_2 \cos\theta_2(t) \\ h_{21}(t) &= I_{zz2} + m_2 r_{c2}^2 + 0.5 m_2 l_1 l_2 \cos\theta_2(t) \\ h_{22}(t) &= I_{zz2} + m_2 r_{c2}^2, \end{aligned}$$

where I_{zzi} with $i = 1, 2$ denotes the moment of the inertia around the ith joint, l_i with $i = 1, 2$ represents the length of the ith link, r_{ci} with $i = 1, 2$ means the centroid vector of the ith link, $r_{ci} = -0.5l_i$, m_i with $i = 1, 2$ is the mass of the ith link. Note that $C(\theta(t), \dot{\theta}(t))$ is a 2×2 centrifugal and Coriolis matrix as given below:

$$C(\theta(t), \dot{\theta}(t)) = \begin{bmatrix} c_{11}(t) & c_{12}(t) \\ c_{21}(t) & c_{22}(t) \end{bmatrix} \tag{7.3}$$

with

$$\begin{aligned} c_{11}(t) &= -0.5 m_2 l_1 l_2 \dot{\theta}_2(t) \sin\theta_2(t) \\ c_{12}(t) &= -0.5 m_2 l_1 l_2 (\dot{\theta}_1(t) + \dot{\theta}_2(t)) \sin\theta_2(t) \\ c_{21}(t) &= 0.5 m_2 l_1 l_2 \dot{\theta}_1(t) \sin\theta_2(t) \\ c_{22}(t) &= 0. \end{aligned}$$

Moreover, $g(\theta(t))$ is a 2×1 gravitational force vector which is shown as

$$g(\theta(t)) = \begin{bmatrix} g_1(t) \\ g_2(t) \end{bmatrix} \tag{7.4}$$

with

$$\begin{aligned} g_1(t) &= (r_{c2} + l_2) m_2 g \cos(\theta_1(t) + \theta_2(t)) + l_1 m_2 g \cos\theta_1(t) \\ &\quad + (r_{c1} + l_1) m_1 g \cos\theta_1(t) \\ g_2(t) &= (r_{c2} + l_2) m_2 g \cos(\theta_1(t) + \theta_2(t)). \end{aligned}$$

Furthermore, $\tau(t)$ is a 2×1 input torque vector

$$\tau(t) = J^T K(\theta(t)) p \tag{7.5}$$

in which J is a Jacobian matrix to describe the tendon transmission routing mechanism and this routing is reasonable and controllable [2]. It is written as

$$J = \begin{bmatrix} r_1 & -r_1 & -r_1 & r_1 \\ 0 & 0 & r_2 & -r_2 \end{bmatrix}^T, \tag{7.6}$$

where r_i denote the radius of the ith joint. Note that $K(\theta(t))$ is a matrix which is demonstrated as

$$K(\theta(t)) = \begin{bmatrix} k_1 & \cdots & 0 \\ \vdots & \ddots & \vdots \\ 0 & \cdots & k_4 \end{bmatrix}, \tag{7.7}$$

where k_j with $j = 1, 2, 3, 4$ is the equivalent coefficient of sectional area of all four pneumatic artificial muscles [3, 4]. It is written as

$$k_j = \zeta \frac{3l_j^2 - b^2}{4\pi N^2}, \tag{7.8}$$

where b denotes the length of wire mesh fiber, N is the enwinding number of fiber, ζ is treated as a compensation factor of the structure parameters, and l_j denotes the real-time length of each pneumatic muscle. Since l_j is the element of length vector $l(\theta(t))$, we get l_j by calculating the vector

$$l(\theta(t)) = l_0 - J\theta(t), \tag{7.9}$$

where l_0 is the initial length vector as

$$l_0 = \left[l_b \; l_b \; l_b \; l_b, \right]^T$$

where l_b means the initial length of each PAM. Moreover, $p(t)$ denotes the air pressure vector of each pneumatic artificial muscle. It is given as

$$p(t) = \left[k_u u_1(t) \; -k_u u_1(t) \; k_u u_2(t) \; -k_u u_2(t) \right]^T, \tag{7.10}$$

where k_u is the gain of proportional control valve and $u_i(t)$ is the control signal for each pneumatic artificial muscle with $i = 1, 2$. Thereby, $u_1(t)$ is the control signal changing for Muscle 1 and Muscle 2, $u_2(t)$ is the control signal changing for Muscle 3 and Muscle 4 in Fig. 7.6. The control input signal changing for Muscle 1 is positive and the input signal variation of Muscle 2 is negative to realize inflation of Muscle 1 and deflation of Muscle 2 in the same pair of PAMs, so do as for both Muscle 3 and Muscle 4. A coupling problem has to be considered for the reason of transmission routing for the two tendons in Fig. 7.6. In order to solve the coupling problem, system (7.1) is rewritten as follows:

$$H(\theta(t))\ddot{\theta}(t) + C(\theta(t), \dot{\theta}(t))\dot{\theta}(t) + g(\theta(t)) = D(t)\begin{bmatrix} k_u u_1(t) \\ k_u u_2(t) \end{bmatrix} \tag{7.11}$$

with

$$D(t) = \begin{bmatrix} A(t) & B(t) \\ 0 & C(t) \end{bmatrix},$$

where

$$A(t) = r_1\zeta\left(\frac{3l_b^2 - b^2}{2\pi N^2} + \frac{3r_1^2\theta_1(t)^2}{2\pi N^2}\right)$$

$$B(t) = -r_1\zeta\left(\frac{3l_b^2 - b^2}{2\pi N^2} + \frac{3(r_1\theta_1(t) - r_2\theta_2(t))^2}{2\pi N^2}\right)$$

$$C(t) = r_2\zeta\left(\frac{3l_b^2 - b^2}{2\pi N^2} + \frac{3(r_1\theta_1(t) - r_2\theta_2(t))^2}{2\pi N^2}\right).$$

Setting $x_1(t) = \theta(t)$, system (7.11) is rewritten as follows:

$$\begin{cases} \dot{x}_1(t) = x_2(t) \\ \dot{x}_2(t) = \tau_u(t) + f(\theta(t), \dot{\theta}(t), t) \\ y(t) = x_1(t), \end{cases} \tag{7.12}$$

where $\tau_u(t)$ is regarded as a vector which consists of elements corresponding to $u_1(t)$ and $u_2(t)$, and it is written as follows:

$$\tau_u(t) = H(\theta(t))^{-1} D k_u U(t), \tag{7.13}$$

where

$$U(t) = \begin{bmatrix} u_1(t) & u_2(t) \end{bmatrix}^T.$$

The nonlinear function $f(\theta(t), \dot{\theta}(t), t)$ indicates a complicated uncertain nonlinear part vector for each joint.

7.4 Conclusion

In this chapter, a pneumatic dexterous hand platform has been introduced. The components of the pneumatic dexterous hand platform have been shown and listed in this chapter. Pneumatic circuit and control circuit for the pneumatic dexterous hand system have been given to show structure of the pneumatic dexterous hand platform. Finally, a dynamic system for the pneumatic dexterous hand system has been derived.

References

1. Yang H, Yu Y, Yuan Y, Fan X (2015) Back-stepping control of two-link flexible manipulator based on an extended state observer. Adv Space Res 56(10):2312–2322
2. Ozawa R, Kobayashi H, Hashirii K (2014) Analysis, classification and design of tendon-driven mechanisms. IEEE Trans Robot 30(2):396–410
3. Klute GK, Czerniecki JM, Hanaford B (1999) Mckibben artificial muscles: Pneumatic actuators with biomechanical intelligence. In: IEEE/ASME international conference on advanced intelligent mechantronics, pp 221–226
4. Tondu B, Lopez P (2000) Modeling and control of Mckibben artificial muscle robot actuators. IEEE Control Syst Mag 20(2):15–38

Chapter 8
Backstepping Control

8.1 Introduction

Personified pneumatic robots have drawn increasing attention in biomimetic fields in recent years. PAMs are widely used as muscle-like actuators of hominine robotics due to large output force–weight/force–volume ratio, high-tension force, long durability, low cost, cleanliness, and compliance [1]. There exists an obvious advantage for pneumatic actuators over rigid actuators when robots need to interact with human beings [2, 3]. However, compressibility of air and nonlinear elasticity of rubber tube bring many properties for PAMs, such as high nonlinearities and time-varying behavior [4]. Hence, it is difficult to get a satisfactory control performance for a system driven by PAMs. The ADRC approach tracks given input signals swiftly with reducing noise signals by TD [5]. Note that ADRC is suitable to solve the multi-input and multi-output decoupling problem without that much complicated calculations. Especially, an ADRC controller shows a good robust property for estimating and compensating uncertainties. A nonlinear robust controller for horizontal motor-tendon-driven joints and a PID controller for vertical cylinder-tendon-driven joints are given in [6, 7]. To the best of our knowledge, there are few researches on two-joint robot finger driven by PAMs in a coupling power transmission way via ADRC, which motivates us to do this work.

In this paper, a pneumatic dexterous hand system with specific tendon transmission route driven by PAMs is designed. For the pneumatic dexterous hand with two degree of freedom (DoF), control shortcomings not only lie in nonlinearities of PAMs but also lie in unmodeled dynamics and coupling which is caused by tendon transmission route. That is, there exist severe uncertain nonlinearities in the two-joint robot finger system. An active disturbance rejection nonlinear controller is proposed based on a backstepping method, which is used to solve nonlinearities in the trajectory tracking process of pneumatic dexterous hand system. Some comparison results with this proposed method are presented in experiment section. The main contributions of this chapter are summarized as follows:

L. Zhao et al., *Pneumatic Servo Systems Analysis*, Advances in Industrial Control,
https://doi.org/10.1007/978-981-16-9515-5_8

i A two-input and two-output decoupling problem is solved via active disturbance rejection control without complicated calculations.
ii An extended state observer is designed to estimate nonlinearities for the pneumatic dexterous hand.
iii Stability of the pneumatic dexterous hand system is shown by a backstepping method.

8.2 Main Results

8.2.1 Schematic Diagram of Control Method

A schematic diagram of the backstepping method for the pneumatic dexterous hand system is shown in Fig. 8.1.

In order to obtain continuous smooth signal and its differential signal, a TD is introduced. A second-order TD is designed for system (7.12). Give an error vector as $\eta(t) = \left[\, \eta_1(t)\ \eta_2(t) \,\right]^T$, where $\eta_1(t)$ and $\eta_2(t)$ are the angle errors assigned for joints. The error vector $\eta(t)$ is calculated from $\eta(t) = v_1(t) - v_0(t)$, where

$$v_0(t) = \left[\, \theta_{1d}(t)\ \ \theta_{2d}(t) \,\right]^T \tag{8.1}$$

is the desired angle signal vector. The TD is expressed as

$$\begin{cases} \dot{v}_1(t) = v_2(t) \\ \dot{v}_2(t) = F_{han}(t), \end{cases} \tag{8.2}$$

where $v_1(t)$ is an output signal vector tracking $v_0(t)$ and $v_2(t)$ is the differential signal vector of $v_1(t)$. That is, $v_2(t)$ is the angular velocity vector.

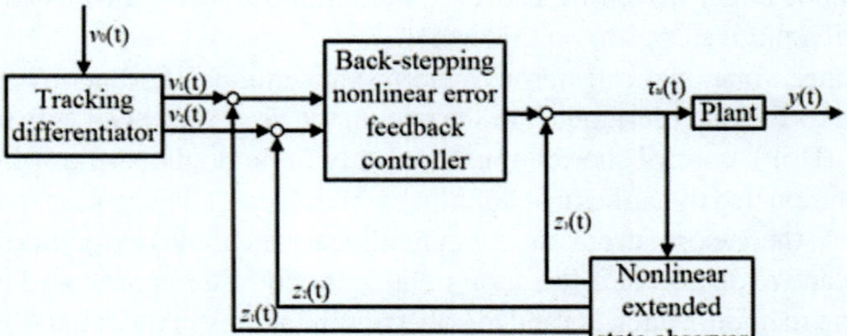

Fig. 8.1 Schematic diagram of the backstepping method

8.2.2 Nonlinear Extended State Observer

In this chapter, a nonlinear ESO is designed to deal with the nonlinear interior disturbance part of the two-DoF two-joint finger system. The nonlinear term is extended as an extended system state, i.e., $f(\theta(t), \dot{\theta}(t), t) = x_3(t)$, for system (7.12). There may be some finite non-differentiable parts in $f(\theta(t), \dot{\theta}(t), t)$ in practice. In the case, some differentiable curves can be used to fit the non-differentiable parts [8]. So function $f(\theta(t), \dot{\theta}(t), t)$ is given to be continuously differentiable and bounded in this paper. System (7.12) is rewritten as follows

$$\begin{cases} \dot{x}_1(t) = x_2(t) \\ \dot{x}_2(t) = \tau_u(t) + x_3(t) \\ \dot{x}_3(t) = \varrho(t), \end{cases} \tag{8.3}$$

where $\varrho(t)$ is the derivative vector of $x_3(t)$. Define an error $e_1(t) = \left[\, e_{11}(t)\ e_{12}(t) \,\right]^T$ and it is calculated from $e_1(t) = z_1(t) - y(t)$. The concrete expression of extended state observer is written as follows:

$$\begin{cases} \dot{z}_1(t) = z_2(t) - \beta_1 e_1(t) \\ \dot{z}_2(t) = z_3(t) - \beta_2 Fal_1 + \tau_u(t) \\ \dot{z}_3(t) = -\beta_3 Fal_2, \end{cases} \tag{8.4}$$

where $z_i(t) = \left[\, z_{i1}(t)\ z_{i2}(t) \,\right]^T$ is the observation value of state $x_i(t) = \left[\, x_{i1}(t)\ x_{i2}(t) \,\right]^T$ with $i = 1, 2, 3$. Note that the subscripts 1 and 2 denote the first joint and the second joint, respectively. β_1, β_2, and β_3 are gain matrices. Fal_1 and Fal_2 are two different function vectors with

$$\begin{aligned} Fal_1 &= \left[\, fal_{11}(e_{11}(t), \sigma_1, \delta)\ fal_{12}(e_{12}(t), \sigma_1, \delta) \,\right]^T \\ Fal_2 &= \left[\, fal_{21}(e_{11}(t), \sigma_2, \delta)\ fal_{22}(e_{12}(t), \sigma_2, \delta) \,\right]^T . \end{aligned}$$

The nonlinear ESO (8.4) is a vector expression which consists of two sub-ESOs. The two sub-ESOs share a similar expression structure. Only the convergence of one sub-ESO is needed to verify and the other sub-ESO is obtained by the same way. One sub-ESO state reconstructed error system is established based on the nonlinear ESO (8.4) and system (8.3). Let $\tau_u(t)$ in the nonlinear ESO (8.4) be the same as the one in system (8.3). The sub-ESO state reconstructed error system is given as follows:

$$\begin{cases} \dot{e}_{11}(t) = e_{21}(t) - \beta_{11} e_{11}(t) \\ \dot{e}_{21}(t) = e_{31}(t) - \beta_{21} fal_{11}(e_{11}(t)) \\ \dot{e}_{31}(t) = -\varrho_1(t) - \beta_{31} fal_{21}(e_{11}(t)), \end{cases} \tag{8.5}$$

where $e_{11}(t) = z_{11}(t) - x_{11}(t)$, $e_{21}(t) = z_{21}(t) - x_{21}(t)$, $e_{31}(t) = z_{31}(t) - x_{31}(t)$, fal_{11}, and fal_{21} denote $fal_{11}(e_{11}(t), \sigma_1, \delta)$ and $fal_{21}(e_{11}(t), \sigma_2, \delta)$, respectively. A self-stable region (SSR) method is introduced to verify the effectiveness of the nonlinear ESO (8.4). That is, $z_{i1}(t)$ is converged to $x_{i1}(t)$ with $i = 1, 2, 3$.

Definition 8.1 Let region R_e be a self-stable region of system (8.5) and its vertex is the origin. Then all error trajectories in R_e will converge to the origin after a certain moment.

Two functions are set based on the following state error: reconstructed system

$$\begin{aligned}
\hbar_1(e_{11}(t)) &= \dot{e}_{11}(t) + \Re_1 h_1(e_{11}(t))\text{sign}(e_{11}(t)) \\
\hbar_2(e_{11}(t), e_{21}(t)) &= \dot{e}_{21(t)} - \beta_{11}\dot{e}_{11}(t) \\
&\quad + \Re_2 h_2(e_{11}(t), e_{21}(t))\text{sat}\left(\frac{\hbar_1(e_{11}(t))}{h_1(e_{11}(t))}\right),
\end{aligned}$$

where both $h_1(e_{11}(t))$ and $h_2(e_{11}(t), e_{21}(t))$ are two arbitrary continuous positive definite functions, with $h_1(0) = 0$ and $h_2(0, 0) = 0$. Note that $\Re_1$ and $\Re_2$ are two constants, both of which are larger than 1.

Give two regions

$$\begin{aligned}
R_{e3} &= \{|\hbar_2(e_{11}(t), e_{21}(t))| \le h_2(e_{11}(t), e_{21}(t))\} \\
R_{e2} &= \{|\hbar_1(e_{11}(t))| \le h_1(e_{11}(t))\}
\end{aligned}$$

and two Lyapunov functions

$$\begin{aligned}
V_{\hbar_1 h_1}(t) &= \frac{1}{2}\left\{\hbar_1^2(e_{11}(t)) - h_1^2(h_{11})\right\} \\
V_{\hbar_2 h_2}(t) &= \frac{1}{2}\left\{\hbar_2^2(e_{11}(t), e_{21}(t)) - h_2^2(e_{11}(t), e_{21}(t))\right\}.
\end{aligned}$$

To relax notation, $h_1(e_{11}(t))$, $h_1\text{sign}(e_{11}(t))$, and $h_2(e_{11}(t), e_{21}(t))$ are simply denoted as h_1, $h_1 s$, and h_2. Both $\hbar_1(e_{11}(t))$ and $\hbar_2(e_{11}(t), e_{21}(t))$ are simply written as $\hbar_1$ and $\hbar_2$, respectively.

Theorem 8.2 *If there exists an error trajectory $(e_{11}(t), e_{21}(t), e_{31}(t))$ which is outside region R_{e3} initially in system (8.5), it will be drawn by the self-stable region R_{e3} and get into the self-stable region R_{e3}.*

Proof The trajectory satisfies that $V_{\hbar_2 h_2}(t) \ge 0$ initially. A Lyapunov function $V_{\hbar_2 h_2}(t)$ is adopted as

$$V_{\hbar_2 h_2}(t) = \frac{1}{2}\left\{\hbar_2^2(e_{11}(t), e_{21}(t)) - h_2^2(e_{11}(t), e_{21}(t))\right\}, \tag{8.6}$$

where $h_2(t)$ in $V_{\hbar_2 h_2}(t)$ is designed as

$$h_2(t) = \Re_3|\hbar_1(t)|, \tag{8.7}$$

where $|\hbar_1(t)|$ is satisfied with $|\hbar_1(t)| \geq h_1(t)$, and $\Re_3$ is a constant which is satisfied with

$$\Re_3 > \frac{(\Re_1+1)^2}{(\Re_2-1)}\left|\frac{d_{h_1(t)}}{d_{e_{11}(t)}}\right| > 0. \tag{8.8}$$

It is obtained that

$$\begin{aligned}
\dot{e}_{11}(t) &= e_{21}(t) - \beta_{11}e_{11}(t) = \hbar_1(t) - \Re_1 h_1(t)s \\
\dot{e}_{21}(t) &= e_{31}(t) - \beta_{21} fal_{11}(e_{11}(t)) \\
&\qquad = \hbar_2(t) + \beta_{11}(\hbar_1(t) - \Re_1 h_1(t)s) - \Re_2 h_2(t)\text{sat} \\
h_2(t)\text{sat} &= h_2\text{sat}(\hbar_1(t)/h_1(t)) = \Re_3|\hbar_1(t)| \\
\frac{\partial \hbar_1(t)}{\partial e_{11}(t)} &= -\beta_{11} + \Re_1\frac{dh_1(t)}{de_{11}(t)}s,\ \frac{\partial(t)\hbar_1(t)}{\partial e_{21}(t)} = 1.
\end{aligned}$$

The derivative of $V_{\hbar_2 h_2}(t)$ is given as follows:

$$\begin{aligned}
\dot{V}_{\hbar_2 h_2}(t) &= \hbar_2(t)\dot{\hbar}_2(t) - h_2(t)\dot{h}_2(t) \\
&= \hbar_2^2(t)(-\beta_{11} + \Re_2\Re_3) + \hbar_2(t)\hbar_1(t)[\Re_1\Re_2\Re_3\frac{dh_1(t)s}{de_{11}(t)} \\
&\quad -\Re_2^2\Re_3^2 - \beta_{21} fal'_{11} + \beta_{11}\Re_2\Re_3] + h_1(t)\hbar_2(t)s[-\beta_{31}\frac{|fal_{21}|}{h_1}(t) \\
&\quad +\Re_1\beta_{21} fal'_{11} - \Re_1^2\Re_2\Re_3\frac{dh_1(t)s}{de_{11}(t)}] - \hbar_2(t)h_2(t)\frac{\partial h_2(t)}{\partial e_{21}(t)} \\
&\quad -\hbar_2(t)\varrho_1(t) - \hbar_1(t)h_2(t)(-\Re_2\Re_3\frac{\partial h_2(t)}{\partial e_{21}(t)} + \frac{\partial h_2(t)}{\partial e_{11}(t)} \\
&\quad +\beta_{11}\frac{\partial h_2(t)}{\partial e_{21}(t)}) + \Re_1 h_1(t)h_2(t)s\left(\beta_{11}\frac{\partial h_2(t)}{\partial e_{21}(t)} + \frac{\partial h_2(t)}{\partial e_{11}(t)}\right).
\end{aligned} \tag{8.9}$$

The derivative of extended state $x_3(t)$ is bounded, i.e., $|\varrho_1(t)| \leq Q$ with $Q > 0$. One has that $|\hbar_2(t)| > h_2(t) > |\hbar_1(t)| > h_1(t)$ is deduced from $V_{\hbar_2(t)h_2(t)}$ and equation (8.7), then $\dot{V}_{\hbar_2 h_2}(t)$ is amplified as follows:

$$
\begin{aligned}
\dot{V}_{\hbar_2 h_2}(t) \le{}& \hbar_2^2(t)(-\beta_{11} + \Re_2\Re_3) + \hbar_2^2(t)\left|-\frac{\beta_{21}}{\Re_3} fal'_{11} + \beta_{11}\Re_2 - \Re_2^2\Re_3\right| \\
&+ \frac{\hbar_2^2(t)}{\Re_3}\left|-\beta_{31}\frac{fal_{21}}{h_1(t)} - \Re_1\beta_{21} fal'_{11}\right| + \hbar_2^2(t)\Re_1\Re_2\left|\frac{dh_1(t)s}{de_{11}(t)}\right| \\
&+ \Re_1^2\Re_2\hbar_2^2(t)\left|\frac{dh_1(t)s}{de_{11}(t)}\right| + |\hbar_2(t)Q| + \hbar_2^2(t)\left|\frac{\partial h_2(t)}{\partial e_2(t)}\right| \\
&+ \frac{\hbar_2^2(t)}{\Re_3}\left|\frac{\partial h_2(t)}{\partial e_{11}(t)} + \beta_{11}\frac{\partial h_2(t)}{\partial e_{21}(t)} - \Re_2\Re_3\frac{\partial h_2(t)}{\partial e_{21}(t)}\right| \\
&+ \frac{\Re_1\hbar_2^2(t)}{\Re_3}\left|\frac{\partial h_2(t)}{\partial e_{11}(t)} + \beta_{11}\frac{\partial h_2(t)}{\partial e_{21}(t)}.\right| \qquad (8.10)
\end{aligned}
$$

Choose parameters β_{11}, β_{21}, and β_{31} such that

$$
\begin{aligned}
\hbar_2^2(t)\left|-\frac{\beta_{21}}{\Re_3} fal'_{11} + \beta_{11}\Re_2 - \Re_2^2\Re_3\right| &= 0 \\
\frac{\hbar_2^2(t)}{\Re_3}\left|-\beta_{31}\frac{fal_2}{h_1(t)} - \Re_1\beta_{21} fal'_1\right| &= 0
\end{aligned}
$$

and $\dot{V}_{\hbar_2 h_2}(t) \le 0$ hold. Then trajectory $(e_{11}(t), e_{21}(t), e_{31}(t))$ outside region R_{e3} is attracted by R_{e3} and it will converge into the region eventually.

Theorem 8.3 *If there exists a trajectory* $(e_{11}(t), e_{21}(t), e_{31}(t))$ *which belongs in region* R_{e3} *but out of* R_{e2} *in system (8.5), it will converge to origin directly or get into* R_{e2} *in a finite time if the following inequation holds:*

$$
h_2(t) > \frac{(\Re_1 + 1)^2}{(\Re_2 - 1)}\left|\frac{d_{h_1}(t)}{d_{e_{11}}(t)}\right| |\hbar_1(t)|. \qquad (8.11)
$$

Proof According to the position information of the trajectory $(e_{11}(t), e_{21}(t), e_{31}(t))$, the trajectory satisfies that $V_{\hbar_2 h_2}(t) < 0$ and $V_{\hbar_1 h_1}(t) \ge 0$. Then an inequality is given as follows:

$$
\hbar_1(t)\left(\hbar_2(t) - \Re_2 h_2(t)\mathrm{sat}(\hbar_1(t)/h_1(t))\right) \le -|\hbar_1(t)|(\Re_2 - 1)h_2(t) < 0. \qquad (8.12)
$$

A Lyapunov function $V_{\hbar_1 h_1}(t)$ is given as follows:

$$
V_{\hbar_1 h_1}(t) = \frac{1}{2}\left\{\hbar_1^2(t)(e_{11}(t), e_{21}(t)) - h_1^2(t)(e_{11}(t))\right\}.
$$

The derivative of Lyapunov function $V_{\hbar_1 h_1}(t)$ is deduced as follows:

$$\begin{aligned}\dot{V}_{\hbar_1 h_1}(t) &= \hbar_1(t)\dot{\hbar}_1(t) - h_1(t)\dot{h}_1(t) \\ &= \hbar_1(t)\left[\hbar_{2(t)} - \Re_2 h_2(t)\mathrm{sat}(\hbar_1(t)/h_1(t)) + \Re_1 \frac{dh_1(t)s}{de_{11}(t)}(\hbar_1(t)\right. \\ &\quad -\Re_1 h_1(t)s)] - h_1(t)\frac{dh_1(t)}{de_{11}(t)}(\hbar_1(t) - \Re_1 h_1(t)s). \end{aligned} \tag{8.13}$$

According to inequation (8.12), $\dot{V}_{\hbar_1 h_1}(t)$ is amplified as follows:

$$\begin{aligned}\dot{V}_{\hbar_1 h_1}(t) &\leq -|\hbar_1(t)|(\Re_2 - 1)h_2(t) + \hbar_1(t)\Re_1 \frac{dh_1(t)s}{de_{11}}(t)(\hbar_1(t) - \Re_1 h_1(t)s) \\ &\leq -h_1(t)\frac{dh_1(t)}{de_{11}}(t)(\hbar_1(t) - \Re_1 h_1(t)s) - |\hbar_1(t)|(\Re_2 - 1)h_2(t) \\ &\leq +\Re_1 \left|\hbar_1(t)\frac{dh_1(t)s}{de_{11}}(t)(\hbar_1(t) - \Re_1 h_1(t)s)\right| \\ &\quad +h_1(t)|\frac{dh_1(t)}{de_{11}}(t)(\hbar_1(t) - \Re_1 h_1(t)s)| - |\hbar_1(t)|(\Re_2 - 1)h_2(t) \\ &\leq +|\frac{dh_1(t)}{de_{11}(t)}||\hbar_1(t) \\ &\quad -\Re_1 h_1 s|(\Re_1|\hbar_1(t)| + h_1(t)) - |\hbar_1(t)|(\Re_2 - 1)h_2(t) \\ &\quad + \left|\frac{dh_1(t)}{de_{11}}(t)\right| |\hbar_1(t) \\ &\leq \Re_1 \hbar_1(t)|(\Re_1|\hbar_1(t)| + |\hbar_1(t)|) - |\hbar_1(t)|(\Re_2 - 1)h_2(t) \\ &\quad +|\hbar_1|^2(t)(\Re_1 + 1)^2 \left|\frac{dh_1(t)}{de_{11}}(t)\right|. \end{aligned} \tag{8.14}$$

There exists $\dot{V}_{\hbar_1 h_1}(t) < 0$ when h_2 satisfies inequation (8.11). The trajectory which belongs in region R_{e3} but out of R_{e2} will converge into R_{e2} or it will converge into region R_{e2} after certain time.

Theorem 8.4 *A trajectory $(e_{11}(t), e_{21}(t), e_{31}(t))$, which belongs in region R_{e2}, will converge to the origin eventually in system (8.5).*

Proof Let a trajectory $(e_{11}(t), e_{21}(t), e_{31}(t))$ stay in R_{e2} after time T, i.e., $V_{\hbar_1 h_1}(t) < 0, \forall t > T$. The following inequation is obtained from region R_{e2} as

$$-\Re_1 h_1(t)s - h_1(t) < e_{21}(t) - \beta_{11}e_{11}(t) < -\Re_1 h_1(t)s + h_1(t)$$

. Design a Lyapunov function

$$V_{e_{11}}(t) = \frac{1}{2}e_{11}^2(t), \tag{8.15}$$

then the derivative of $V_{e_{11}}(t)$ is shown as

$$\begin{aligned}\dot{V}_{e_{11}}(t) &= e_{11}(t)\dot{e}_{11}(t)\\ &= e_{11}(t)(e_{21}(t) - \beta_{11}e_{11}(t))\\ &\leq -(\Re_1 s - 1)e_{11}(t)h_1(t).\end{aligned} \tag{8.16}$$

There exist $\dot{V}_{e_{11}}(t) \leq -(\Re_1 - 1)e_{11}(t)h_1(t) < 0$ when $e_{11}(t) > 0$ and $\dot{V}_{e_{11}}(t) \leq -(\Re_1 + 1)|e_{11}(t)|h_1(t) < 0$ when $e_{11}(t) < 0$, then $\dot{V}_{e_{11}}(t) < 0$ for all $e_{11}(t) \neq 0$. It is shown that $e_{11}(t)$ converges to zero from $\dot{V}_{e_{11}}(t) < 0$, then $e_{21}(t)$ and $e_{31}(t)$ will also converge to zero, which is obtained from region R_{e2} and R_{e3}. So trajectory $(e_{11}(t), e_{21}(t), e_{31}(t))$ belonging in region R_{e2} will eventually converge to the origin.

Theorem 8.5 *If a trajectory arrives at a point of the boundary of one certain region in system (8.5), such as R_{e2} or R_{e3}, after some time, then the point is the origin.*

Proof Taking the boundary of region R_{e2}, for example, if the trajectory $(e_{11}(t), e_{21}(t))$ arrives at a point of the boundary of region R_{e2}, then this point is called the utmost point $(\bar{e}_{11}(t), \bar{e}_{21}(t))$ of the R_{e2}. The utmost point is satisfied with $\dot{V}_{\hbar_1 h_1}(t) = 0$. It is known that only the origin satisfies that $\dot{V}_{\hbar_1 h_1}(t) = 0$. Hence, the utmost point is the origin. The similar conclusion is also obtained for the boundary of region R_{e3}.

Taking all theorems in this chapter into consideration comprehensively, no matter which region a trajectory belongs initially, every error trajectory will converge to the origin directly, or it will be attracted by a certain region and got into it, or it will arrive to the utmost point after some time. It is obtained that R_{e3} is a SSR of the error reconstructed system (8.5).

8.2.3 Backstepping Nonlinear Error Feedback Controller

In this chapter, a backstepping nonlinear error feedback controller is designed to make the rotary movement of pneumatic dexterous hand track the desired angles precisely. The error variable system is shown as

$$\begin{cases} s_1(t) = \upsilon_1(t) - x_1(t)\\ s_2(t) = \phi_1(t) - x_2(t), \end{cases} \tag{8.17}$$

where $s_1(t)$ and $s_2(t)$ are error variables, $\upsilon_1(t)$ is the given reference angle vector, and $\phi_1(t)$ is a virtual control variable as

$$\phi_1(t) = \upsilon_2(t) + k_1 Fal_{s1} \tag{8.18}$$

with

$$Fal_{s1} = \left[\, fal(s_{11}(t), \sigma_3, \delta)\ fal(s_{12}(t), \sigma_3, \delta)\,\right]^T.$$

Taking the derivative of the error variable system (8.17), it is rewritten as follows:

$$\begin{cases} \dot{s}_1(t) = \dot{v}_1(t) - x_2(t) \\ \dot{s}_2(t) = \dot{\phi}_1(t) - x_3(t) - H^{-1}(\theta(t))Dk_u U(t). \end{cases} \tag{8.19}$$

The backstepping nonlinear error feedback controller is designed as

$$U(t) = \frac{1}{k_u} D^{-1} H(\theta(t)) \left(\dot{\phi}_1 - z_3(t) + s_1(t) + k_2 Fal_{s2}\right), \tag{8.20}$$

where

$$Fal_{s2} = \left[\, fal(s_{21}(t), \sigma_4, \delta) \;\; fal(s_{22}(t), \sigma_4, \delta)\right]^T.$$

The backstepping nonlinear error feedback controller (8.20) is substituted into system (8.19), and the closed-loop system is obtained as follows:

$$\begin{cases} \dot{s}_1(t) = \dot{v}_1(t) - x_2(t) \\ \dot{s}_2(t) = z_3(t) - x_3(t) - s_1(t) - k_2 Fal_{s2}. \end{cases} \tag{8.21}$$

Theorem 8.6 *The pneumatic dexterous hand system (8.21) is stable if the elements of the gain matrices* k_1 *and* k_2 *in the backstepping nonlinear error feedback controller (8.20) are tuned as sufficiently large as possible, i.e.,* $\dot{v}_1(t) - x_2(t) \to 0$*, then the movement of both joints track the input angles precisely.*

Proof Step 1: Design a Lyapunov function V_1 as follows:

$$V_1(t) = \frac{1}{2} s_1^T(t) s_1(t). \tag{8.22}$$

Taking the derivative of the Lyapunov function $V_1(t)$, the following equation is obtained:

$$\begin{aligned} \dot{V}_1(t) &= \frac{1}{2}\dot{s}_1^T(t)s_1(t) + \frac{1}{2}s_1^T(t)\dot{s}_1(t) \\ &= \dot{s}_1^T(t)s_1(t) \\ &= (\dot{v}_1(t) - \dot{x}_1(t))^T(v_1(t) - x_1(t)) \\ &= (v_2(t) - x_2(t))^T(v_1(t) - x_1(t)) \\ &= (v_2(t) + s_2(t) - \phi_1)^T(v_1(t) - x_1(t)) \\ &= [-k_1 Fal_{s_1}]^T s_1(t) + s_2^T(t)s_1(t). \end{aligned} \tag{8.23}$$

To relax notation, $fal(s_{ij}(t), \sigma_o, \delta)$ with $i, j = 1, 2,\ o = 3, 4$ is noted as $fal(*)$. Note that $fal(*)$ is an odd function, so there exists $(Fal_{s_i})^T s_i(t) \geq 0, (Fal_{s_i})^T s_i(t) =$

0 if and only if $s_i(t) = 0$ holds. If the first system is stable, there exists $\dot{V}_1(t) \leq 0$ when $s_2 = 0$.

Step 2: A Lyapunov function is given as follows:

$$V_2(t) = V_1(t) + \frac{1}{2}s_2^T(t)s_2(t). \tag{8.24}$$

Taking the derivative of $V_2(t)$, one has that

$$\begin{aligned}
\dot{V}_2(t) &= \dot{V}_1(t) + \frac{1}{2}\dot{s}_2^T(t)s_2(t) + \frac{1}{2}s_2^T(t)\dot{s}_2(t) \\
&= \dot{V}_1(t) + \dot{s}_2^T(t)s_2(t) \\
&= \dot{V}_1(t) + (\dot{\phi}_1(t) - \dot{x}_2^T(t))s_2(t) \\
&= (\dot{\phi}_1(t) - H(\theta(t))^{-1}Dk_uU(t) - x_3(t))^T s_2(t) \\
&\quad + s_2^T(t)s_1(t) - [k_1 Fal_{s_1}]^T s_1(t).
\end{aligned} \tag{8.25}$$

Considering the backstepping nonlinear error feedback controller (8.20), there exists

$$\dot{V}_2(t) = -[k_1 Fal(s_1)(t)]^T s_1(t) - [k_2 Fal(s_2)]^T s_2(t) + \Delta^T s_2(t), \tag{8.26}$$

where $\Delta = z_3(t) - x_3(t)$ which is small enough because $z_3(t)$ converges to $x_3(t)$. Both k_1 and k_2 are positive gains matrix. There exists $\dot{V}_2(t) < 0$ if we tune the parameters in k_1 and k_2 to appropriate ones. That is, the closed-loop system is convergent and stable by letting k_1 and k_2 be large enough.

Remark 8.7 The backstepping nonlinear error feedback controller (8.20) contains the variable $\dot{\phi}_1(t)$ which is obtained by the TD (8.2) as follows:

$$\begin{cases} v_{\phi_1}(t) = v_{\phi_1}(t) + h\dot{\phi}_1(t) \\ \dot{\phi}_1(t) = \dot{\phi}_1(t) + hF_{han1} \end{cases} \tag{8.27}$$

with

$$\begin{aligned}
v_{\phi_1}(t) &= \left[v_{\phi_{11}}(t)\ v_{\phi_{12}}(t)\right]^T,\ \phi_1(t) = \left[\phi_{11}(t)\ \phi_{12}(t)\right]^T \\
F_{han1} &= \begin{bmatrix} f_{han}^T(v_{\phi_{11}}(t) - \phi_{11}(t), \dot{\phi}_{11}(t), r_{03}, h_{03}) \\ f_{han}^T(v_{\phi_{12}}(t) - \phi_{12}(t), \dot{\phi}_{12}(t), r_{04}, h_{04}) \end{bmatrix},
\end{aligned}$$

where $v_{\phi_1}(t)$ is used to estimate the tracking signal of $\phi_1(t)$. Hence, $\dot{\phi}_1(t)$ is used to express the differential signal of $v_{\phi_1}(t)$, where $\phi_{11}(t)$ and $\phi_{12}(t)$ are two variables, and r_{03}, r_{04}, h_{03}, and h_{04} are four parameters.

Remark 8.8 The contribution of this chapter is that an active disturbance rejection controller is proposed for a coupling two-joint system actuated by PAMs. A special mechanical design has been proposed to solve a coupling problem of the two-joint system actuated by DC motors in [9]. Rigid DC motors are chosen in [9] as actuators, while the actuators in this chapter are compliant PAMs. The DC motors have a fast dynamic performance than PAMs [3], so the response time in [9] is better than this chapter. However, PAMs have many properties such as high force-to-weight ratios and inherent compliances. Hence, the PAMs have been used in many occasions where DC motors can't be used [3]. An elaborately designed mechanism of the two-joint system is introduced in [9] to tackle the coupling problem. In this chapter, the coupling problem is processed by proposing an effective decoupling control algorithm. The decentralized SISO controller in [9] is not suitable for the coupling problem in this chapter. The controller in this chapter can simplify the mechanical design, saving cost in manufacturing and maintaining.

8.3 Experiments and Results

In this chapter, some experimental results are given to demonstrate the performance of the backstepping nonlinear error feedback controller (8.20) under desired step signal. Note that the experiment is carried out in a maximal air pressure of 0.6 Mpa, and the input–output function of the electrical proportional valve is expressed as follows:

$$P = 0.09E + 0.001,$$

where P is the output air pressure (Mpa) and E is input voltage (v).

Firstly, the coupling effect is measured. One joint is controlled to track a step signal while the other one is under passive control, respectively. Coupling effect results are recorded in Fig. 8.2. Figure 8.2a depicts the tracking result of the first joint when it is given a 15° reference signal, Fig. 8.2b shows that the second joint is subject to the disturbance of the first joint in a large extent, and the second joint will rotate 8° clockwise passively when the first joint tracks the reference signal. The first joint gets little effect from the second joint which is depicted in Fig. 8.2c, and it will rotate 0.05° counterclockwise when the second joint tracks a 15° reference signal in Fig. 8.2d.

Set the step length h as 0.01. The gain matrix parameters for the nonlinear ESO (8.4) are tuned by trial-and-error method, which are demonstrated as follows:

$$\beta_1 = \begin{bmatrix} 50 & 0 \\ 0 & 50 \end{bmatrix}, \beta_2 = \begin{bmatrix} 150 & 0 \\ 0 & 150 \end{bmatrix}, \beta_3 = \begin{bmatrix} 500 & 0 \\ 0 & 600 \end{bmatrix}.$$

The gain matrices of the backstepping nonlinear error feedback controller (8.20) k_1 and k_2 are given as follows, which is tuned by trial-and-error method, too.

$$k_1 = \begin{bmatrix} 510 & 0 \\ 0 & 510 \end{bmatrix}, k_2 = \begin{bmatrix} 600 & 0 \\ 0 & 600 \end{bmatrix}.$$

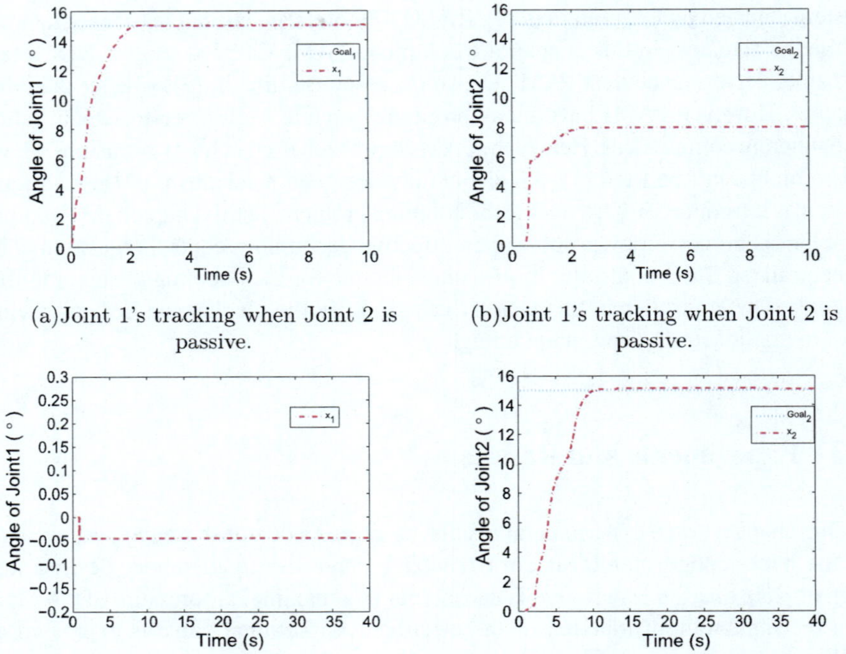

(a) Joint 1's tracking when Joint 2 is passive.

(b) Joint 1's tracking when Joint 2 is passive.

(c) Joint 1's tracking when Joint 1 is passive.

(d) Joint 2's tracking when Joint 1 is passive.

Fig. 8.2 Coupling effect of the two joints

A step reference signal with 15° is set for each joint. The experiment results of the backstepping nonlinear error feedback controller (8.20) and some PID controller results are demonstrated in Fig. 8.3. Figure 8.3a, b depicts the position-tracking results of two joints, respectively. In the figures, the reference line is the given step signal, PID line is the joint's angle which tracks the given signal by PID controller, and ADRC line shows the situation in which the joint is controlled by the backstepping nonlinear error feedback controller (8.20). The control signals of two control technology are depicted in Fig. 8.4.

As shown in Fig. 8.3b, there exists an overshoot with 6°-8° and response time with 5s–7s by using the PID controller in controlling joints. In contrast, the backstepping nonlinear error feedback controller (8.20) can overcome the coupling effect in controlling the system, which is demonstrated from the low overshoot in Fig. 8.3. Response time with 3s–4s and steady error with 0.02° are depicted in Fig. 8.3, which shows the advantage over PID controller.

Remark 8.9 As an actuator, a similar tendon-driven two-joint mechanism is designed using DC motors in [6]. Nonlinear robust controller is introduced for the system. However, only simulation results are demonstrated in that paper and no experiment results to support the validity of the backstepping nonlinear error feed-

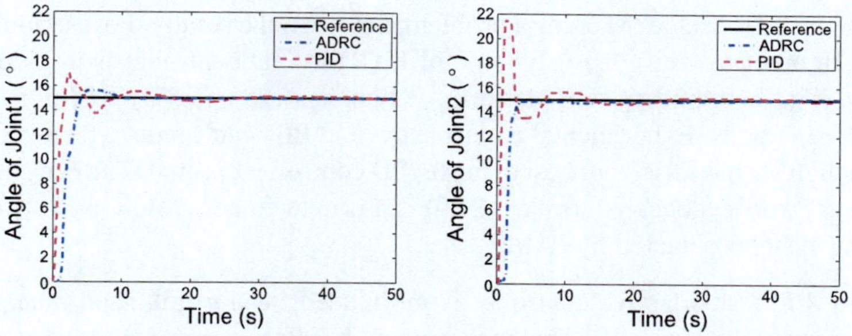

(a) Position tracking of the first joint. (b) Position tracking of the second joint.

Fig. 8.3 Comparison performance of two joints under two control methods

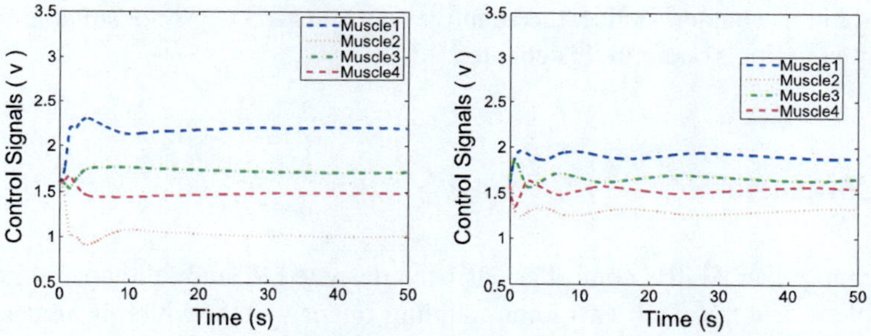

(a) Control signals of ADRC method. (b) Control signals of PID method.

Fig. 8.4 Control signals of both control methods

Table 8.1 Compared results with other literature

Control methods	Response time	Overshoot
Controller in [7]	15–17 s	18°–20°
PID controller	5–7 s	6°–8°
Controller in this paper	3–4 s	0°–0.5°

back controller (8.20). In [7], a PID feedback control method is used to control a tendon-driven manipulator system. There are some differences in the mechanical design, pneumatic cylinders have been utilized as actuators instead of pneumatic artificial muscles. In Fig. 8.4, the experiment results in this chapter are compared with the ones in [7] and a PID controller, which is depicted in Table 8.1.

Remark 8.10 Note that PID control is widely used in control systems. Many researchers apply PID controllers in pneumatic systems, so we choose PID as a comparison control method. The controller in [7] is chosen as a comparison due to its similar two-joint coupling mechanism which is powered by pneumatic actuators.

PID is adopted to tackle the control problem; however, the results show serious overshoots. In order to verify the validation of PID, the PID controller is introduced in our two-joint coupling system platform. Parameters are tuned by traditional trial-and-error methods. Experimental results show that PID can't reduce the overshoots even though we have tried our best to make PID controller optimal. The backstepping nonlinear error feedback controller (8.20) can provide a new solution for two-joint coupling systems actuated by PAMs.

Remark 8.11 An adaptive controller is introduced for a n-link rigid manipulator with uncertain loads in [10]. The backstepping nonlinear error feedback controller (8.20) is actuated by PAMs in this chapter. The uncertain parameters are assumed to be constant in [10]. However, the unknown disturbances in our paper are time varying. Only simulation results are provided in [10], but the simulation requires that two links are identical and parameters are known. Reliable experiment verifications are provided in this chapter. Hence, the adaptive controller in [10] is not suitable for the two-joint coupling system in this chapter.

8.4 Conclusion

In this chapter, an ADRC controller has been designed to study the position tracking problem of a two-DoF two-joint coupling finger system which is actuated by PAMs. A desired trajectory has been tracked via TD with low overshoot by arranging transitional process. All state variables have been estimated via a designed ESO effectively. The closed-loop control system is stable if the gain matrices are tuned large sufficiently, which is proved by a backstepping method. Experiment results show the validity of proposed ADRC method. The response time of the closed-loop system is 3s–4s and the steady error is demonstrated as 0.02°.

References

1. Ahn KK, Nguyen HTC (2007) Intelligent switching control of a pneumatic muscle robot arm using learning vector quantization neural network. Mechatronics 17(4–5):255–262
2. Nuchkrua T, Leephakpreeda T (2013) Fuzzy self-tuning PID control of hydrogen-driven pneumatic artificial muscle actuator. J Bionic Eng 10(3):329–340
3. Shin D, Yeh X, Khatib O (2014) A new hybrid actuation scheme with artificial pneumatic muscles and a magnetic particle brake for safe human-robot collaboration. Int J Robot Res 33(4):507–518
4. Chang MK, Liou JJ, Chen ML (2011) T-S fuzzy model-based tracking control of a one-dimensional manipulator actuated by pneumatic artificial muscles. Control Eng Pract 19(12):1442–1449
5. Xia Y, Shi P, Liu G-P, Rees D, Han J (2007) Active disturbance rejection control for uncertain multivariable systems with time-delay. IET Control Theory Appl 1(1):75–81
6. Okur B, Aksoy O, Zergeroglu E, Tatlicioglu E (2015) Nonlinear robust control of tendon-driven robot manipulators. J Intell Robot Syst 80(1):3–14

7. Nakamura S, Oyama O, Yoshimitsu T (2007) Development of tendon-driven assistance manipulator system by air pressure control. In: SICE annual conference, pp 2857–2860
8. Yang H, Fan X, Shi P, Hua C (2016) Nonlinear control for tracking and obstacle avoidance of a wheeled mobile robot with nonholonomic constraint. IEEE Trans Control Syst Technol 24(2):741–746
9. Grossard M, Martin J, Pacheco G (2015) Control-oriented design and robust decentralized control of the CEA dexterous robot hand. IEEE-ASME Trans Mechatron 20(4):1809–1821
10. Slotine JJE, Li W (1987) On the adaptive control of robot manipulators. Int J Robot Res 6(3):49–59

Chapter 9
Sliding Mode Control

9.1 Introduction

Pneumatic artificial muscles are widely applied as robotic actuators. As environment-friendly products, PAMs have strong nonlinearities which are caused by compressibility of air and elasticity of embedded tube. The control difficulties not only come from PAMs actuators' nonlinearities but also lie in tendon transmission coupling. Tendon driving mechanisms are widely used in lightweight humanoid robots; however, coupling tendon manner degrades the merits of control performance in PAM-actuator systems [1]. As a robust control strategy, SMC is widely applied and developed by a lot of researchers to a variety of systems [2]. The SMC is applied to solve against model errors and uncertain disturbances, which are close to pneumatic system's time-varying and strong nonlinearity characteristics. In [3], an indirect sliding mode controller is presented to control pneumatic artificial muscles, which have a good position-tracking performance. The SMC is an appropriate control method to be applied in pneumatic systems due to its robustness characteristic [4]. ESO is a centerpiece of ADRC theory which was proposed by Jingqing Han. ADRC proposes that all nonlinearities and disturbances in control system are considered as a total disturbance, which can be observed and estimated by ESO; furthermore, the observation value is eliminated in controller to resist the disturbances [5, 6]. To the best of our knowledge, a good disturbance observer can enhance the performance of sliding mode controller, which motivates us to do this work.

In this chapter, a pneumatic dexterous hand system driven by four PAMs is designed, and an active disturbance rejection nonlinear controller is proposed to study the trajectory tracking control of a two-joint system driven by pneumatic artificial muscles. Furthermore, a nonlinear sliding mode controller based on ESO is used to deal with nonlinearities and coupling disturbances in the trajectory tracking procedure for improving the control performance of the pneumatic dexterous hand system. Experiment results show that the nonlinear sliding mode controller can effectively reject the disturbances and coupling nonlinearities. The main contributions of this paper are summarized as follows:

L. Zhao et al., *Pneumatic Servo Systems Analysis*, Advances in Industrial Control,
https://doi.org/10.1007/978-981-16-9515-5_9

i. A sliding mode controller with extended state observer is proposed to cope with nonlinearities and disturbances for the pneumatic dexterous hand system.
ii. Convergence of the extended state observer is presented and stability analysis of the closed-loop system is also demonstrated with the sliding mode controller.
iii. Experimental results show that the nonlinear sliding mode controller can effectively reject the disturbances and coupling nonlinearities.

9.2 Main Results

9.2.1 Schematic Diagram of Control Method

A sliding mode control technique is proposed to process the coupling disturbances and strong nonlinearities due to its robustness. To reduce the interferences of the uncertainties from nonlinearities and coupling disturbances, an ESO is designed to estimate the disturbance from the actuators and pneumatic dexterous hand. A schematic diagram of the sliding control method is shown in Fig. 9.1.

It is shown from Fig. 9.1 that the active disturbance rejection control consists of a nonlinear ESO and a sliding mode controller.

9.2.2 Nonlinear Extended State Observer

In this chapter, a nonlinear ESO is designed to deal with the nonlinear interior disturbance item of the pneumatic dexterous hand system. Not only the observer value of system state but also the estimated value of total disturbance term is obtained in the nonlinear ESO. From an idea of active disturbance rejection control theory, the total disturbance term is extended as an extended state, i.e., $f(\theta(t), \dot{\theta}(t), t) = x_3(t)$ for system (7.12). $f(\theta(t), \dot{\theta}(t), t)$ is assumed to be continuously differentiable and bounded. System (7.12) is expressed as follows:

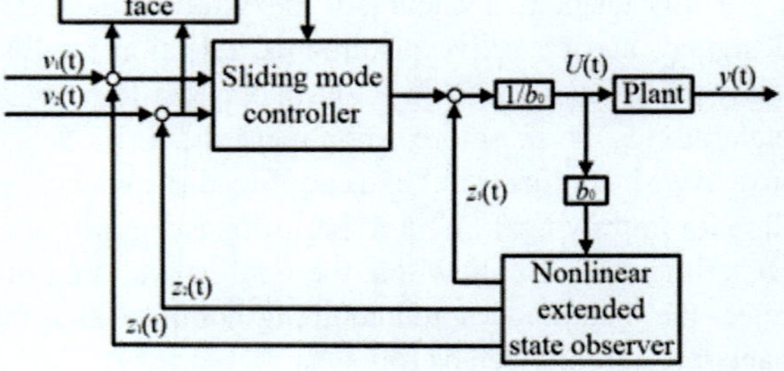

Fig. 9.1 A schematic diagram of the sliding control method

$$\begin{cases} \dot{x}_1(t) = x_2(t) \\ \dot{x}_2(t) = H(\theta(t))^{-1} Dk_u U(t) + x_3(t) \\ \dot{x}_3(t) = \varrho(t), \end{cases} \tag{9.1}$$

where $\varrho(t)$ is the derivative vector of $x_3(t) \in R^2$. Define an observation error vector $e_1(t) \in R^2$ and it is calculated from $e_1(t) = z_1(t) - y(t)$. The nonlinear ESO is shown as follows:

$$\begin{cases} \dot{z_1}(t) = z_2(t) - \beta_1 e_1(t) \\ \dot{z_2}(t) = z_3(t) - \beta_2 Fal_1 + H(\theta(t))^{-1} Dk_u U(t) \\ \dot{z_3}(t) = -\beta_3 Fal_2, \end{cases} \tag{9.2}$$

where $z_i(t) \in R^2$ are observed value vectors of state $x_i(t) \in R^2$ with $i = 1, 2, 3$. $\beta_1 \in R^{2\times 2}$, $\beta_2 \in R^{2\times 2}$, and $\beta_3 \in R^{2\times 2}$ are gain matrices. $Fal_1 \in R^2$ and $Fal_2 \in R^2$ are two different function vectors with

$$Fal_1 = \left[\, fal_{11}(e_{11}(t), \sigma_1, \delta) \;\; fal_{12}(e_{12}(t), \sigma_1, \delta) \,\right]^T$$

and

$$Fal_2 = \left[\, fal_{21}(e_{11}(t), \sigma_2, \delta) \;\; fal_{22}(e_{12}(t), \sigma_2, \delta) \,\right]^T$$

.

The convergence of the nonlinear ESO (9.2) is obtained from the error system and the pneumatic dexterous hand system (9.1). In order to facilitate the calculation, τ_u in the nonlinear ESO (9.2) is regarded to be unchanged with the pneumatic dexterous hand system (9.1). The error system is demonstrated as follows:

$$\begin{cases} \dot{e}_{11}(t) = e_{21}(t) - \beta_{11} e_{11}(t) \\ \dot{e}_{21}(t) = e_{31}(t) - \beta_{21} fal_{11}(e_{11}(t)) \\ \dot{e}_{31}(t) = -\varrho_1(t) - \beta_{31} fal_{21}(e_{11}(t)). \end{cases} \tag{9.3}$$

Theorem 9.1 *Considering the estimation error dynamic system (9.3), if the gains $\beta_{11}, \beta_{21}, \beta_{31}$ of the nonlinear ESO (9.2) are tuned larger efficiently, then the trajectories of estimation errors $e_{11}(t), e_{21}(t), e_{31}(t)$ will converge to zero asymptotically.*

Proof The Lyapunov function for the stability proof of the nonlinear ESO (9.2) is designed as follows:

$$\begin{aligned} V_{eso}(t) = {} & \frac{1}{2} q_1 (e_{11}(t) - e_{21}(t))^2 + \frac{1}{2} q_2 (e_{11}(t) - e_{31}(t))^2 + \frac{1}{2} q_3 (e_{21}(t) \\ & - e_{31}(t))^2 + \frac{1}{2} q_4 e_{11}^2(t) + \frac{1}{2} q_5 e_{21}^2(t) + \frac{1}{2} q_6 e_{31}^2(t), \end{aligned} \tag{9.4}$$

where $e_{ij}(t)$ with $i = 1, 2, 3,\ j = 1$ indicates the observation error, q_1, q_2, and q_3 are nonnegative, q_4, q_5, and q_6 are positive. To simplify the expression, $V_{eso}(t)$ is rewritten as follows:

$$
\begin{aligned}
V_{eso}(t) = &\frac{1}{2}(q_1 + q_2 + q_4)e_{11}^2(t) + \frac{1}{2}(q_1 + q_3 + q_5)e_{21}^2(t) \\
&+\frac{1}{2}(q_2 + q_3 + q_6)e_{31}^2(t) - q_1 e_{11}(t)e_{21}(t) \\
&-q_2 e_{11}(t)e_{31}(t) - q_3 e_{21}(t)e_{31}(t).
\end{aligned} \tag{9.5}
$$

To simplify the expression, let

$$
\begin{aligned}
o_1 &= q_1 + q_2 + q_4 \\
o_2 &= q_1 + q_3 + q_5 \\
o_3 &= q_2 + q_3 + q_6.
\end{aligned} \tag{9.6}
$$

One have that

$$
\begin{aligned}
V_{eso}(t) = &\frac{1}{2}o_1 e_{11}^2(t) + \frac{1}{2}o_2 e_{21}^2(t) + \frac{1}{2}o_3 e_{31}^2(t) - q_1 e_{11}(t)e_{21}(t) \\
&-q_2 e_{11}(t)e_{31}(t) - q_3 e_{21}(t)e_{31}(t).
\end{aligned} \tag{9.7}
$$

Differentiating $V_{eso}(t)$ along with (9.3), we have that

$$
\begin{aligned}
\dot{V}_{eso}(t) = &o_1 e_{11}(t)\dot{e}_{11}(t) + o_2 e_{21}(t)\dot{e}_{21}(t) + o_3 e_{31}(t)\dot{e}_{31}(t) - q_1 \dot{e}_{11}(t)e_{21}(t) \\
&-q_1 e_{11}(t)\dot{e}_{21}(t) - q_2 \dot{e}_{11}(t)e_{31}(t) - q_2 e_{11}(t)\dot{e}_{31}(t) \\
&-q_3 \dot{e}_{21}(t)e_{31}(t) - q_3 e_{21}(t)\dot{e}_{31}(t) \\
= &o_1 e_{11}(t)(e_{21}(t) - \beta_{11}e_{11}(t)) + o_2 e_{21}(t)(e_{31}(t) - \beta_{21} fal_{11}(e_{11}(t))) \\
&+o_3 e_{31}(t)(-\beta_{31} fal_{21}(e_{11}(t)) - \varrho_1(t)) - q_1 e_{21}(t)(e_{21}(t) \\
&-\beta_{11}e_{11}(t)) - q_1 e_{11}(t)(e_{31}(t) - \beta_{21} fal_{11}(e_{11}(t))) \\
&-q_2 e_{31}(t)(e_{21}(t) - \beta_{11}e_{11}(t)) - q_2 e_{11}(t)\,(-\beta_3 fal_{21}(e_{11}(t)) \\
&-\varrho_1(t)) - q_3 e_3(t)(e_3(t) - \beta_2 fal_{11}(e_{11}(t))) \\
&-q_3 e_{21}(t)(-\beta_{31} fal_{21}(e_{21}(t)) - \varrho_1(t)).
\end{aligned} \tag{9.8}
$$

Letting $fal_{11}(e_{11}(t)) = fal_{21}(e_{11}(t))$, $\dot{V}_{eso}(t)$ is further derived as follows:

$$
\begin{aligned}
\dot{V}_{eso}(t) = &-o_1\beta_1 e_{11}^2(t) + (q_2\beta_{31} + q_1\beta_{21})e_{11}(t) fal_{11}(e_{11}(t)) - q_1 e_{21}^2(t) \\
&-q_3 e_{31}^2(t) + (o_1 + q_1\beta_{11})e_{11}(t)e_{21}(t) + (q_3\beta_{31} \\
&-o_2\beta_{21})\, e_{21}(t) fal_{11}(e_{11}(t)) - (q_1 - q_2\beta_{11})e_{11}(t)e_{31}(t) \\
&+(q_3\beta_{21} - q_3\beta_{31})e_{31}(t) fal_{11}(e_{11}(t)) \\
&+(o_2 - q_2)e_{21}(t)e_{31}(t) + (q_2 e_{11}(t) \\
&+q_3 e_{21}(t) - o_3 e_{31}(t))\, \varrho_1(t).
\end{aligned} \tag{9.9}
$$

Letting $q_1 = 0$, $q_2 = 0$, and $q_3 = 0$, $\dot{V}_{eso}(t)$ is simplified as follows:

$$\begin{aligned}\dot{V}_{eso}(t) = &-o_1\beta_{11}e_{11}^2(t) + o_1e_{11}(t)e_{21}(t) - o_2\beta_{21}e_{21}(t)fal_{11}(e_{11}(t)) \\ &-o_3\beta_{31}e_{31}(t)fal_{11}(e_{11}(t)) + o_2e_{21}(t)e_{31}(t) \\ &-o_3e_{31}\varrho_1(t). \end{aligned} \tag{9.10}$$

The derivative of the nonlinearities $\varrho_1(t)$ is bounded when the reality is taken into consideration. There exists $\dot{V}_{eso}(t) < 0$ if the parameter β_{11} is chosen as large as possible. In this situation, the error trajectories converge to zero, and the stability of the nonlinear ESO (9.2) is verified.

Remark 9.2 Adaptive is a good method to compensate disturbance to make system robust, but it can't estimate relevant state information. ESO and disturbance observer (DOB) are two different disturbance/uncertainty estimation and attenuation (DUEA) methods, and they are widely used in disturbance-observer-based control [7]. Taking a nonlinear disturbance observer (NDOB) proposed in [8], for example, the NDOB need to design the observer gain function to keep the disturbance estimation error dynamics asymptotically stable, which show that the design process is more complicated than ESO. In the design process of DOB, velocity information is needed, which means that more sensors are fabricated. In contrast, the expression of ESO is certain when the dynamic model is written as a state-space expression including total disturbance. Not limited by the design simplicity, ESO not only observes the total disturbance as an extended state, but also estimates all states in the system, i.e., $z_i(t) \rightarrow x_i(t), i = 1, 2, 3$. The obvious characteristic of ESO is that it needs minimum information of the dynamic system, estimating both the model dynamics and external disturbance [7].

9.2.3 *Sliding Mode Controller*

A sliding mode controller based on the nonlinear ESO (9.2) is proposed to make the rotary movement of both joints of pneumatic dexterous hand system (9.1) track the reference trajectory precisely with tolerable time delay.

In order to design the sliding mode controller based on the nonlinear ESO (9.2), a sliding surface is considered as follows:

$$s(t) = c_1e_t(t) + \dot{e}_t(t), \tag{9.11}$$

where $s(t) \in R^2$, it represents the sliding surface vectors for joint 1 and joint 2, respectively. $c_1 > 0$ is a vector where its elements are tunable. $e_t(t) \in R^{2\times 1}$ expresses the tracking error vector, it is calculated as follows:

$$e_t(t) = \theta(t) - \theta_r(t),$$

where $\theta_r(t) \in R^2$ is the reference trajectory vector.

Then the sliding mode controller with the nonlinear ESO (9.2) is designed as follows:

$$U = \frac{1}{k_u} D^{-1} H(\theta) \left(-c_1 \dot{e}_t(t) - z_3(t) + \ddot{\theta}_r(t) - \lambda s(t) - \sigma \text{sign}(s(t))\right), \quad (9.12)$$

where $\lambda \in R^{2\times2}$ is a positive matrix. $\sigma > |z_{\max}(t)|$, $|z_{\max}(t)| \in R^{2\times2}$ is used to indicate the upper bound of observation error of nonlinearities and disturbances, which are treated as a total disturbance. The stability of the pneumatic dexterous hand system (9.1) is analyzed in the following.

Theorem 9.3 *Considering the pneumatic dexterous hand system (9.1) and the sliding mode controller (9.12) based on the nonlinear ESO (9.2), if the parameter λ in the controller and gains β_1, β_2, and β_3 are tuned as large as possible, the sliding surface (9.11) will be driven onto zero, meanwhile the pneumatic dexterous hand system (9.1) is asymptotically stable.*

Proof Taking one joint into consideration, sliding surface system (9.11) and observer error system (9.3) are used to design a Lyapunov function as follows:

$$V(t) = \frac{1}{2} s_1^2(t) + V_{eso}(t). \quad (9.13)$$

To simplify the notification, the Lyapunov function is expressed as follows:

$$V_1(t) = \frac{1}{2} s_1^2(t).$$

Differentiating $V_1(t)$ along with (7.12), one noted that

$$\begin{aligned}
\dot{V}_1(t) &= s_1(t)\dot{s}_1(t) \\
&= s_1(t)(c_1 \dot{e}_{r1}(t) + \ddot{e}_{r1}(t)) \\
&= s_1(t)[c_1(\dot{x}_1(t) - \dot{\theta}_r(t)) + (\dot{x}_2(t) - \ddot{\theta}_r(t))] \\
&= s_1(t)[f(\theta(t), \dot{\theta}(t), t) - z_{31}(t) - \lambda_1 s_1(t) - \sigma_1 \text{sign}(s_1(t))] \\
&= -\lambda_1 s_1^2(t) - s_1(t)(\tilde{z}(t) - \sigma(s_1(t))) \\
&\le 0. \qquad (9.14)
\end{aligned}$$

Then considering the $\dot{V}_1(t)$ and $\dot{V}_{eso}(t)$ simultaneously, one have that

$$\dot{V}(t) = \dot{V}_1(t) + \dot{V}_{eso}(t) \le 0. \quad (9.15)$$

So the pneumatic dexterous hand system (9.1) is converged to be stable if the parameter λ in the controller and parameters in the nonlinear ESO (9.2) are going to be chosen as large as possible. So the stable method is suitable for another joint.

Remark 9.4 ESO is used to observe the value of total disturbance in the system, then the disturbance is compensated in the sliding mode controller proposed, and this idea is based on disturbance-observer-based control (DOBC) method. If the controller is only used without ESO, the system will be easily chattered. In contrast, if the ESO is only applied, the system will lack the robustness for disturbance that sliding mode controller could realize. To support these opinions aforementioned, some comparison experiments are carried out in experiment section.

9.3 Experiments and Results

In this chapter, some experiment results are given to demonstrate the validity of the sliding mode controller proposed. Note that the experiments are carried out in a maximal air pressure 0.6Mpa, which keeps the electrical proportional valves have sufficient upstream pressure, and the input–output function of electrical proportional valve is expressed as follows:

$$p = 0.09\Delta_u + 0.001,$$

where p is the output air pressure (Mpa) and Δ_u is input voltage (v).

The coupling effect significantly impacts the control performance, which should be measured in reality before the control experiment is processed. The coupling effect measurement results are demonstrated in Fig. 9.2. Figure 9.2a shows the results that joint 1 is given a 15° tracking signal while Fig. 9.2b depicts that joint 2 is subjected to the disturbance caused by coupling effect, as shown in the picture, joint 2 has suffered an 8° rotating clockwise passively. However, joint 1 accepts little impact from joint 2 when joint 2 is given a same 15° tracking reference signal, Fig. 9.2c, d describes the coupling effect results when joint 1 executes the passive control, joint 1 only rotates 0.05° counterclockwise when joint 2 tracks the reference signal.

According to the aforementioned theory analysis of the nonlinear ESO (9.2) and the pneumatic dexterous hand system (9.1), all parameters are tuned by the empirical trial-and-error way in the experiment procedure. Set the step length h as 0.01. The gain matrix parameters for the nonlinear ESO (9.2) are demonstrated as follows:

$$\beta_1 = \begin{bmatrix} 100 & 0 \\ 0 & 100 \end{bmatrix}, \ \beta_2 = \begin{bmatrix} 250 & 0 \\ 0 & 250 \end{bmatrix}, \ \beta_3 = \begin{bmatrix} 500 & 0 \\ 0 & 600 \end{bmatrix}.$$

The gain matrix λ of the controller is given as follows, which is tuned by the trial-and-error method too.

$$\lambda = \begin{bmatrix} 1200 & 0 \\ 0 & 1000 \end{bmatrix}.$$

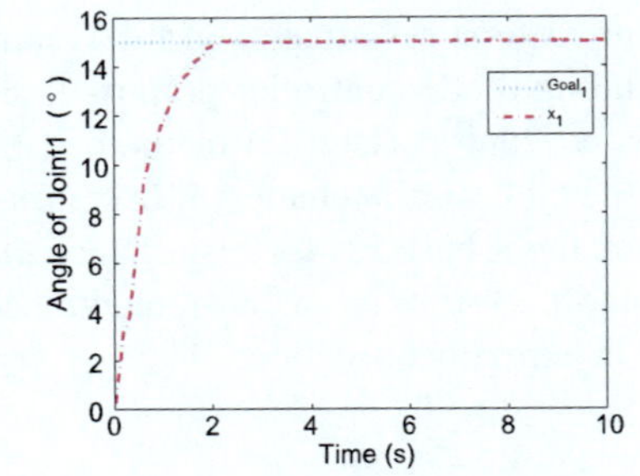

(a) Joint 1's tracking when Joint 2 is passive.

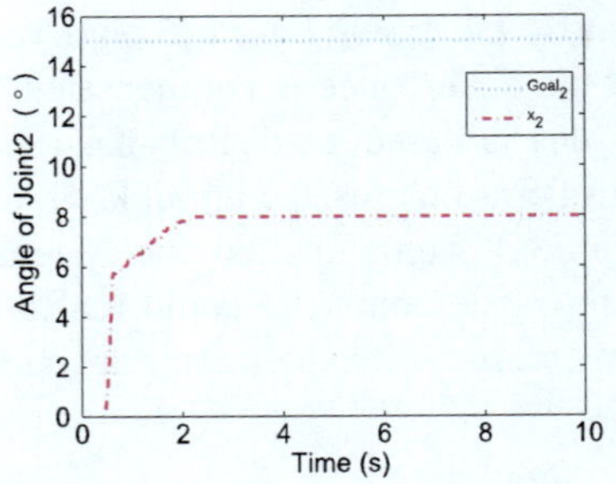

(b) Joint 2's tracking when Joint 2 is passive.

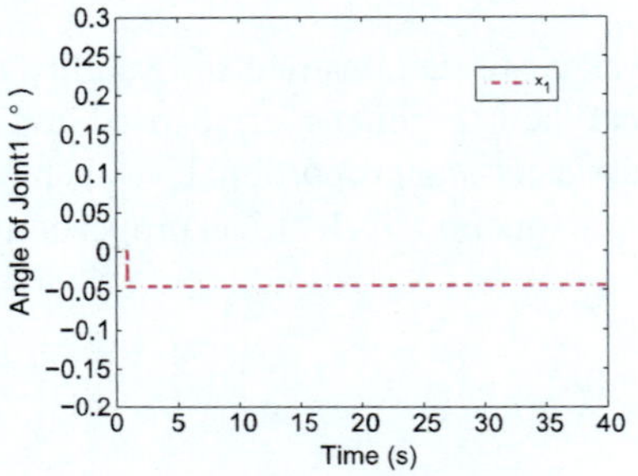

(c) Joint 1's tracking when Joint 1 is passive.

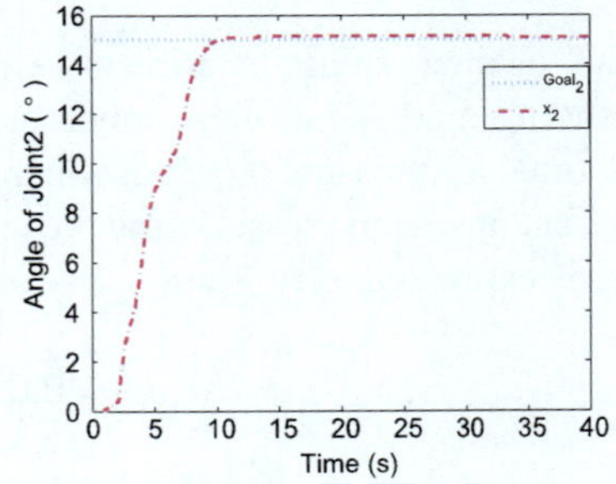

(d) Joint 2's tracking when Joint 1 is passive.

Fig. 9.2 Coupling effect to both joints

Table 9.1 The parameters of the pneumatic dexterous hand system (9.1)

Parameters	First link	Second link
Rotation inertia	86.54 kg mm^2	61.23 kg mm^2
Mass of joint	0.158 kg	0.027 kg
Length of link	50 mm	41 mm
Radius of joint	14 mm	11.5 mm

Other parameters with respect to the pneumatic dexterous hand system (9.1) are presented in Tables 9.1 and 9.2.

A step reference tracking signal of amplitude 15° is set for both joints. The experimental results of proposed controller and contrastive experimental results of PID controller for the pneumatic dexterous hand are demonstrated in Fig. 9.3. Figure 9.3a clearly describes the tracking performance of joint 1; furthermore, control consequence of joint 2 is depicted in Fig. 9.3b.

As vividly demonstrated in Fig. 9.3, sliding mode control based on the nonlinear ESO (9.2) is robust to high nonlinearities and coupling disturbances, which make the pneumatic dexterous hand system (9.1) have a control advantage over the one

Table 9.2 The parameters of PAMs

Constant	Value	Constant	Value
η	0.75	b	164
N	2.7	l_0	150

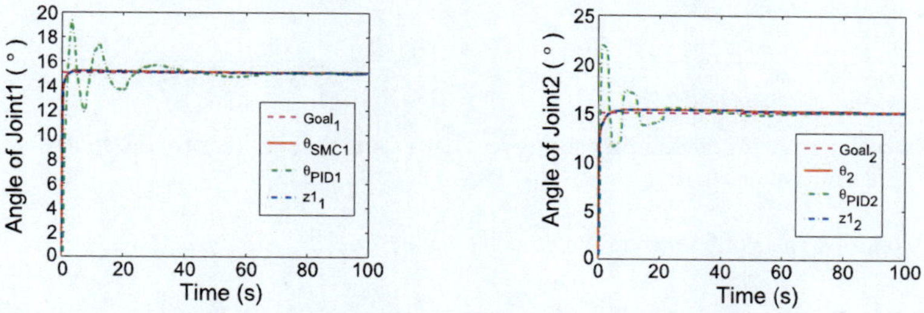

(a) Position tracking of the first joint. (b) Position tracking of the second joint.

Fig. 9.3 Comparison performance of pneumatic dexterous hands

Fig. 9.4 The changing trend of sliding surface

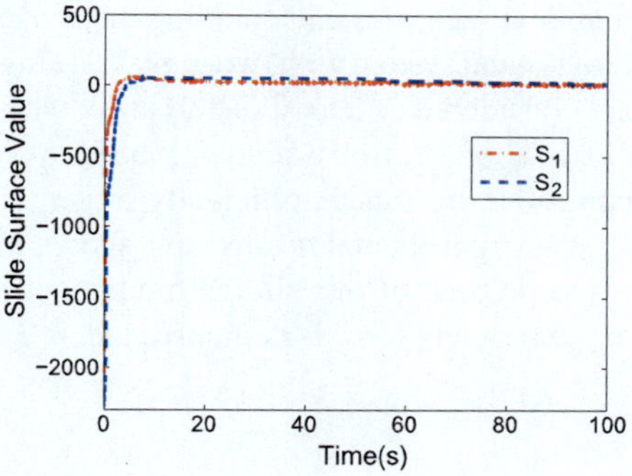

based on PID controller. Figure 9.3 shows that the sliding mode controller (9.12) can track the reference signal swiftly with tiny overshoot; however, PID has a serious overshoot and vibration characteristics, which indicates that PID controller cannot process the nonlinearities and coupling disturbance of joint 2 in an effective way. The changing trend of sliding surface is shown in Fig. 9.4, which indicates the pneumatic dexterous hand system (9.1) can converge in a short time. Control signals for each muscles of the sliding mode controller (9.12) and PID controller are demonstrated in Fig. 9.5.

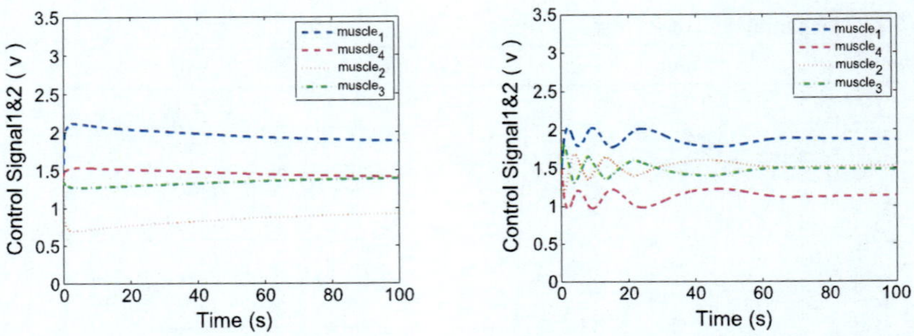

(a) Control signals of the sliding mode controller (9.12).

(b) Control signals of PID controller.

Fig. 9.5 Control signals of both control methods

9.4 Conclusion

In this chapter, we have designed a sliding mode controller based on ESO to tackle high nonlinearities and coupling disturbance control problem of the pneumatic dexterous hand system with two joints. The high nonlinearities and coupling disturbances are considered as a total disturbance, which is extended as a new state by ESO. The closed-loop control system will be converged if the gain matrices λ in the nonlinear controller are tuned sufficiently large, which is derived from Lyapunov principle. Lastly, experimental results have shown the validity of this nonlinear controller. The response time of this sliding mode controller closed-loop control system is 3 s–4 s and the steady error is demonstrated as $0.05°$.

References

1. Ozawa R, Kobayashi H, Hashirii K (2014) Analysis, classification, and design of tendon driven mechanisms. IEEE Trans Robot 30:396–410
2. Robison RM, Kothera CS, Sanner RM, Wereley NM (2016) Nonlinear control of robotic manipulators driven by pneumatic artificial muscles. IEEE/ASME Trans Mechatron 21:55–68
3. Ba DX, Ahn KK (2015) Indirect sliding mode control based on gray-box identification method for pneumatic artificial muscle. Mechatronics 32:1–11
4. Carneiro JF, Almeida FG (2015) Accurate motion control of a servopneumatic system using integral sliding mode control. Int J Adv Manuf Technol 77:1533–1548
5. Pan H, Sun W, Gao H, Hayat T, Alsaadi F (2015) Nonlinear tracking control based on extended state observer for vhicle active suspensions with performance constrains. Mechatronics 30:363–370
6. Xia Y, Zhu Z, Fu M (2011) Back-stepping sliding mode control for missile systems based on extended state observer. IET Control Theory Appl 5(1):93–102
7. Chen W-H, Yang J, Guo L, Li S (2016) Disturbance-observer-based control and related method-an overview. IEEE Trans Ind Electron 63(2):1083–1095
8. Chen W-H, Ballance DJ, Gawthrop PJ, O'Reilly J (2000) A nonlinear disturbance observer for robotic manipulators. IEEE Trans Ind Electron 47(4):932–938

Chapter 10
Nonlinear Feedback Control

10.1 Introduction

Rehabilitation robots call for flexible interaction with people. As a supple actuator, PAMs are widely implemented in rehabilitation robots in recent years. There are more desirable advantages of PAMs than conventional electrical and hydraulic actuators, such as high force–volume/power–mass ratio, inherent compliance, and low cost [1–3]. These characteristics endow PAMs the abilities to intimate the behaviors of biological muscles, which are essential for bionic robots. However, it is difficult to achieve the good position control performance due to the asymmetrical hysteresis nonlinear property, compressibility of air, and friction forces [2, 4]. Besides, coupling problems are caused by tendons that connect PAMs and manipulator [2, 5]. In order to solve these negative problems of the system, many control methods are proposed in existing literatures. A nonlinear PID controller based on neural network, which has the ability of learning, adaptation, and disturbance rejection, has been used to improve the control performance of two-axes pneumatic artificial muscle manipulator [6, 7]. However, above control methods require highly complete mathematical model which are obtained hardly in the PAMs. Therefore, ADRC is proposed to deal with the problems for the pneumatic dexterous hand system.

In this chapter, TD is used to handle input signal to obtain a corresponding smooth signal and its differential signal. Besides, a nonlinear error feedback controller is proposed to improve precision for the pneumatic dexterous hand system. Furthermore, an ESO is introduced to estimate strong nonlinear owing to coupling and disturbances. Then, both convergence of the ESO and stability of the pneumatic dexterous hand are analyzed. Finally, experimental results verify the effectiveness of the proposed control strategy. The main contributions of this paper are summarized as follows:

i. A nonlinear error feedback controller is designed to improve position accuracy by solving strong nonlinearities in a pneumatic dexterous hand system.
ii. A nonlinear ESO is presented to estimate strong nonlinearities from coupling and disturbances in the pneumatic dexterous hand system.

L. Zhao et al., *Pneumatic Servo Systems Analysis*, Advances in Industrial Control,
https://doi.org/10.1007/978-981-16-9515-5_10

iii. Both effectiveness of the ESO and stability of the pneumatic dexterous hand system are analyzed using Lyapunov approaches.

10.2 Main Results

10.2.1 Schematic Diagram of Control Method

An ADRC method is introduced to deal with strong nonlinearities which come from coupling and disturbances between two joints. A TD is designed to arrange a transition process to curb overshot. A nonlinear ESO is introduced to estimate the strong nonlinearities in the pneumatic dexterous hand system (7.12). A nonlinear error feedback controller is designed to ensure good position performance for the pneumatic dexterous hand system (7.12). A control strategy block diagram of ADRC is shown in Fig. 10.1.

A TD is designed for the pneumatic dexterous hand system (7.12); furthermore, a tracking error vector $\eta(t) \in R^2$ is given as follows:

$$\eta(t) = \left[\, \tilde{\eta_1}(t)\ \tilde{\eta_2}(t) \,\right], \tag{10.1}$$

where $\tilde{\eta_i}(t)$ with $i = 1, 2$ means the tracking error of TD for the ith joint. Tracking error vector $\tilde{\eta_i}(t)$ is calculated in the equation as follows:

$$\tilde{\eta_i}(t) = v_{1i}(t) - v_{0i}(t), \tag{10.2}$$

where $v_{0i}(t) \in R^2$ is the given position signal of both joints:

$$v_{0i}(t) = \left[\, v_{01}(t)\ v_{02}(t) \,\right]^T. \tag{10.3}$$

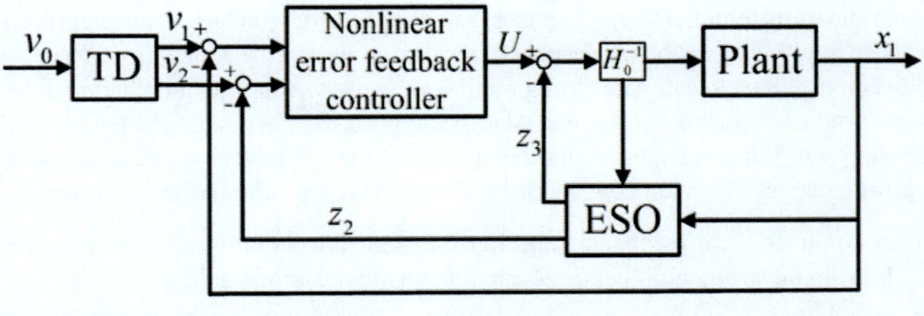

Fig. 10.1 The control strategy block diagram of ADRC

The TD is expressed as follows:

$$\begin{cases} F_{han}(t) = \left[\, f_{han}(\tilde{\eta_2}, v_{22}, r_{01}, h_{01}) \;\; f_{han}(\tilde{\eta_1}, v_{21}(t), r_{01}, h_{01}) \,\right]^T \\ \dot{v_1}(t) = v_2(t) \\ \dot{v_2}(t) = F_{han}(t) \\ v_1(t) = \left[\, v_{11}(t) \;\; v_{12}(t) \,\right]^T \\ v_2(t) = \left[\, v_{21}(t) \;\; v_{22}(t) \,\right]^T , \end{cases} \tag{10.4}$$

where $v_{1i}(t)$ is the position-tracking signal of given $v_{0i}(t)$ and $v_{2i}(t)$ is the differential signal of $v_{0i}(t)$.

10.2.2 Nonlinear Extended State Observer

From an idea of active disturbance rejection control theory, the total disturbance term is extended as an extended state, i.e., $f(\theta(t), \dot{\theta}(t), t) = x_3(t)$. $f(\theta(t), \dot{\theta}(t), t)$ is assumed to be continuously differentiable and bounded in practice. System (7.12) is expressed as follows:

$$\begin{cases} \dot{x}_1(t) = x_2(t) \\ \dot{x}_2(t) = \tau_u(t) + x_3(t) \\ \dot{x}_3(t) = \varrho(t), \end{cases} \tag{10.5}$$

where $\varrho(t)$ is the derivative vector of $x_3(t) \in R^2$. Define an observation error vector $e_1(t) \in R^2$ and it is calculated from $e_1(t) = z_1(t) - y(t)$. The concrete expression of extended state observer is given below:

$$\begin{cases} \dot{z}_1(t) = z_2(t) - \beta_1 e_1(t) \\ \dot{z}_2(t) = z_3(t) - \beta_2 Fal_1 + \tau_u(t) \\ \dot{z}_3(t) = -\beta_3 Fal_2, \end{cases} \tag{10.6}$$

where $z_i(t) \in R^2$ is observed value vectors of state $x_i(t) \in R^2$ with $i = 1, 2, 3$. $\beta_1 \in R^{2\times2}$, $\beta_2 \in R^{2\times2}$, and $\beta_3 \in R^{2\times2}$ are gain matrices. $Fal_1 \in R^2$ and $Fal_2 \in R^2$ are two different function vectors with

$$Fal_1 = \left[\, fal_{11}(e_{11}(t), \sigma_1, \delta) \;\; fal_{12}(e_{12}(t), \sigma_1, \delta) \,\right]^T$$
$$Fal_2 = \left[\, fal_{21}(e_{11}(t), \sigma_2, \delta) \;\; fal_{22}(e_{22}(t), \sigma_2, \delta) \,\right]^T .$$

Then an observation error system is demonstrated as follows:

$$\begin{cases} \dot{e}_{11}(t) = e_{21}(t) - \beta_{11} e_{11}(t) \\ \dot{e}_{21}(t) = e_{31}(t) - \beta_{21} fal_{11}(e_{11}(t)) \\ \dot{e}_{31}(t) = -\varrho_1(t) - \beta_{31} fal_{21}(e_{11}(t)). \end{cases} \tag{10.7}$$

Theorem 10.1 *Considering the observation error system (10.7), there exist appropriate parameters* β_{11}, β_{21}, *and* β_{31} *such that variables* $e_{11}(t)$, $e_{21}(t)$, *and* $e_{31}(t)$ *converge to zero. That is, the extended state observer (10.6) estimates state variables of system effectively.*

Proof A Lyapunov function is designed as follows:

$$
\begin{aligned}
V_{eso}(t) = &\frac{1}{2}q_1e_{11}^2(t) + \frac{1}{2}q_2e_{21}^2(t) + \frac{1}{2}q_3e_{31}^2(t)\\
&-\frac{1}{2}q_4(e_{11}(t) - e_{21}(t) - e_{31}(t))^2,
\end{aligned}
\tag{10.8}
$$

where $e_{ij}(t)$ indicates the observation error with $i = 1, 2, 3$, $j = 1$, q_1, q_2, q_3, and q_4 are positive numbers to keep $V_{eso}(t)$ positive definition. To simplify the expression (10.8), $V_{eso}(t)$ is rewritten as follows:

$$
\begin{aligned}
V_{eso}(t) = &\frac{1}{2}(q_1 - q_4)e_{11}^2(t) + \frac{1}{2}(q_2 - q_4)e_{21}^2(t) + \frac{1}{2}(q_3 - q_4)e_{31}^2(t)\\
&+q_4(e_{11}(t)e_{21}(t) + e_{11}(t)e_{31}(t) + e_{21}(t)e_{31}(t)).
\end{aligned}
\tag{10.9}
$$

Let

$$
\begin{cases}
o_1 = q_1 - q_4\\
o_2 = q_2 - q_4\\
o_3 = q_3 - q_4.
\end{cases}
\tag{10.10}
$$

Substituting (10.10) into (10.9), one has that

$$
\begin{aligned}
V_{eso}(t) = &\frac{1}{2}o_1e_{11}^2(t) + \frac{1}{2}o_2e_{21}^2(t) + \frac{1}{2}o_3e_{31}^2(t) + q_4[e_{11}(t)e_{21}(t)\\
&+e_{11}(t)e_{31}(t) + e_{21}(t)e_{31}(t)].
\end{aligned}
\tag{10.11}
$$

The derivative of $V_{eso}(t)$ is obtained as follows:

$$
\begin{aligned}
\dot{V}_{eso}(t) = &o_1e_{11}(t)\dot{e}_{11}(t) + o_2e_{21}(t)\dot{e}_{21}(t) + o_3e_{31}(t)\dot{e}_{31}(t)\\
&+q_4(\dot{e}_{11}(t)e_{21}(t) + e_{11}(t)\dot{e}_{21}(t) + \dot{e}_{11}(t)e_{31}(t)\\
&+e_{11}(t)\dot{e}_{31}(t) + \dot{e}_{21}(t)e_{31}(t) + e_{21}(t)\dot{e}_{31}(t))\\
= &o_1e_{11}(t)(e_{21}(t) - \beta_{11}e_{11}(t)) + o_2e_{21}(t)[e_{31}(t)]\\
&-\beta_{21}fal_{11}(e_{11}(t)) + o_3e_{31}(t)[-\beta_{31}fal_{21}(e_{11}(t)) - \varrho_1(t)]\\
&+q_4e_{21}(t)(e_{21}(t) - \beta_{11}e_{11}(t)) + q_4e_{11}[e_{31}(t) - \beta_{21}fal_{11}(e_{11}(t))]\\
&+q_4e_{31}(e_{21}(t) - \beta_{11}e_{11}) + q_4e_{11}(t)[-\beta_{31}fal_{21}(e_{11}) - \varrho_1(t)]\\
&+q_4e_{31}(t)[e_{31}(t) - \beta_{21}fal_{11}(e_{11}(t))]\\
&+q_4e_{21}(t)[-\beta_{31}fal_{21}(e_{11}(t)) - \varrho_1(t)].
\end{aligned}
\tag{10.12}
$$

Letting $fal_{11}(e_{11}(t)) = fal_{21}(e_{11}(t)) = fal(e_{11}(t))$, Eq. (10.12) is rewritten as follows:

$$\begin{aligned}
\dot{V}_{eso}(t) &= -q_4(\beta_{21}+\beta_{31})e_{11}(t)fal(e_{11}(t)) + q_4(e_{21}(t)^2 + e_{31}(t)^2) \\
&\quad +(o_1 - q_4\beta_{11})e_{11}(t)e_{21}(t) + (o_2 + q_4)e_{21}(t)e_{31}(t) - (o_2\beta_{21}e_{21}(t) \\
&\quad +o_3\beta_{31}e_{31}(t) + q_4\beta_{21}e_{31}(t) + q_4\beta_{31}e_{21}(t))fal(e_{11}(t)) \\
&\quad +q_4(1-\beta_{11})e_{11}(t)e_{31}(t) - o_1\beta_{11}e_{11}^2(t) \\
&\quad -(o_3e_{31}(t) + q_4e_{11}(t) + q_4e_{21}(t))\varrho_1(t) \\
&= -o_1\beta_{11}e_{11}^2(t) - q_4(\beta_{21}+\beta_{31})e_{11}(t)fal_{11}(e_{11}(t)) + \omega(t).
\end{aligned} \tag{10.13}$$

The derivative of the nonlinearities $\varrho_1(t)$ is bounded when the reality is taken into consideration [8]. There exists $\dot{V}_{eso}(t) < 0$ if the parameters β_{11}, β_{21}, and β_{31} are chosen as large as possible. Therefore, the observation error system (10.6) converges to zero.

10.2.3 Nonlinear Error Feedback Controller

In this section, a nonlinear error feedback controller is considered to compensate the disturbance of the system. A second-order tracking error system is introduced by TD (10.4) and system (10.5):

$$\begin{cases} \epsilon_{1i}(t) = v_{1i}(t) - x_{1i}(t) \\ \epsilon_{2i}(t) = v_{2i}(t) - z_{2i}(t), \end{cases} \tag{10.14}$$

where $\epsilon_{1i}(t) \in R^2$ represents the position error vector of both joints with $\epsilon_{1i}(t) = \left[\epsilon_{11}(t)\ \epsilon_{12}(t)\right]^T$, $\epsilon_{2i}(t) \in R^2$ is the velocity error vector with $\epsilon_{2i}(t) = \left[\epsilon_{21}(t)\ \epsilon_{22}(t)\right]^T$.

Remark 10.2 In this situation, $z_{2i}(t)$ in the ESO (10.6) is used to substitute x_{2i} because the velocity can't be obtained in reality. The velocity signal is not suitable for the controller although it could be measured.

The nonlinear error feedback controller is expressed as follows:

$$U(t) = \mathbb{H}^{-1}(k_3Fal_3 + k_4Fal_4 - z_3(t)), \tag{10.15}$$

where $k_j \in R^{2\times 2}$ $(j = 3, 4)$ are the control gain matrices. The control gain matrices are given as follows:

$$k_3 = \begin{bmatrix} k_{31} & 0 \\ 0 & k_{32} \end{bmatrix},\ k_4 = \begin{bmatrix} k_{41} & 0 \\ 0 & k_{42} \end{bmatrix}$$

and $Fal_j \in R^2\,(j = 3, 4)$ is nonlinear function vectors based on feedback error with

$$Fal_3 = \left[\, fal(\epsilon_{11}(t), \sigma_3, \delta) \;\; fal(\epsilon_{12}(t), \sigma_3, \delta) \,\right]^T$$
$$Fal_4 = \left[\, fal(\epsilon_{21}(t), \sigma_4, \delta) \;\; fal(\epsilon_{22}(t), \sigma_4, \delta) \,\right]^T$$

in which $0 < \sigma_3 < 1 < \sigma_4$.

Theorem 10.3 *Considering the pneumatic dexterous hand system (7.12) and the nonlinear error feedback controller (10.15), the tracking errors of the closed-loop control system converge to zero asymptotically if the gain matrices k_3 and k_4 are large sufficiently.*

Proof In stability proof process of the pneumatic dexterous hand system, the coupling effect of each joint subsystem is considered as the disturbance for the relevant sub-joint system.

The derivative of tracking error system is rewritten as follows:

$$\begin{cases} \dot{\epsilon}_{11}(t) = v_{21}(t) - x_{21}(t) \\ \dot{\epsilon}_{21}(t) = v_{31}(t) - \mathbb{H}_{11}U_1(t) - f_1(t), \end{cases} \tag{10.16}$$

where $v_{31}(t)$ is the derivative of $v_{21}(t)$. Note that $v_{21}(t)$ is derivable and its derivative is bounded. A Lyapunov function is designed for stability analysis of the closed-loop control system as follows:

$$V_a(t) = \frac{1}{2}(\epsilon_{11}(t) + \epsilon_{21}(t))^2 + \frac{1}{2}\epsilon_{11}^2(t). \tag{10.17}$$

The derivative of $V_a(t)$ is shown as follows:

$$\begin{aligned} \dot{V}_a(t) = {} & (\epsilon_{11}(t) + \epsilon_{21}(t))(v_{21}(t) - x_{21}(t) + v_{31}(t) - \mathbb{H}_{11}U_1(t) \\ & - f_1(t)) + \epsilon_{11}(t)(v_{21}(t) - x_{21}(t)). \end{aligned} \tag{10.18}$$

Substituting $U(t)$ into the derivative of $\dot{V}_a(t)$, one has that

$$\begin{aligned} \dot{V}_a(t) = {} & (\epsilon_{11}(t) + \epsilon_{21}(t))(v_{21}(t) - x_{21}(t) + v_{31}(t) - k_{31}Fal_{31} - k_{41}Fal_{41} \\ & + z_{31}(t) - f_1(t)) + \epsilon_{11}(t)(v_{21}(t) - x_{21}(t)), \end{aligned} \tag{10.19}$$

where $v_{21}(t)$, $v_{31}(t)$, and $x_{21}(t)$ are bounded in practice. According to Theorem 10.1, $z_{31}(t) - f_1 \to \delta_0$, where δ_0 converges to zero. Considering Eq. (10.19), it follows that

$$\begin{aligned}\dot{V}_a(t) &= (\epsilon_{11}(t) + \epsilon_{21}(t))[B_v(t) - k_{31} fal(\epsilon_{11}(t)) - k_{41} fal(\epsilon_{21}(t)) + \delta_0(t)] \\ &\quad + \epsilon_{11}(t)(v_{21}(t) - x_{21}(t)) \\ &= -(\epsilon_{11}(t) + \epsilon_{21}(t))[k_{31} fal(\epsilon_{11}(t)) + k_{41} fal(\epsilon_{21}(t))] \\ &\quad + (\epsilon_{11}(t) + \epsilon_{21}(t))(B_v(t) + \delta_0(t)) + \epsilon_{11}(t)(v_{21}(t) - x_{21}(t)),\end{aligned} \tag{10.20}$$

where $B_v(t)$ indicates the sum of $v_{21}(t) - x_{21}(t) + v_{31}(t)$, and it is bounded, $fal(\epsilon, \sigma, \delta)$ is the monotonous odd function. Let $k_{31} = k_{41} = k$ and given $\sigma_3 < \sigma_4$. If both $\epsilon_{11}(t)$ and $\epsilon_{21}(t)$ are positive or negative, then there exist that $k(fal(\epsilon_{11}(t)) + fal(\epsilon_{21}(t))$ is positive or negative. If $\epsilon_{11}(t)$ is positive and $\epsilon_{21}(t)$ is negative, then four situations are shown as follows:
(1) If $\epsilon_{11}(t) > 1$ and $\epsilon_{21}(t) < -1$, one has that

$$\begin{aligned}&-k(\epsilon_{11}(t) + \epsilon_{21}(t))(fal(\epsilon_{11}(t)) + fal(\epsilon_{21}(t))) \\ &< -k[fal(\epsilon_{11}(t)) + fal(\epsilon_{21}(t))]^2.\end{aligned} \tag{10.21}$$

(2) If $\epsilon_{11}(t) > 1$ and $-1 < \epsilon_{21}(t) < 0$, one has that

$$\begin{aligned}&-k(\epsilon_{11}(t) + \epsilon_{21}(t))(fal(\epsilon_{11}(t)) + fal(\epsilon_{21}(t))) \\ &< -k[fal(\epsilon_{11}(t)) + \epsilon_{21}(t)]^2.\end{aligned} \tag{10.22}$$

(3) If $1 > \epsilon_{11}(t) > 0$ and $\epsilon_{21}(t) < -1$, one has that

$$\begin{aligned}&-k(\epsilon_{11}(t) + \epsilon_{21}(t))(fal(\epsilon_{11}(t)) + fal(\epsilon_{21}(t))) \\ &< -k[\epsilon_{11}(t) + fal(\epsilon_{21}(t))]^2.\end{aligned} \tag{10.23}$$

(4) If $1 > \epsilon_{11}(t) > 0$ and $-1 < \epsilon_{21}(t) < 0$, one has that

$$\begin{aligned}&-k(\epsilon_{11}(t) + \epsilon_{21}(t))(fal(\epsilon_{11}(t)) + fal(\epsilon_{21}(t))) \\ &< -k[\epsilon_{11}(t) + \epsilon_{21}(t)]^2.\end{aligned} \tag{10.24}$$

No matter what $\epsilon_{11}(t)$ and $\epsilon_{21}(t)$ is, if k as large as possible, there exists

$$\begin{aligned}\dot{V}_a(t) &= (\epsilon_{11}(t) + \epsilon_{21}(t))(B_v(t) - k_{31} fal(\epsilon_{11}(t)) - k_{41} fal(\epsilon_{21}(t)) + \delta_0(t)) \\ &= -k(\epsilon_{11}(t) + \epsilon_{21}(t))(fal(\epsilon_{11}(t)) + fal(\epsilon_{21}(t))) + (\epsilon_{11}(t) \\ &\quad + \epsilon_{21}(t))(B_v(t) + \delta_0(t)) + \epsilon_{11}(t)(v_{21}(t) - x_{21}(t)) < 0.\end{aligned} \tag{10.25}$$

The derivative of $V_a(t)$ is negative definite, so the closed-loop control system is globally asymptotically stable when the control gains k_{31} and k_{41} are large enough. So the same as another joint.

10.3 Experiments and Results

In this section, some experimental results are demonstrated to show the effectiveness of the nonlinear error feedback controller (10.15). This experiment is carried out at a 0.6 Mpa upstream pressure of compressed air before inputting electrical proportional valves. In order to control the electrical proportional valves, the input–output mathematical model of proportional valve is obtained as follows:

$$p = 0.09e + 0.001, \tag{10.26}$$

where p represents the air pressure (Mpa) of outlet and e is input voltage (v) of control signal. The strong coupling caused by the way of tendons twining for this pneumatic dexterous hand system means great disturbances for each joint. An experiment is carried out to give a phenomenon of coupling disturbance. A reference signal is given for a joint to track, and the other one is passive to rotate.

The experiment results are shown in Fig. 10.2. Figure 10.2a shows the first joint's tracking result when it is given a 15° signal, and Fig. 10.2b shows the second joint's response for coupling disturbance. Moreover, it can be seen that joint 2 has a 8° clockwise passively. However, Fig. 10.2c, d demonstrates the first joint suffers little impact from joint 2 due to coupling. The joint 1 moves 0.05° when joint 2 tracks the reference signal.

Set the step length of the ESO (10.6) h as 0.01 and gains of the ESO (10.6) are demonstrated as follows:

$$\beta_1 = \begin{bmatrix} 150 & 0 \\ 0 & 150 \end{bmatrix}, \ \beta_2 = \begin{bmatrix} 450 & 0 \\ 0 & 450 \end{bmatrix}, \ \beta_3 = \begin{bmatrix} 1200 & 0 \\ 0 & 1200 \end{bmatrix}.$$

The control gains of the nonlinear error feedback controller (10.15) are given as follows:

$$k_3 = \begin{bmatrix} 1000 & 0 \\ 0 & 1500 \end{bmatrix}, \ k_4 = \begin{bmatrix} 1000 & 0 \\ 0 & 1000 \end{bmatrix}.$$

Other parameters with respect to the pneumatic dexterous hand system are presented in Table 10.1.

Figure 10.3 shows the validity of the TD (10.4) and the ESO (10.6). The v_2 and z_2 are synchronized which means that the parameters and gains in the TD (10.4) and the ESO (10.6) are reasonable. The signals obtained by TD is more smoother than ESO's z_2 signal, but the trend of z_2 is within the vicinity of v_2.

Control signals of the nonlinear error feedback controller (10.15) and PID controller for all muscles are demonstrated in Fig. 10.4a and b, respectively. It is shown that the control signal of the nonlinear error feedback controller (10.15) is more stable than the control signal of PID controller.

The decoupling ability and effectiveness of the nonlinear error feedback controller (10.15) are verified by a step signal track experiment. The advantage of the nonlinear

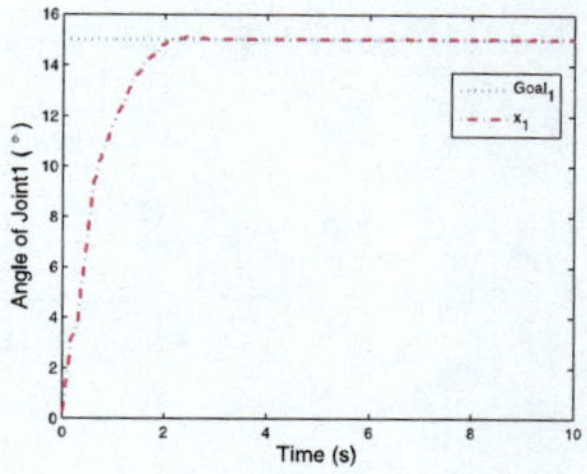

(a) Joint1's tracking when joint2 is passive.

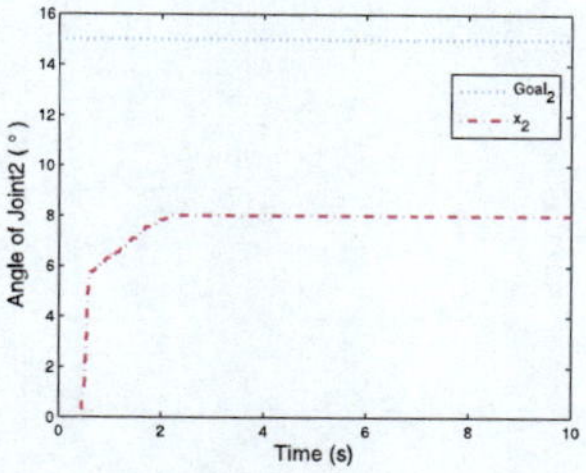

(b) Joint2's tracking when joint2 is passive

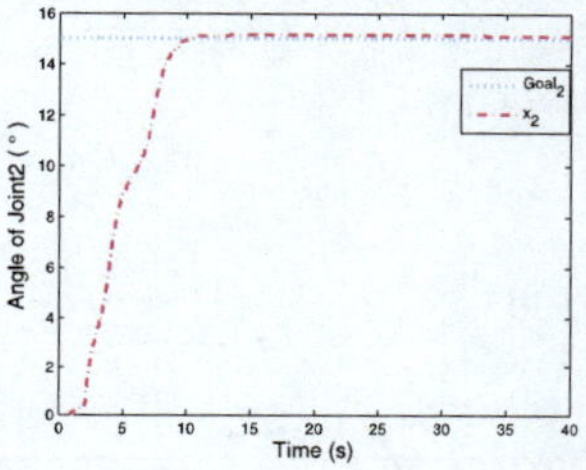

(c)Joint2's tracking when joint1 is passive

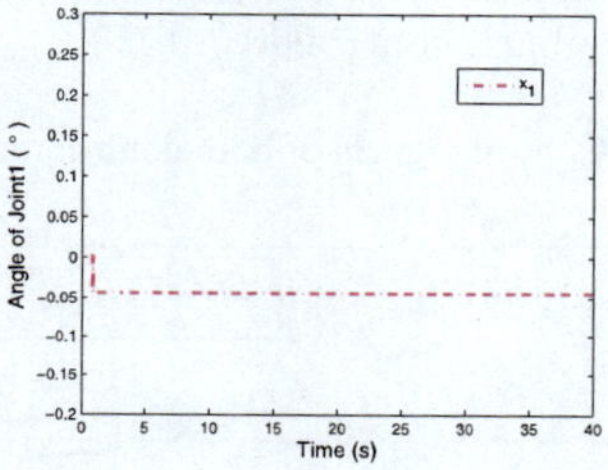

(d)Joint1's tracking when joint1 is passive

Fig. 10.2 Coupling effect to both joints

Table 10.1 The parameters of the pneumatic dexterous hand system

Parameters	First link	Second link	Constants	Value
Rotation inertia	$86.54\,\mathrm{kg\,mm^2}$	$61.23\,\mathrm{kg\,mm^2}$	η	0.75
Mass of joint	0.158 kg	0.027 kg	N	2.7
Length of link	50 mm	41 mm	b	164
Radius of joint	14 mm	11.5 mm	l_0	150

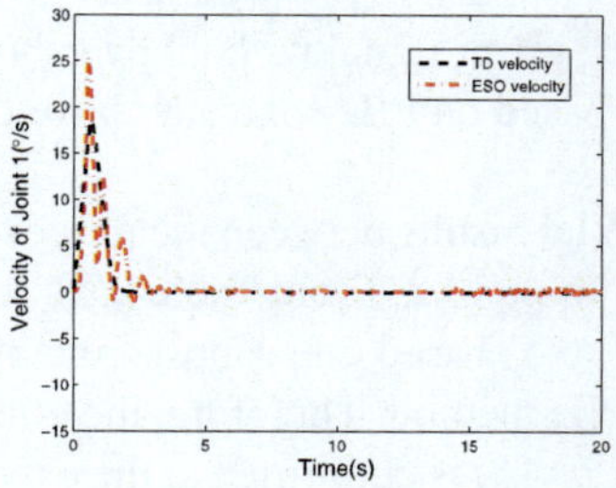

(a) Velocity estimation of the joint 1.

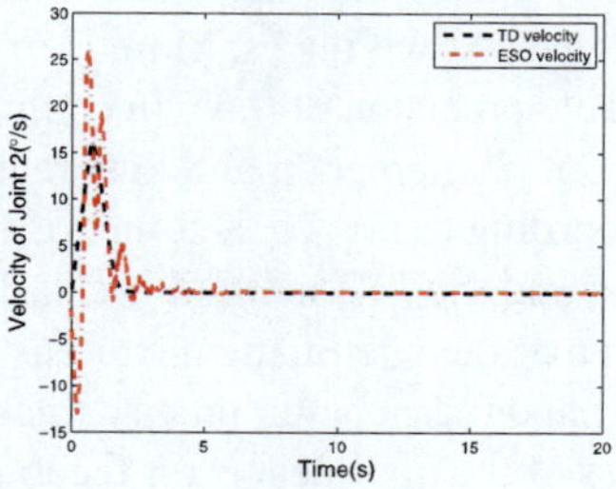

(b) Velocity estimation of the joint 2.

Fig. 10.3 Velocity estimation of both joints

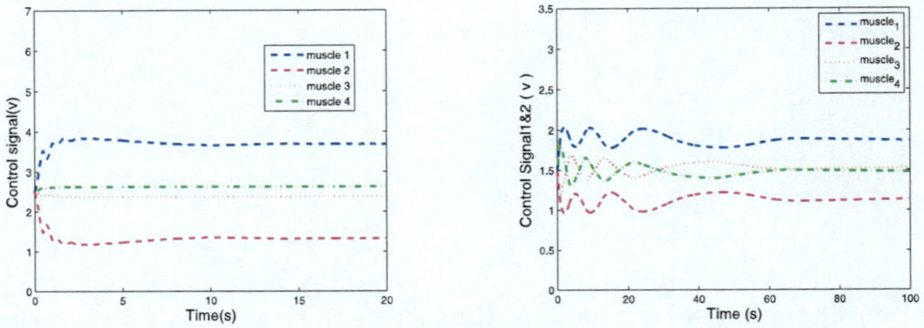

(a) Control signals of the nonlinear error feedback controller (10.15). (b) Control signals of PID controller.

Fig. 10.4 Control signals of both control methods

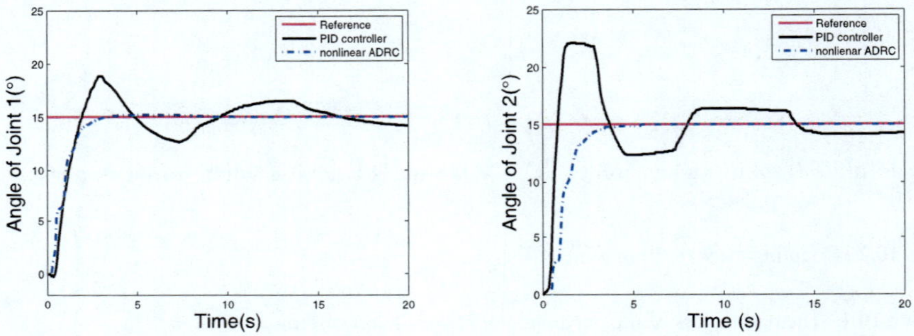

(a) Position tracking of the joint 1. (b) Position tracking of the joint 2.

Fig. 10.5 Comparison performance of the pneumatic dexterous hand under two control methods

error feedback controller (10.15) and the ability of PID controller are shown in Fig. 10.5a and b, respectively. It is shown from Fig. 10.5 that two joints track the given step signal 15° within 3 s. Moreover, two joints track given signal without overshoot at a stable precision of 0.5°. In contrast, two joints based on PID controller have the overshoot phenomenon and low tracking precision.

According to Fig. 10.5, comparison of experimental results between the proposed method and the PID method is clearly arranged in Table 10.2. From Table 10.2, it is known that the pneumatic dexterous hand system (10.5) based on proposed method in this paper has a better precise tracking than the PID method. Therefore, the effectiveness of the nonlinear error feedback controller (10.15) is confirmed in the experimental platform of the pneumatic dexterous hand system.

Table 10.2 Comparison of experimental result

Joint	Control method	Overshoot	Convergence time	Error
Joint1	PID	4°	12 s	1°
Joint1	ADRC	0°	3 s	0.5°
Joint2	PID	7°	8 s	1°
Joint2	ADRC	0°	3 s	0.5°

10.4 Conclusion

In this chapter, a nonlinear error feedback controller is introduced to deal with the strong nonlinearities and disturbances for the pneumatic dexterous hand system. All nonlinearities and disturbances are estimated by ESO and compensated by a nonlinear error feedback controller. Furthermore, experimental results have shown the validity of the proposed controller. The response time and the steady error of the closed-loop control system are less than 3 s and 0.5°, respectively.

References

1. Serres JL, Reynolds DB, Phillips CA, Gerschutz MJ, Repperger DW (2009) Characterisation of a phenomenological model for commercial pneumatic muscles actuators. Comput Methods Biomech Biomed Eng 12(4):423–430
2. Tsai Y-C, Huang A-C (2008) Multiple-surface sliding controller design for pneumatic servo systems. Mechatronics 18(9):506–512
3. Tondu B, Ippolito S, Guiochet J, Daidie A (2005) A seven-degrees-of-freedom robot-arm driven by pneumatic artificial muscles for humanoid robot. Int J Robot Res 24(4):257–274
4. Vo-Minh T, Tjahjowidodo T, Ramon H, Brussel HV (2011) A new approach to modeling hysteresis in a pneumatic artificial muscle using the maxwell-slip model. IEEE-ASME Trans Mechatron 16(1):177–186
5. Linares-Flores J, Barahona-Avalos J, Sira-Ramirez H, Contreras-Ordaz MA (2012) Robust passivity-based control of a buckcboost-converter/dc-motor system: an active disturbance rejection approach. IEEE Trans Ind Appl 48(6):2362–2371
6. Vinagre BM, Monje CA, Calderon AJ, Suarez JI (2007) Fractional PID controllers for industry application. J Vib Control 13(9–10):1419–1429
7. Tu DCT, Ahn KK (2006) Nonlinear PID control to improve the control performance of 2 axes pneumatic artificial muscle manipulator using neural network. Mechatronics 16(9):577–587
8. Tang Z, Ge SS, Lee KP, He W (2016) Adaptive neural control for an uncertain robotic manipulator with joint space constraints. Int J Control 89(7):1428–1446

Part III
Pneumatic Motion Simulation System

Chapter 11
Platform Introduction

11.1 Application Background

Motion simulation systems are widely used in military, such as flight simulation training, chariot road test, and submarine swaying test. Nowadays, a class of similar motion simulation systems is also applied in rehabilitation treatment, theme park dynamic cinemas, science and technology museums, and so on. Hydraulic and pneumatic servo systems are widely used to control the motion simulation system at home and abroad. Most of the motion simulation systems achieve the desired motion by reasonable control strategies on position or force control. With the development of pneumatic technology, PAMs are widely used in motion simulation systems. As shown in Fig. 11.1a, a motion simulation system driven by PAMs is presented, and it is used to simulate the movement of cars, ships, and aircrafts. Therefore, the motion simulation system can be used to train drivers of cars, ships, and aircrafts. Moreover, a vehicle motion simulation system is researched by BIT University which is shown in Fig. 11.1b.

11.2 Platform Structure

11.2.1 Platform Components

As shown in Fig. 11.2, a pneumatic motion simulation platform which is an experimental platform with a two-DoF parallel mechanism of PAMs is introduced in this chapter.

Two deflection movements around X-axis and Y-axis for the two-DoF parallel mechanism of PAMs are shown in Figs. 11.3 and 11.4, respectively.

Moreover, motion parameters on the pneumatic motion simulation platform are shown in Table 11.1.

L. Zhao et al., *Pneumatic Servo Systems Analysis*, Advances in Industrial Control,
https://doi.org/10.1007/978-981-16-9515-5_11

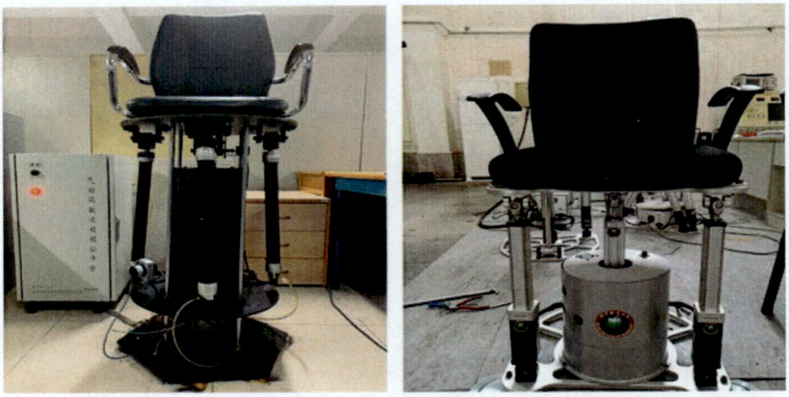

(a) Motion simulation platform driven by PAMs (b) Vehicle motion simulation platform

Fig. 11.1 Motion simulation platform

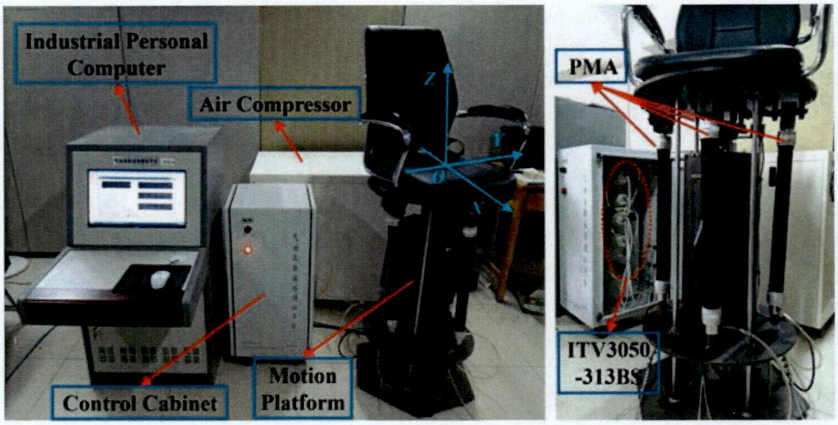

Fig. 11.2 The pneumatic motion simulation platform

The pneumatic motion simulation platform consists of an air compressor (DA7003CS), an industrial personal computer (Advantech, 610H), a control cabinet, and a motion platform. The motion platform contains four PAMs (Festo, DMSP40) and a cylinder (SMC, CE2G100-400). Four pressure proportional valves (SMC, ITV3050-313BS) and a proportional directional control valve (Festo, MPYE-5-3/8-010-B) are installed in the control cabinet. In addition, there are two angle sensors with 2000 P/R resolution (OMRON, E6B2-CWZ3E), a counting card (Advantech, PCL-833), and a D/A card (Advantech, PCL-726). Main components of the pneumatic motion simulation platform are shown in Table 11.2.

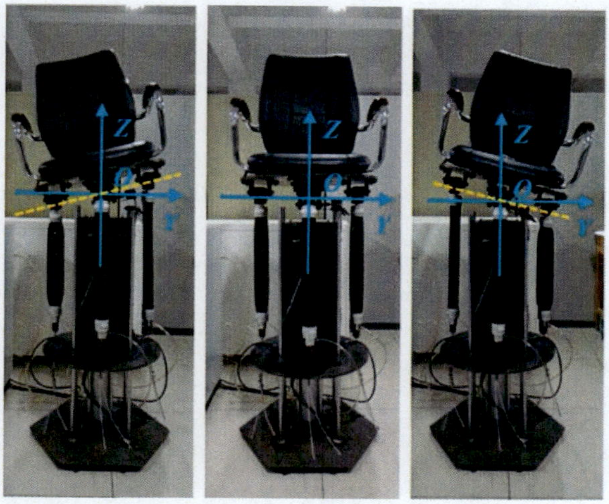

Fig. 11.3 The deflection motion around X-axis for the two-DoF parallel mechanism

Fig. 11.4 The deflection motion around Y-axis for the two-DoF parallel mechanism

Table 11.1 Motion parameters of pneumatic motion simulation platform

DoF	Range of motion
X-axis	$-15° \sim 15°$
Y-axis	$-15° \sim 15°$
Z-axis	$0 \sim 400$ mm

The motion platform is designed as a series–parallel platform with three DoF. It is composed of a two-DoF parallel mechanism of PAMs and a cylinder with measurable stroke in series. It is also driven by the cylinder and the four PAMs. The motion platform is composed of an upper platform and a lower platform. The lower platform is designed as straw hat shape, and the upper platform is connected with the

Table 11.2 Main components of pneumatic motion simulation platform

Component name	Component model	Main performance indicators
PAM	DMSP40(Festo)	Initial diameter: 40 mm
		Initial length: 450 mm
		Working pressure: 0 ~ 1.0 MPa
Cylinder	CE2G100-400 (SMC)	Working stroke: 400 mm
		Working pressure: 0 ~ 1.0 MPa
Pressure proportional valve	ITV3050 (SMC)	Output voltage: 0 ~ 10VDC
		Working pressure: 0 ~ 1.0 MPa
Proportional directional control valve	MPYE-5-3/8-010-B (Festo)	Output voltage: 0 ~ 10VDC
		Working pressure: 0 ~ 1.0 MPa
		Rate of flow: 100 ~ 200 L/min
D/A input	PCL-726	6 path D/A output
		Output range: 0 ~ 10 V
Counter card	PCL-833	Orthogonal frequency: 1.0 Mpa
		Pulse frequency: 2.4 Mpa
		24-bit counter

lower platform by a Hooke hinge. The four PAMs are connected between the upper platform and the lower platform by four flexible steel wires which prevent the PAMs from bending when they are not inflated. Furthermore, some structure sizes of the motion platform are shown in Table 11.3.

11.2.2 Control Circuit

The pneumatic motion simulation platform is driven by four PAMs and one cylinder. A pneumatic circuit of the pneumatic motion simulation platform is shown in Fig. 11.5.

The proportional directional control valve used in the pneumatic circuit is shown in Fig. 11.6, which is a MPYE-type valve produced by Festo.

In the proportional directional control valve, supply voltage is 24 V, setpoint value input is analogue voltage signal from 0 V to 10 V with 5 V as mid-position voltage and standard nominal flow rate is 2000 L/min. As shown in Fig. 11.6 b, a component symbol of the proportional directional control valve is given. Control characteristics of the proportional directional control valve are shown in Fig. 11.8.

As shown in Fig. 11.8, control voltage of the proportional directional control valve is 5 V when the spool of valve is in the middle position. Moreover, dead-zone phenomenon of the proportional directional control valve exists when the control voltage is around 5 V.

Table 11.3 Structure sizes of the experimental platform

Radius of upper platform		250 mm
Radius of lower platform		250 mm
Lower platform sleeve	Radius	220 mm
	Height	540 mm
PAM	Diameter	40 mm
	Length of joint part	2 × 90 mm
	Length of rubber part	450 mm
Cylinder	Diameter	100 mm
	Stroke	400 mm
Distance between upper platform and lower platform	700 mm	
Distance between pedestal and lower platform	80 mm	
Side length of regular hexagon pedestal	300 mm	
Guide length	780 mm	
Radius of position for the PAM installation	200 mm	
Radius of position for the Guide installation	160 mm	
Length of flexible steel wire	60 mm	

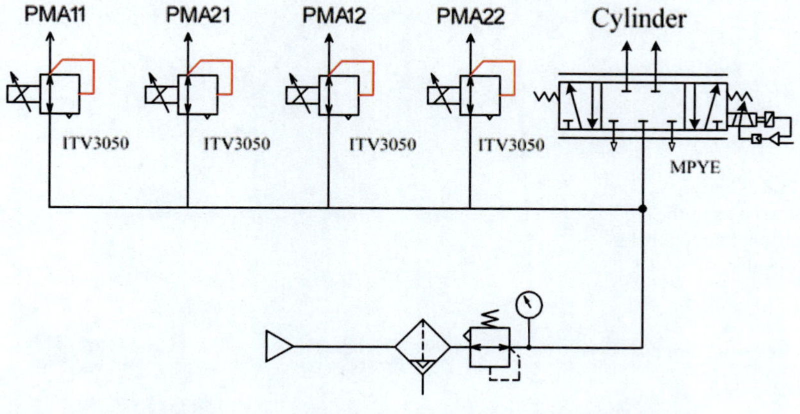

Fig. 11.5 The pneumatic circuit of the pneumatic motion simulation platform

The pressure proportional valve used in the pneumatic circuit is shown in Fig. 11.7, which is produced by SMC company.

Control characteristics of the pressure proportional valve are shown in Fig. 11.9. According to Fig. 11.9, good control characteristics are owned by the pressure proportional valve.

A control process for the two-DoF parallel mechanism of PAMs is shown in Fig. 11.10.

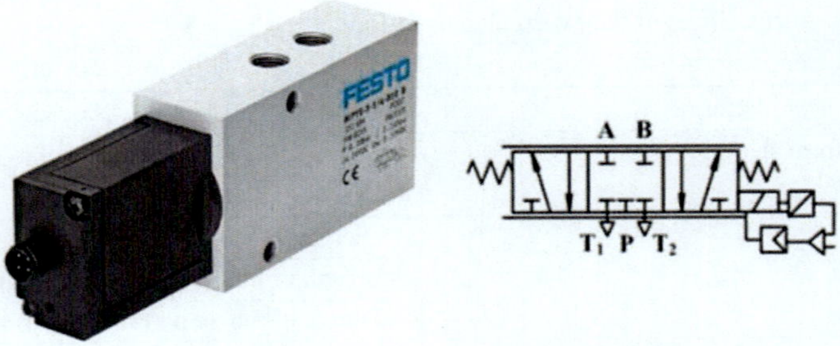

(a) Real prototype (b) Component symbol

Fig. 11.6 Proportional directional control valve

Fig. 11.7 Real prototype of the pressure proportional valve

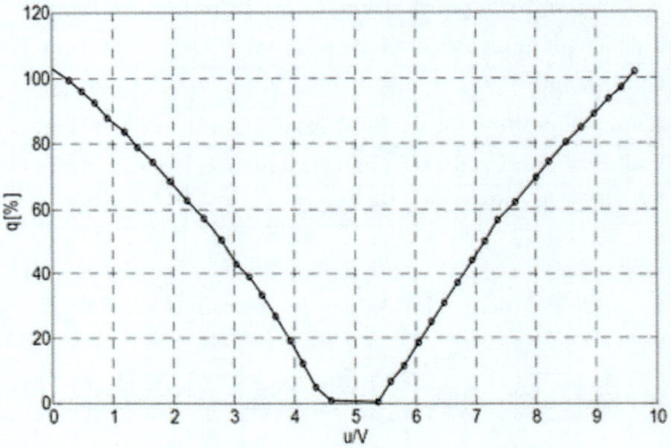

Fig. 11.8 Control characteristics of the proportional directional control valve

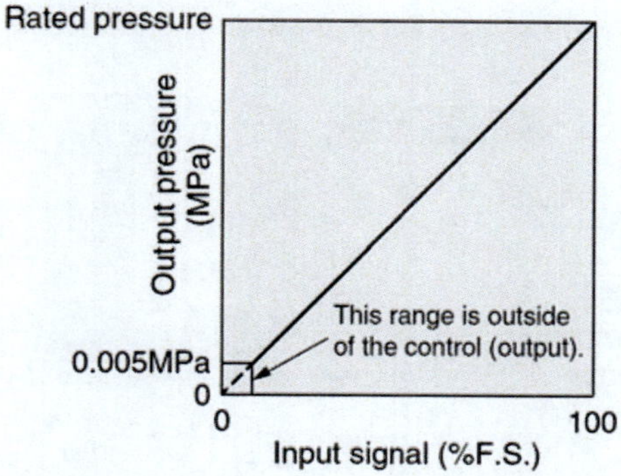

Fig. 11.9 Real prototype of the pressure proportional valve

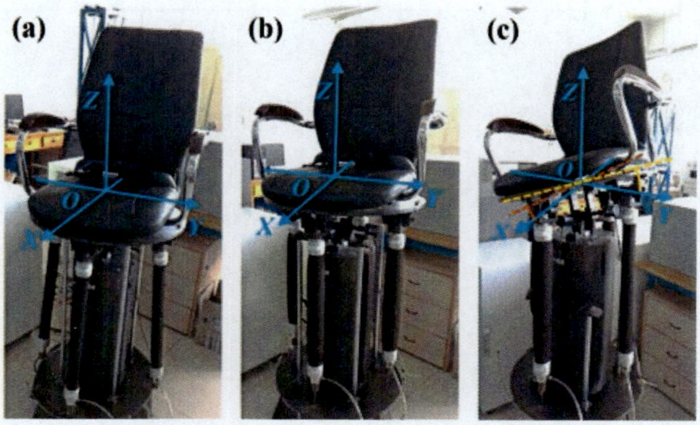

Fig. 11.10 The control process for the parallel mechanism of PAMs

In Fig. 11.10a, the four PAMs and the cylinder are without the compressed air, all input control voltages of the four pressure proportional valves and the proportional directional control valve are 0V. In Fig. 11.10b, the two-DoF parallel mechanism of PAMs rises to a specified position by the cylinder in the direction of Z-axis, and four preloaded internal pressures of PAMs are adjusted by the four pressure proportional valves. Then four preloaded pulling forces of PAMs are produced by the four preloaded internal pressures. In Fig. 11.10c, deflection movements around X-axis and Y-axis for the two-DoF parallel mechanism of PAMs are implemented by four pulling forces of PAMs. Each pulling force is changed with PMA's internal pressure which is adjusted by the pressure proportional valve. According to the control process for the two-DoF parallel mechanism of PAMs, a control circuit for the two-DoF parallel mechanism of PAMs is shown in Fig. 11.11.

In this chapter, single-DoF angle tracking control for the motion mechanism of PAMs is studied. A schematic diagram on control structure for the motion mechanism of PAMs is shown in Fig. 11.12.

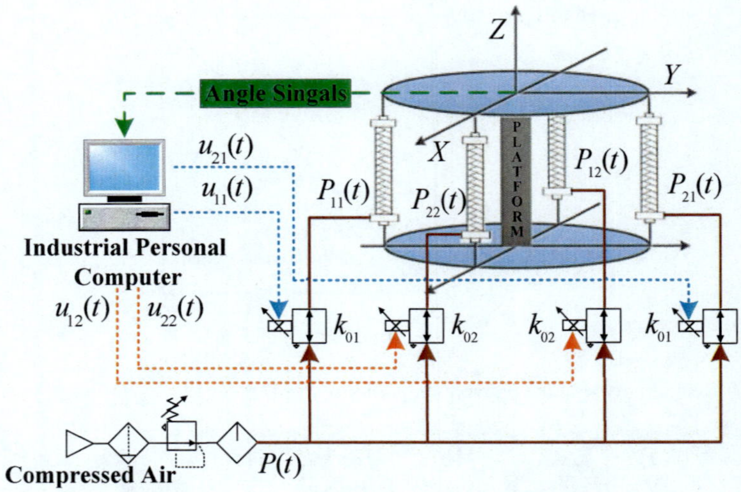

Fig. 11.11 The control structure for the two-DoF parallel mechanism of PAMs

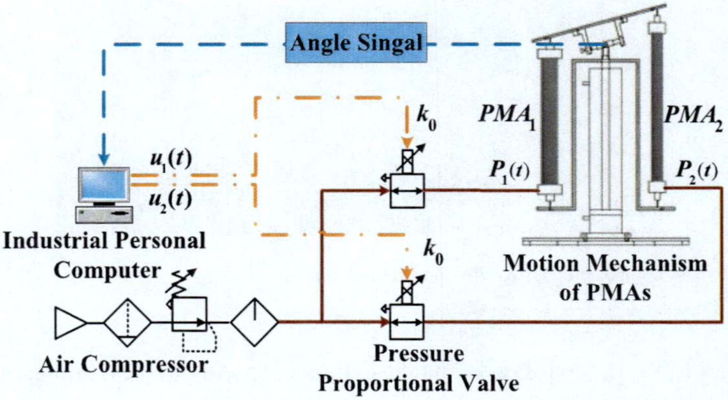

Fig. 11.12 Control structure for the motion mechanism of PAMs

In Fig. 11.12, $u_1(t)$ and $u_2(t)$ are two control voltages of the pressure proportional valves, $P_1(t)$ and $P_2(t)$ are two internal pressures of PAMs, and k_0 is a proportional coefficient of the control voltages and output pressures for the pressure proportional valves. Note that the output pressures for the pressure proportional valves are the internal pressures of PAMs.

11.3 System Model

Considering the control circuit as shown in Fig. 11.12, two control variables for the mechanism of PAMs are presented as follows:

$$\begin{cases} u_1(t) = u_0 + k_u u(t) \\ u_2(t) = u_0 - k_u u(t), \end{cases} \tag{11.1}$$

where k_u is a coefficient of voltage distribution, u_0 is a preloaded voltage, $u(t)$ is a control law, and $u_1(t)$ and $u_2(t)$ are two input control voltages of pressure proportional valves. Then two internal pressures of PAMs are given as follows:

$$\begin{cases} P_1(t) = P_0 + \Delta P(t) = k_0 u_1(t) \\ P_2(t) = P_0 - \Delta P(t) = k_0 u_2(t), \end{cases} \tag{11.2}$$

where k_0 is a proportional coefficient of the input control voltages and the output pressures for pressure proportional valves, P_0 is a preload internal pressure of PAMs, $\Delta P(t)$ is a variation of pressure, and $P_1(t)$ and $P_2(t)$ are two internal pressures of PAMs. According to [1], the following mathematical models of PAMs are obtained as

$$\begin{cases} F_1(t) = P_1(t)(k_1\varepsilon_1^2(t) + k_2\varepsilon_1(t) + k_3) + k_4 \\ F_2(t) = P_2(t)(k_1\varepsilon_2^2(t) + k_2\varepsilon_2(t) + k_3) + k_4, \end{cases} \tag{11.3}$$

where $F_1(t)$ and $F_2(t)$ are two pulling forces of PAMs; k_1, k_2, k_3, and k_4 are four parameters for the mathematical models of PAMs; and $\varepsilon_1(t)$ and $\varepsilon_2(t)$ are two shrinking rates which are ratios of contraction lengths and initial lengths for PAMs. The deflection movement for the mechanism of PAMs is driven by the two pulling forces of PAMs. A schematic force for the mechanism of PAMs is shown in Fig. 11.13.

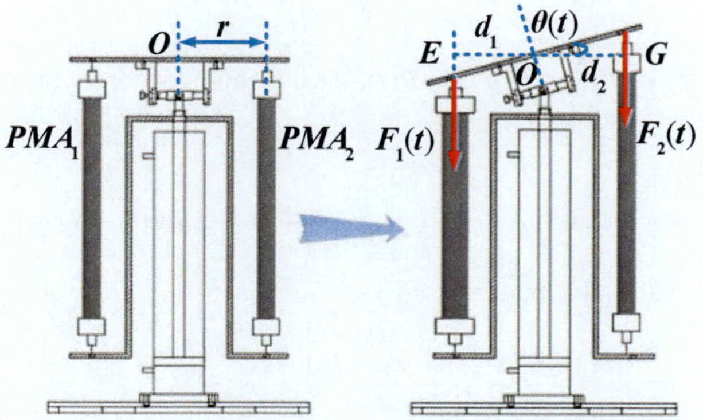

Fig. 11.13 Schematic force for the mechanism of PAMs

In Fig. 11.13, $\theta(t)$ is a deflection angle for the mechanism of PAMs, r is a radius of upper platform for the mechanism of PAMs, both $OE = d_1$ and $OG = d_2$ express two values for arm of force. Because the deflection angle $\theta(t)$ is small enough, let $d_1 = r$ and $d_2 = r$ in modeling the dynamical system for the mechanism of PAMs. The relationships between the shrinkage rates of PAMs and the deflection angle for the mechanism of PAMs are given as follows:

$$\begin{cases} \varepsilon_1(t) = \varepsilon_0 + rL_0^{-1}\theta(t) \\ \varepsilon_2(t) = \varepsilon_0 - rL_0^{-1}\theta(t), \end{cases} \tag{11.4}$$

where ε_0 and L_0 are the initial shrinking rate and initial length for PAMs, respectively. By law of rotation, the moment $T(t)$ for the mechanism of PAMs is obtained as follows:

$$T(t) = J\ddot{\theta}(t) + b_v\dot{\theta}(t) = F_1(t)d_1 - F_2(t)d_2 + \vartheta(t), \tag{11.5}$$

where J is a moment of inertia for the mechanism of PAMs, b_v is a coefficient of damping, and $\vartheta(t)$ indicates an unknown term such as external disturbances and unmodeled dynamics for the mechanism of PAMs. Substituting (11.2), (11.3), and (11.4) into (11.5), the expression of moment $T(t)$ is derived as follows:

$$\begin{aligned} T(t) = {} & k_0u_0r(4k_1\varepsilon_0 rL_0^{-1} + 2k_2rL_0^{-1})\theta(t) + k_0k_ur(2k_1\varepsilon_0^2 \\ & +2k_1(r\theta(t)L_0^{-1})^2 + 2k_2\varepsilon_0 + 2k_3)u(t) + \vartheta(t), \end{aligned} \tag{11.6}$$

where $2k_1(r\theta(t)L_0^{-1})^2u(t)$ is considered in $\vartheta(t)$. Then the dynamic model for the mechanism of PAMs is obtained from (11.6) as follows:

$$\begin{aligned} \ddot{\theta}(t) = {} & -\frac{b_v}{J}\dot{\theta}(t) + \frac{2k_0u_0r^2(2k_1\varepsilon_0 + k_2)L_0^{-1}}{J}\theta(t) \\ & + \frac{2k_0k_ur(k_1\varepsilon_0^2 + k_2\varepsilon_0 + k_3)}{J}u(t) + \nu(t). \end{aligned} \tag{11.7}$$

Letting $x_1(t) = \theta(t)$ and $x_2(t) = \dot{\theta}(t)$, the dynamic system for the mechanism of PAMs is obtained as follows:

$$\begin{cases} \dot{x}_1(t) = x_2(t) \\ \dot{x}_2(t) = b_1x_1(t) + b_2x_2(t) + \nu(t) + (b - b_0)u(t) + b_0u(t), \end{cases} \tag{11.8}$$

where b_0 is an adjustable parameter,

$$b = \frac{2k_0k_ur(k_1\varepsilon_0^2 + k_2\varepsilon_0 + k_3)}{J},$$

$$b_1 = \frac{2k_0u_0r^2(2k_1\varepsilon_0 + k_2)L_0^{-1}}{J}, \quad b_2 = -\frac{b_v}{J}.$$

Table 11.4 Parameters in a PAM mechanism model

Parameters		
$k_0 = 0.09$	$\varepsilon_0 = 0.111$	$r = 200$ mm
$k_1 = 0.0501$	$k_2 = -0.0472$	$k_3 = 0.0106$
$L_0 = 450$ mm	$u_0 = 2$ V	$k_u = 0.45$

Note that the uncertain term $\nu(t)$ in the dynamic system for the mechanism of PAMs (11.8) is continuously differentiable and bounded as in [1]. Furthermore, the item $\nu(t) + (b - b_0)u(t)$ is treated as an extended state $x_3(t)$, then the dynamic system for the mechanism of PAMs (11.8) is rewritten as follows:

$$\begin{cases} \dot{x}_1(t) = x_2(t) \\ \dot{x}_2(t) = b_1 x_1(t) + b_2 x_2(t) + x_3(t) + b_0 u(t) \\ \dot{x}_3(t) = h(t, X) \\ y(t) = x_1(t), \end{cases} \tag{11.9}$$

where $h(t, X)$ is a derivative of $x_3(t)$, X is a variable associated with states $x_1(t)$, $x_2(t)$, and $x_3(t)$.

11.4 Simulation and Results

In this section, a simulation is given to show the effectiveness of the dynamic system for the mechanism of PAMs. All constant parameters of the dynamic system (11.9) for the mechanism of PAMs are given in Table 11.4.

According to the dynamic system (11.9) for the mechanism of PAMs, a simulation is carried out. A sinusoidal signal with 0.5Hz frequency and 10° amplitude is given as a given desired signal $v_0(t)$. A PID control method is used to control the dynamic system (11.9) for the mechanism of PAMs. Simulation results are shown in Figs. 11.14 and 11.15.

According to the simulation results shown in Figs. 11.14 and 11.15, the mechanism of PAMs can be represented by the dynamic nonlinear system (11.9).

11.5 Conclusion

In this chapter, a three-DoF pneumatic motion simulation platform has been introduced. Platform components of the motion simulation platform have been shown and listed in this chapter. Pneumatic circuit and control circuit for the motion simulation platform have been given to show structure of the motion simulation plat-

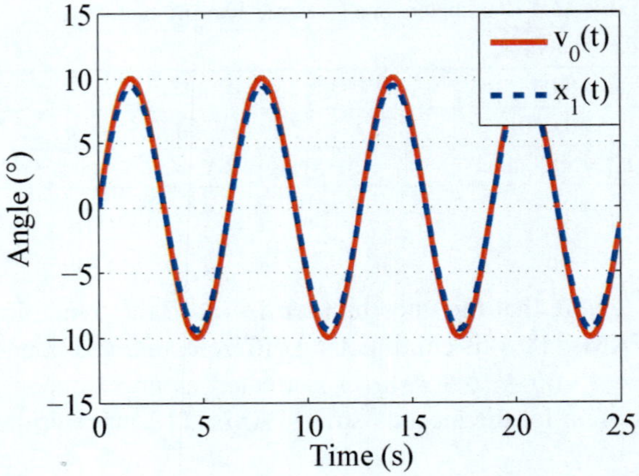

Fig. 11.14 Angle tracking control of the dynamic system for the mechanism of PAMs

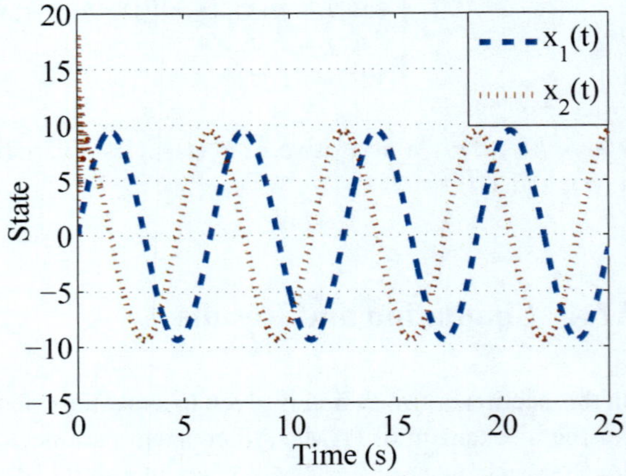

Fig. 11.15 States of the dynamic system for the mechanism of PAMs

form. Moreover, a dynamic nonlinear system for the motion simulation platform has been derived to express the motion simulation platform. Finally, effectiveness of the dynamic nonlinear system has been proved by simulation.

Reference

1. Yang H, Yu Y, Zhang J (2017) Angle tracking of a pneumatic muscle actuator mechanism under varying load conditions. Control Eng Pract 61:1–10

Chapter 12
Nonlinear Feedback Control

12.1 Introduction

Due to the advantages of rapid response, low cost, and high power–weight ratio, PAMs are used in the pneumatic motion simulation platform. Moreover, PAMs are also applied widely in other various situations. A two-axis planar articulated robot driven by four PAMs has been introduced [1]. PAMs have also been used in a specially designed hand rehabilitation device [2]. A sliding mode controller has been designed for a linear driver with an antagonistic pair of PAMs in [3]. Three strategies for compensation on hysteresis in force characteristics of PAMs have been shown in [4]. An adaptive wearable ankle robot which is manufactured by PAMs has been presented in [5]. In [6], a practical approach to deal with variant payloads has been proposed for a two-DoF parallel manipulator. A friction observer has been employed for friction compensation control by considering load changes [7]. Since it was proposed, active disturbance rejection control has been successfully used in a lot of nonlinear systems [8–10]. The active disturbance rejection control has been applied to achieve precise position control for magnetic rodless cylinders [11]. In [12], the active disturbance rejection control has been adopted to solve a robot trajectory tracking problem. As an important part of the active disturbance rejection control, ESOs are widely used to estimate uncertain terms in nonlinear systems [13, 14].

In this chapter, a pneumatic motion simulate system driven by PAMs is studied. In order to control the mechanism motion as accurately as desired, particular attention is taken onto the ADRC. The varying load conditions are treated as external disturbances which are estimated by a linear ESO. A nonlinear error feedback controller is presented to compensate negative impacts induced by the varying loads. Convergence analysis of the linear ESO and stability analysis of the closed-loop pneumatic system are performed, respectively. Finally, experimental results indicate that the proposed active disturbance rejection control approach has distinct advantage in dealing with varying loads for the mechanism with PAMs.

L. Zhao et al., *Pneumatic Servo Systems Analysis*, Advances in Industrial Control,
https://doi.org/10.1007/978-981-16-9515-5_12

The main contributions of this chapter are summarized as follows:

i. ADRC is proposed for a PMA mechanism to achieve angle tracking precisely under varying load conditions.
ii. The varying load conditions are treated as external disturbances which are estimated by a linear extended state observer.
iii. An active disturbance rejection controller is presented to compensate negative impacts induced by the varying loads.

12.2 Main Results

12.2.1 Schematic Diagram of Control Method

The ADRC is developed to guarantee control precision in face of varying loads, unknown disturbances, and nonlinear terms for system (11.9). A schematic diagram of the ADRC for the PMA mechanism is shown in Fig. 12.1.

It is shown from Fig. 12.1 that the ADRC consists of three parts: a tracking differentiator, a linear ESO, and a nonlinear error feedback controller. To prevent the occurrence of saturation in the initial stage of deflection angle $\theta(t)$, a nonlinear tracking differentiator is designed to arrange a transition process. So an abrupt input signal is turned into a continuous smooth signal which avoids damaging to equipment. At the same time, the differential signal of the continuous smooth signal is also obtained as in [15]. For system (11.9), a second-order tracking differentiator is designed as follows:

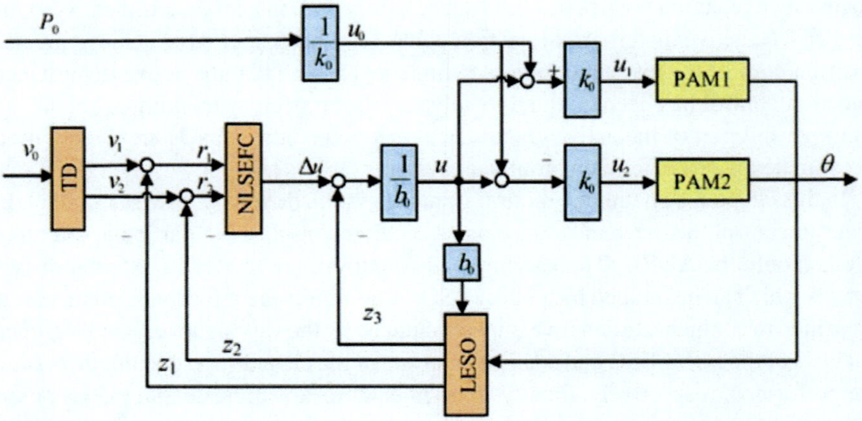

Fig. 12.1 Schematic diagram of ADRC

$$\begin{cases} f_h(t) = f_{han}(v_1(t) - v_0(t), v_2(t), r_0, h_0) \\ \dot{v}_1(t) = v_2(t) \\ \dot{v}_2(t) = f_h(t), \end{cases} \tag{12.1}$$

where $v_0(t)$ is the given trajectory, $v_1(t)$ is the tracking trajectory of $v_0(t)$, $v_2(t)$ is the differential signal of $v_1(t)$, and h_0 and r_0 are the filter factor and velocity factor, respectively. The filtering effect of the tracking differentiator is positive correlation with the filter factor h_0. Along with the increase of r_0, the tracking velocity of the tracking differentiator becomes fast. However, the accurate transient process cannot be obtained if h_0 and r_0 are beyond the range. That is, both $v_1(t)$ and $v_2(t)$ cannot be obtained precisely and quickly.

12.2.2 *Linear Extended State Observer*

Different from general observers, the ESO is not only used to estimate state variables, but also estimates total unknown nonlinear terms. In this section, a linear ESO is introduced to deal with the unknown nonlinearity in system (11.9). Note that an extended state $x_3(t)$ is chosen as $x_3(t) = f(x(t), \omega(t))$. Moreover, the derivative of $f(x(t), \omega(t))$ is set as $h(x(t), u(t), \omega(t), \dot{\omega}(t))$, i.e., $h(x(t), u(t), \omega(t), \dot{\omega}(t)) = \dot{f}(x(t), \omega(t))$. Therefore the system (11.9) is rewritten as

$$\begin{cases} \dot{x}_1(t) = x_2(t) \\ \dot{x}_2(t) = x_3(t) + b_0 u(t) \\ \dot{x}_3(t) = h(x(t), u(t), \omega(t), \dot{\omega}(t)) \\ y(t) = x_1(t). \end{cases} \tag{12.2}$$

The third-order linear ESO is designed as follows:

$$\begin{cases} e_1(t) = z_1(t) - y(t) \\ \dot{z}_1(t) = z_2(t) - \beta_{01} e_1(t) \\ \dot{z}_2(t) = z_3(t) - \beta_{02} e_1(t) + b_0 u(t) \\ \dot{z}_3(t) = -\beta_{03} e_1(t), \end{cases} \tag{12.3}$$

where $z_1(t)$, $z_2(t)$, and $z_3(t)$ are the estimates of $x_1(t)$, $x_2(t)$, and $x_3(t)$, respectively, $\beta_{01} > 0$, $\beta_{02} > 0$, and $\beta_{03} > 0$ are three adjustable parameters.

According to (12.2) and (12.3), the following error system is obtained:

$$\begin{cases} e_1(t) = z_1(t) - y(t) \\ \dot{e}_1(t) = e_2(t) - \beta_{01} e_1(t) \\ \dot{e}_2(t) = e_3(t) - \beta_{02} e_1(t) \\ \dot{e}_3(t) = -h(x(t), u(t), \omega(t), \dot{\omega}(t)) - \beta_{03} e_1(t), \end{cases} \tag{12.4}$$

which is rewritten as

$$\dot{E}(t) = AE(t) + Bh(x(t), u(t), \omega(t), \dot{\omega}(t)), \tag{12.5}$$

where

$$E(t) = \begin{bmatrix} e_1(t) \\ e_2(t) \\ e_3(t) \end{bmatrix}, \; A = \begin{bmatrix} -\beta_{01} & 1 & 0 \\ -\beta_{02} & 0 & 1 \\ -\beta_{03} & 0 & 0 \end{bmatrix}, \; B = \begin{bmatrix} 0 \\ 0 \\ -1 \end{bmatrix}.$$

To deal with the differential equation (12.5), it follows that

$$E(t) = e^{At}(t)E(0) + e^{At}(t)\int_0^t e^{-A\tau}(t)Bh(\tau)d\tau, \tag{12.6}$$

where A is a Hurwitz matrix by choosing parameters β_{0i}, $i = 1, 2, 3$.

Remark 12.1 Note that equality (12.6) is a standard Luenberger observer that estimates both states and a disturbance variable of system (11.9) by measurement x_1. All eigenvalues of the estimation error dynamics can be placed by β_{ij} and guarantee asymptotic stability of the estimation error system (12.4). Therefore, the convergence analysis of the estimation error system (12.4) is dependent on the eigenvalues of matrix A in this chapter.

Before analyzing the convergence of the estimation error system (12.4), the following lemma is given to show main results in this paper.

Lemma 12.2 ([2]) *For* $\forall\lambda < 0$, *the following equation*

$$\lim_{t\to\infty}\int_0^t (t-\tau)^k e^{\lambda(t-\tau)}(t)d\tau = \frac{k!}{(-\lambda)^{k+1}} \tag{12.7}$$

holds.

Proof Setting $t - \tau = s$, one has that

$$\int_0^t (t-\tau)^k e^{\lambda(t-\tau)}(t)d\tau = \int_0^t s^k e^{\lambda s}(t)ds.$$

Then (12.7) is rewritten as

$$\lim_{t\to\infty}\int_0^t s^k e^{\lambda s}(t)ds = \frac{k!}{(-\lambda)^{k+1}}. \tag{12.8}$$

Step 1: When $k = 0$, it is obtained that

$$\lim_{t\to\infty}\int_0^t e^{\lambda s}(t)ds = \lim_{t\to\infty}\frac{1}{\lambda}(e^{\lambda t}(t) - 1).$$

Due to $\lambda < 0$, it follows that $\lim_{t\to\infty} e^{\lambda t}(t) = 0$. One has that

$$\lim_{t\to\infty}\int_0^t e^{\lambda s}(t)ds = -\frac{1}{\lambda},$$

which illustrates that (12.7) holds.

Step 2: Note that Eq. (12.7) is satisfied with $k = p$, i.e.,

$$\lim_{t\to\infty}\int_0^t s^p e^{\lambda s}(t)ds = \frac{p!}{(-\lambda)^{p+1}}.$$

Step 3: When $k = p + 1$, there exists

$$\begin{aligned}\lim_{t\to\infty}\int_0^t s^{p+1}e^{\lambda s}(t)ds &= \lim_{t\to\infty}\frac{1}{\lambda}\int_0^t s^{p+1}de^{\lambda s}(t)\\ &= \lim_{t\to\infty}\frac{1}{\lambda}\left(t^{p+1}e^{\lambda t}(t) - (p+1)\int_0^t s^p e^{\lambda s}(t)ds\right).\end{aligned}$$

For $\lambda < 0$, it follows that

$$\lim_{t\to\infty} t^{p+1}e^{\lambda t}(t) = 0.$$

Moreover, one has that

$$\begin{aligned}\lim_{t\to\infty}\int_0^t s^{p+1}e^{\lambda s}(t)ds &= -\frac{1}{\lambda}\lim_{t\to\infty}(p+1)\int_0^t s^p e^{\lambda s}(t)ds\\ &= -\frac{1}{\lambda}\frac{p!}{(-\lambda)^{p+1}} = \frac{(p+1)!}{(-\lambda)^{p+2}}.\end{aligned}$$

Therefore, Eq. (12.7) holds for $k = p + 1$. The lemma is shown.

In order to analyze the convergence of the error system (12.4), an approach in [2] is adopted in the following theorem.

Theorem 12.3 *Consider the third-order linear ESO (12.3) and the estimation error system (12.4). If the uncertain term $f(x(t), \omega(t))$ is bounded for all t, i.e., $|f(x(t), \omega(t))| \le M,\ t \ge 0$, and $E(0)$ is also bounded, then there exist appropriate parameters β_{01}, β_{02}, and β_{03} such that $\lim_{t\to\infty} E(t)$ is bounded.*

Proof Because $E(0)$ is bounded, the first term of the right side of (12.6) converges to zero, i.e.,

$$\lim_{t\to\infty} e^{At}(t)E(0) = 0. \tag{12.9}$$

Denote the second term of (12.6) as the following function:

$$H(t) = \int_0^t e^{A(t-\tau)}(t)Bh(\tau)d\tau. \tag{12.10}$$

Let A have r different eigenvalues $\lambda_i < 0,\ i = 1, 2, \ldots, r$, where the multiple numbers of λ_i are l_i and

$$\sum_{i=1}^{r} l_i = 3.$$

Then $e^{A(t-\tau)}(t)$ is expressed as follows:

$$e^{A(t-\tau)}(t) = \sum_{i=1}^{r} e^{\lambda_i(t-\tau)}(t) \sum_{j=0}^{l_i-1} P_{ij} \frac{(t-\tau)^j}{j!}, \tag{12.11}$$

where $P_{ij} \in R^{3\times 3}$ is a constant matrix determined by A. Substituting (12.11) into (12.10), one has that

$$H(t) = \int_0^t \sum_{i=1}^{r} \sum_{j=0}^{l_i-1} P_{ij} B \frac{(t-\tau)^j}{j!} e^{\lambda_i(t-\tau)}(t) h(\tau) d\tau. \tag{12.12}$$

By $|f(x(t), \omega(t))| < M$, there exists an upper bound of $|H(t)|$ as follows:

$$\begin{aligned}
|H(t)| &= \left| \int_0^t \sum_{i=1}^{r} \sum_{j=0}^{l_i-1} P_{ij} B \frac{(t-\tau)^j}{j!} e^{\lambda_i(t-\tau)}(t) h(\tau) d\tau \right| \\
&= \left| \int_0^t \sum_{i=1}^{r} \sum_{j=0}^{l_i-1} P_{ij} B \frac{(t-\tau)^j}{j!} e^{\lambda_i(t-\tau)}(t) df(\tau) \right| \\
&\le \left| \sum_{i=1}^{r} \sum_{j=0}^{l_i-1} P_{ij} B \int_0^t \left(\frac{\lambda_i (t-\tau)^j}{j!} + \frac{(t-\tau)^{j-1}}{(j-1)!} \right) e^{\lambda_i(t-\tau)}(t) f(\tau) d\tau \right| \\
&\quad + \left| \sum_{i=1}^{r} \sum_{j=0}^{l_i-1} P_{ij} B \frac{(t-\tau)^j}{j!} e^{\lambda_i(t-\tau)}(t) f(\tau) \Big|_0^t \right| \\
&= \left| \sum_{i=1}^{r} \sum_{j=0}^{l_i-1} P_{ij} B \int_0^t \left(\frac{\lambda_i (t-\tau)^j}{j!} + \frac{(t-\tau)^{j-1}}{(j-1)!} \right) e^{\lambda_i(t-\tau)}(t) f(\tau) d\tau \right| \\
&\quad + \left| \sum_{i=1}^{r} P_{i0} B f(x(t), \omega(t)) - \sum_{i=1}^{r} \sum_{j=0}^{l_i-1} P_{ij} B \frac{t^j}{j!} e^{\lambda_i t}(t) f(0) \right|
\end{aligned}$$

under the supposition that $(t-\tau)^{j-1} = 0$ if $j - 1 < 0$. According to Lemma 12.2 and $|f(x(t), \omega(t))| \le M$, it is obtained that

$$\lim_{t\to\infty}\left|\sum_{i=1}^{r}\sum_{j=0}^{l_i-1}P_{ij}B\int_0^t\left(\frac{\lambda_i(t-\tau)^j}{j!}+\frac{(t-\tau)^{j-1}}{(j-1)!}\right)e^{\lambda_i(t-\tau)}(t)f(\tau)d\tau\right|$$
$$\leq\sum_{i=1}^{r}\sum_{j=0}^{l_i-1}|P_{ij}B|\,M\frac{2}{-(\lambda_i)^j}.$$

Therefore, one has that

$$\lim_{t\to\infty}|H(t)|\leq\sum_{i=1}^{r}|P_{i0}B|M+\sum_{i=1}^{r}\sum_{j=0}^{l_i-1}|P_{ij}B|M\frac{2}{(-\lambda_i)^j}. \tag{12.13}$$

That is, $\lim_{t\to\infty}E(t)$ is bounded when parameters β_{01}, β_{02}, and β_{03} are appropriately designed.

Remark 12.4 The internal model principle has been introduced to guarantee the robustness of system; however, the mathematical models for perturbations and disturbances as an important prerequisite are necessary. That is, only the disturbance model can be obtained accurately and embedded in controllers, and the disturbances can be entirely canceled by the internal model principle [16]. However, the disturbances are unknown in the PMA mechanism system (11.9). Therefore, the linear ESO (12.3) which is used to estimate the disturbances and the convergence of the error system (12.4) is also analyzed in this chapter.

12.2.3 Nonlinear Error Feedback Controller

In this subsection, a nonlinear error feedback controller is designed such that the output angle θ tracks a desired reference input. Error signals between the tracking differentiator (12.1) and the linear ESO (12.3) are given as follows:

$$\begin{cases} r_1(t)=v_1(t)-z_1(t)\\ r_2(t)=v_2(t)-z_2(t),\end{cases} \tag{12.14}$$

where $z_1(t)$ and $z_2(t)$ are the observations of $x_1(t)$ and $x_2(t)$, respectively. The state error feedback controller $u(t)$ is designed as

$$u=\frac{\Delta u(t)-z_3(t)}{b_0}, \tag{12.15}$$

where $\Delta u(t)=\beta_1 fal(r_1(t),\alpha_1,\delta)+\beta_2 fal(r_2(t),\alpha_2,\delta)$, both β_1 and β_2 are two gains of the state error feedback controller (12.15), $z_3(t)$ is the compensation of system (11.9), both $fal(r_1(t),\alpha_1,\delta)$ and $fal(r_2(t),\alpha_2,\delta)$ are two nonlinear functions. To avoid notational overhead, the two variables α and δ are skipped for $fal(r(t), \alpha, \delta)$, when the context is clear.

The two error variables $s_1(t)$ and $s_2(t)$ are set such that

$$\begin{cases} s_1(t) = v_1(t) - x_1(t) \\ s_2(t) = v_2(t) - x_2(t). \end{cases} \tag{12.16}$$

Considering the system (11.9), the differential equation of (12.16) is given as follows:

$$\begin{cases} \dot{s}_1(t) = s_2(t) \\ \dot{s}_2(t) = v_3(t) - x_3(t) - b_0 u(t), \end{cases} \tag{12.17}$$

where $v_3(t)$ is the derivative of $v_2(t)$. From error system (12.4) and equality (12.14), we get the following equation:

$$\begin{cases} r_1(t) = s_1(t) - e_1(t) \\ r_2(t) = s_2(t) - e_2(t). \end{cases} \tag{12.18}$$

By Eqs. (12.17) and (12.18), one has that

$$\begin{cases} \dot{r}_1(t) = r_2(t) - \dot{e}_1(t) + e_2(t) \\ \dot{r}_2(t) = v_3(t) - x_3(t) - \dot{e}_2(t) - b_0 u(t). \end{cases} \tag{12.19}$$

The following theorem is given to show the stabilization of the error system (12.19).

Theorem 12.5 *Considering the closed-loop system (12.19), there exist two appropriate parameters r and h in the tracking differentiator (12.1); parameters β_{01}, β_{02}, and β_{03} in the linear ESO (12.3); and parameters β_1 and β_2 in the nonlinear state error feedback controller (12.15) such that the closed-loop system (12.19) is stable. That is, the actual deflection angle $x_1(t)$ can track the desired angle $v_0(t)$ for the PMA mechanism system (11.9).*

Proof Based on system (12.19), the following candidate Lyapunov function is selected:

$$V(r_1(t), r_2(t)) = \frac{1}{2}r_1^2(t) + \frac{1}{2}r_2^2(t) + |r_1(t)||r_2(t)| \tag{12.20}$$

with $\dot{V}(0, 0) = 0$. Considering the positive and negative of $r_1(t)$ and $r_2(t)$, there exist two cases for the Lyapunov function (12.20) such that

$$V_1(t) = \frac{1}{2}r_1^2(t) + \frac{1}{2}r_2^2(t) + r_1(t)r_2(t)$$

and

$$V_2(t) = \frac{1}{2}r_1^2(t) + \frac{1}{2}r_2^2(t) - r_1(t)r_2(t).$$

The derivative of $V_1(t)$ is computed as

$$\begin{aligned}\dot{V}_1(t) &= r_1(t)\dot{r}_1(t) + r_2(t)\dot{r}_2(t) + \dot{r}_1(t)r_2(t) + r_1(t)\dot{r}_2(t) \\ &= r_1(t)(r_2(t) + \beta_{01}e_1(t)) + r_2(t)(v_3(t) - \beta_1 fal(r_1(t)) \\ &\quad +e_3(t) - \beta_2 fal(r_2(t)) - e_3(t) + \beta_{02}e_1(t)) + (r_2(t) \\ &\quad +\beta_{01}e_1(t))r_2(t) + r_1(t)(v_3(t) + e_3(t) - \beta_1 fal(r_1(t)) \\ &\quad -\beta_2 fal(r_2(t)) - e_3(t) + \beta_{02}e_1(t)) \\ &= r_1(t)r_2(t) + \beta_{01}s_1(t)e_1(t) + r_2(t)(v_3(t) + \beta_{02}e_1(t)) \\ &\quad -\beta_1 r_2(t) fal(r_1(t)) - \beta_2 r_2(t) fal(r_2(t)) + \beta_{01}e_1(t)s_2(t) \\ &\quad -\beta_{01}e_1(t)e_2(t) + r_1(t)(v_3(t) + \beta_{02}e_1(t)) + r_2^2(t) \\ &\quad -\beta_1 r_1(t) fal(r_1(t)) - \beta_2 r_1(t) fal(r_2(t)) - \beta_{01}e_1^2(t).\end{aligned} \tag{12.21}$$

As shown in (12.21), $\dot{V}_1(t)$ is related to the piecewise function $fal(r(t))$. The properties of $fal(r(t))$ are used to analyze the size of $\dot{V}_1(t)$ in the following. For simplify analysis, letting $\beta_1 = \beta_2 = \beta^*$, the derivative of $V_1(t)$ is rewritten as follows:

$$\begin{aligned}\dot{V}_1(t) &= r_1(t)r_2(t) + (r_1(t) + r_2(t))(v_3(t) + \beta_{02}e_1(t)) \\ &\quad +\beta_{01}e_1(t)(s_1(t) + s_2(t)) - \beta_{01}e_1(t)e_2(t) - \beta_{01}e_1^2(t) \\ &\quad -\beta^*(r_1(t) + r_2(t))(fal(r_1(t)) + fal(r_2(t))) + r_2^2(t).\end{aligned}$$

It is shown that $r_1(t)fal(r_1(t)) \geq 0$ and $r_2(t)fal(r_2(t)) \geq 0$. Based on the characteristic curve of the piecewise function $fal(r)$, the following conclusions are presented:

(i) If $r_1(t) \geq 0$ and $r_2(t) \geq 0$, then $fal(r_1(t)) \geq 0$ and $fal(r_2(t)) \geq 0$, i.e., $(r_1(t) + r_2(t))(fal(r_1(t)) + fal(r_2(t))) \geq 0$.

(ii) If $r_1(t) \leq 0$ and $r_2(t) \leq 0$, then $fal(r_1(t)) \leq 0$ and $fal(r_2(t)) \leq 0$, i.e., $(r_1(t) + r_2(t))(fal(r_1(t)) + fal(r_2(t))) \geq 0$.

(iii) If $r_1(t) > 0 > r_2(t)$ and $r_1(t) > |r_2(t)|$, then $fal(r_1(t)) > 0$, $fal(r_2(t)) < 0$ and $fal(r_1(t)) > |fal(r_2(t))|$, i.e., $(r_1(t) + r_2(t))(fal(r_1(t)) + fal(r_2(t))) > 0$.

(iv) If $r_1(t) > 0 > r_2(t)$ and $r_1(t) < |r_2(t)|$, then $fal(r_1(t)) > 0$, $fal(r_2(t)) < 0$ and $fal(r_1(t)) < |fal(r_2(t))|$, i.e., $(r_1(t) + r_2(t))(fal(r_1(t)) + fal(r_2(t))) > 0$.

One has that

$$(r_1(t) + r_2(t))(fal(r_1(t)) + fal(r_2(t))) \geq 0 \tag{12.22}$$

which holds for any $r_1(t)$ and $r_2(t)$. By setting

$$\begin{aligned}\Delta(t) &= |r_1(t)r_2(t) + (r_1(t) + r_2(t))(v_3(t) + \beta_{02}e_1(t)) \\ &\quad +r_2^2(t) + \beta_{01}e_1(t)(s_1(t) + s_2(t)) - \beta_{01}e_1(t)e_2(t)|\end{aligned}$$

inequality

$$\dot{V}_1(t) < -\beta^*(r_1(t)+r_2(t))(fal(r_1(t))+fal(r_2(t))) - \beta_{01}e_1^2(t) + \Delta(t) \quad (12.23)$$

holds. It is shown for β^* large enough that $\dot{V}_1(t) < 0$ is satisfied with. Therefore, there exists $\dot{V}_1(t) < 0$ by choosing appropriate parameters β_1 and β_2.

For the second case, the analysis approach is similar to the first one by setting $\beta_1 = -\beta_2 = \beta^*$ with $\beta^* > 0$. That is, there are appropriate parameters β_1 and β_2 such that $\dot{V}_2(t) < 0$ holds. Based on the analysis above, there exists $\dot{V}(t) < 0$ by choosing appropriate parameters β_1 and β_2. Therefore, the closed-loop system (12.19) is stable when appropriate parameters are given.

If the proper parameters are chosen, then the desired angle $v_0(t)$ is tracked quickly by the tracing signal $v_1(t)$. The output deflection angle $x_1(t)$ is estimated accurately by the estimation $z_1(t)$. Moreover, the two parameters β_1 and β_2 are large enough in the nonlinear state error feedback controller (12.15), i.e., $r_1(t) = v_1(t) - z_1(t) = 0$. That is, the tracking signal $v_1(t)$ is tracked completely by the estimation $z_1(t)$. Therefore, the desired deflection angle $v_0(t)$ of the pneumatic motion simulate platform is tracked fast and effectively by the deflection angle $x_1(t)$ under the action of the state error feedback controller (12.15).

12.3 Experiments and Results

In order to illustrate the disturbance rejection ability and demonstrate the correctness of theoretical analysis of the proposed active disturbance rejection control approach, experiments with different loads are used in this part as shown in Fig. 12.2.

(a) **(b)**

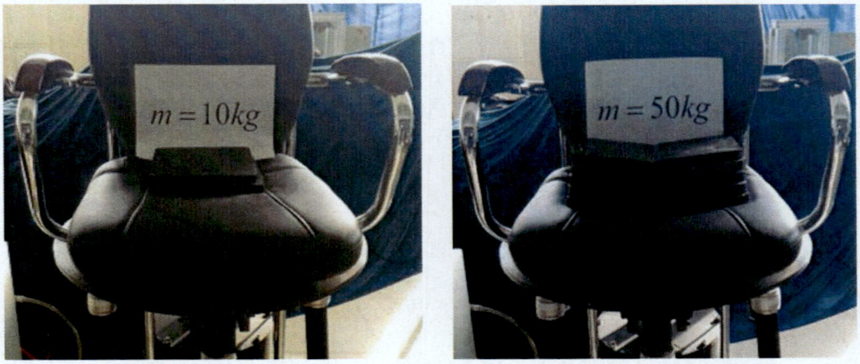

Load mass m = 10 kg. Load mass m = 50 kg.

Fig. 12.2 Experiments of varying load

Table 12.1 Parameters of tracking differentiator and segment-wise function

	Parameters		
Tracking differentiator	$r_0 = 30$	$h_0 = 0.2$	
Segment-wise function	$\alpha_1 = 0.5$	$\alpha_2 = 0.25$	$\delta = 0.05$

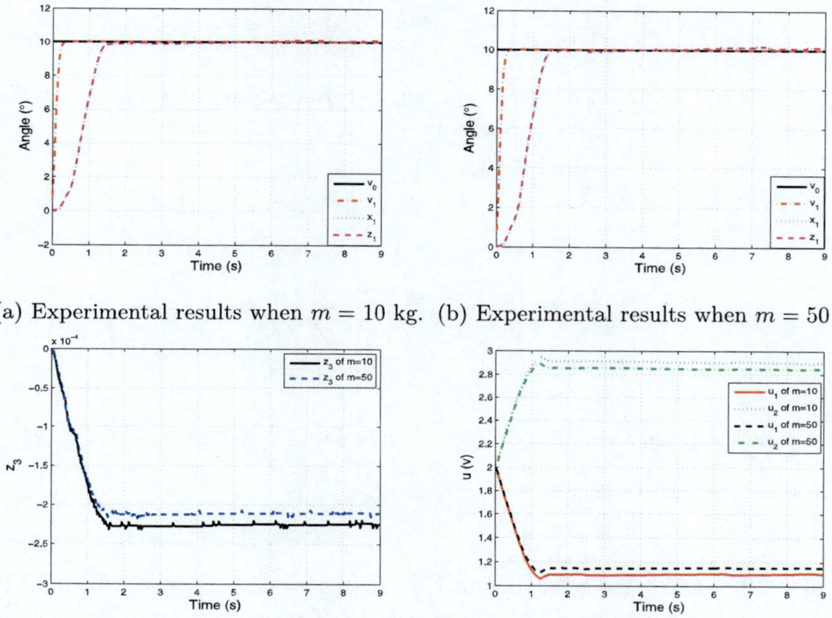

(a) Experimental results when $m = 10$ kg. (b) Experimental results when $m = 50$ kg.

(c) Estimations of nonlinearity. (d) Control voltages.

Fig. 12.3 Experimental results of active disturbance rejection control for a step signal

Note that the load masses are $m = 10$ kg in Fig. 12.2a and $m = 50$ kg in Fig. 12.2b, respectively. Moreover, comparing results with the traditional PID control method is also performed in this experiment. In the first experiment, a step signal at 10° is given and the effectiveness of the developed technique is verified by the comparison with the PID control strategy in this chapter. The two parameters r_0 and h_0 of the tracking differentiator (12.1) and three parameters α_1, α_2, and δ of the nonlinear function are shown in Table 12.1.

Based on the analysis of Theorem 12.3 and trial and error, the parameters of the linear ESO (12.3) are set as $\beta_{01} = 100$, $\beta_{02} = 1000$, and $\beta_{03} = 10$. According to Theorem 12.5, the parameters of the state error feedback controller (12.15) are set as $\beta_1 = 75$, $\beta_2 = 62$, and $b_0 = 6000$. The experimental results obtained by the proposed active disturbance rejection control are shown in Fig. 12.3.

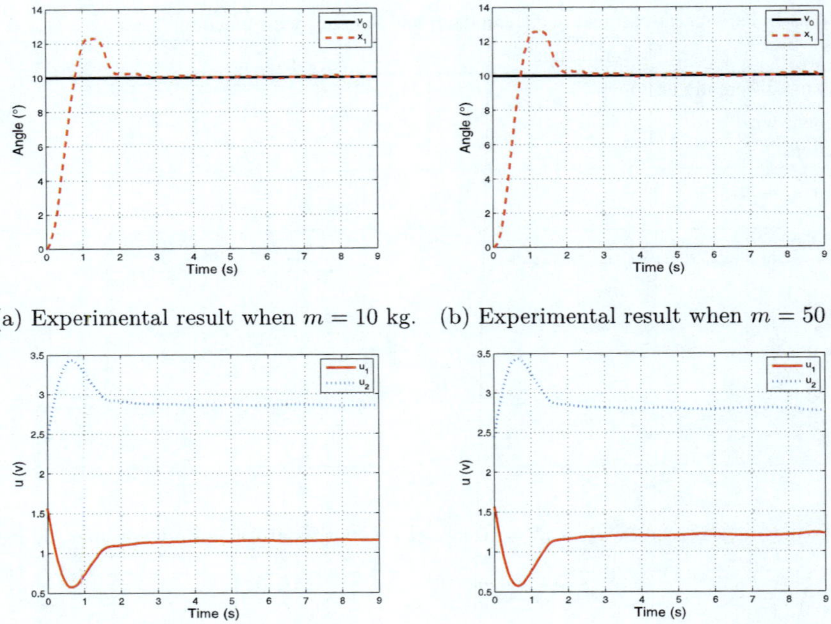

(a) Experimental result when $m = 10$ kg. (b) Experimental result when $m = 50$ kg.

(c) Control voltages when $m = 10$ kg. (d) Control voltages when $m = 50$ kg.

Fig. 12.4 Experimental results of PID for a step signal

In Fig. 12.3a and b, the actual output $x_1(t)$ attains the desired input $v_0(t)$ quickly and accurately no matter $m = 10$ kg or $m = 50$ kg. The output $v_1(t)$ of the tracking differentiator tracks the given signal $v_0(t)$ quickly and steadily. The observed value $z_1(t)$ is also almost coincident with $x_1(t)$. Figure 12.3c shows the estimation of the nonlinear uncertain terms $z_3(t)$ when $m = 10$ kg and $m = 50$ kg, respectively. It is shown that $z_3(t)$ is changed with the mass m, which is consistent with the fact. The voltages of two PAMs $u_1(t)$ and $u_2(t)$ for $m = 10$ kg and $m = 50$ kg are shown in Fig. 12.3d, respectively. From the two figures of Fig. 12.3c and d, it is shown that the changing trend of the control input $u(t)$ is contrary with the nonlinear uncertain term $z_3(t)$.

Based on experience and debugging repeatedly, the parameters of PID are given as $k_p = 0.003$, $k_i = 0.00001$, and $k_d = 0.03$ in the experiment. The experimental results obtained by PID control method are shown in Fig. 12.4.

The results of actual output when $m = 10$ kg and $m = 50$ kg are shown in Fig. 12.4a and b, respectively. It is easy to see that the PID control method leads to serious overshoot in these two curves, and the overshoot increases with the increase of the load mass m. This phenomenon is also seen from the voltage curves u_1 and u_2 with mass $m = 10$ kg and $m = 50$ kg as shown in Fig. 12.4c and d, respectively.

In the second experiment, a sinusoidal signal with 0.5 Hz and 10° amplitude is given. The parameters of tracking differentiator (12.1) and the nonlinear function are

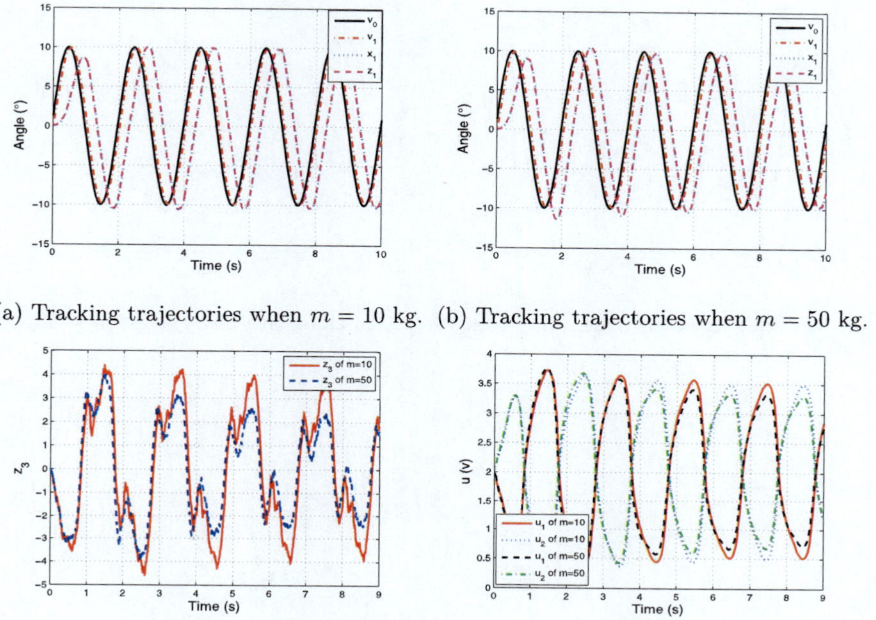

(a) Tracking trajectories when $m = 10$ kg. (b) Tracking trajectories when $m = 50$ kg.

(c) Estimations of nonlinearity (d) Control voltages

Fig. 12.5 Experimental results of active disturbance rejection control for tracking a sinusoidal signal

also applied as shown in Table 12.1. Note that the step signal is related to a (piecewise) constant steady-state reference compared with the sinusoidal signal. Due to a certain frequency of the sinusoidal signal, there exists much inertia when the platform moves with the frequency. To overcome the influence of inertia on stability of system, the parameter β_{03} in the linear ESO (12.3) has been increased to accurately estimate the inertia. By the analysis of Theorem 1, parameters of the linear ESO (12.3) are chosen as $\beta_{01} = 100$, $\beta_{02} = 1500$, and $\beta_{03} = 1500$. Moreover, the parameters of the state error feedback controller (12.15) are given as $\beta_1 = 25$, $\beta_2 = 24$, and $b_0 = 1500$ to ensure stability of the closed-loop system (11.9). The experimental results obtained by applying the proposed active disturbance rejection control are shown in Fig. 12.5.

Figure 12.5a, b shows that the tracking trajectories meet the movement trend mainly when the mass is changing from $m = 10$ kg to $m = 50$ kg although there are certain delays. The estimations of nonlinear and uncertain terms are shown in Fig. 12.5c. The estimated results of nonlinear and uncertain terms are complied with the actual situation that the estimated value becomes smaller with the increase of the load mass m. It is shown that the estimated results are in agreement with the model analyzing of the PMA mechanism system (11.9) for the reason of that the load mass m is presented in the denominator of the nonlinear and uncertain terms $f(x(t), \omega(t)) = b_1 x_1(t) + b_2 x_2(t) + \omega(t)$. Moreover, the controlling input voltage

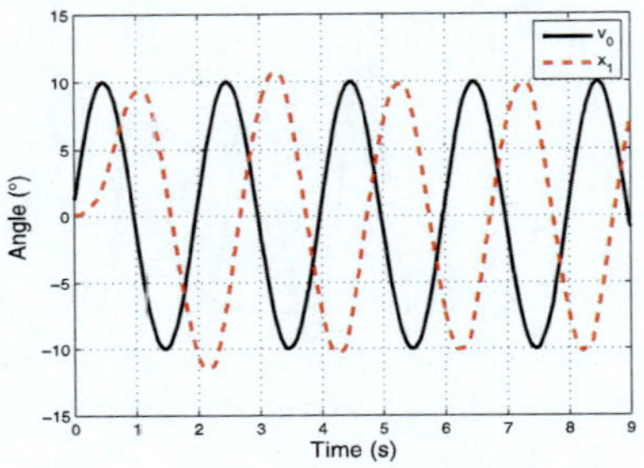

(a) Tracking trajectory when $m = 10$ kg.

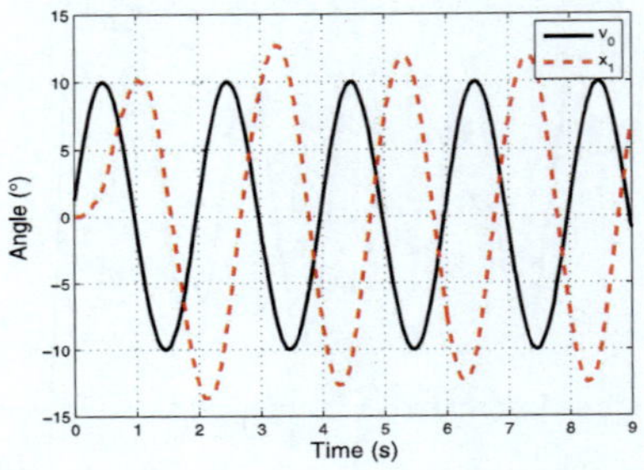

(b) Tracking trajectory when $m = 50$ kg.

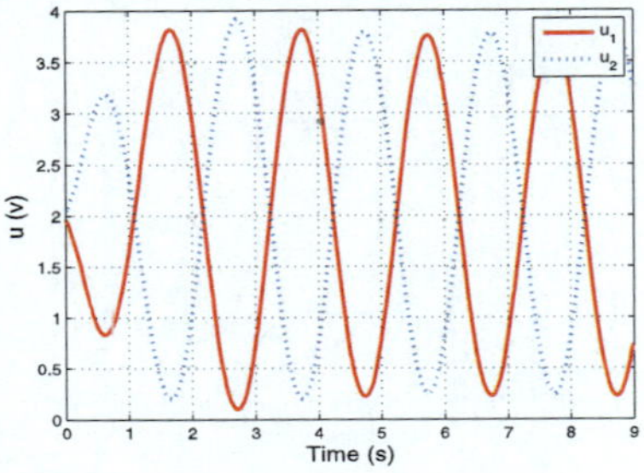

(c) Control voltage when $m = 10$ kg.

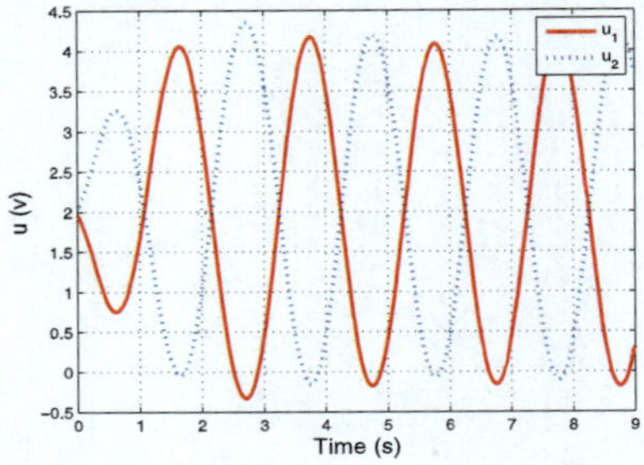

(d) Control voltage when $m = 50$ kg.

Fig. 12.6 Experimental results of PID for tracking a sinusoidal signal

curves $u_1(t)$ and $u_2(t)$ with $m = 10$ kg and $m = 50$ kg are shown in Fig. 12.5d, respectively.

For tracking the sinusoidal signal, the parameters of PID are adjusted to $k_p = 0.0025$, $k_i = 0.00001$, and $k_d = 0.03$. The experimental results obtained by PID control method for sinusoidal signal are shown in Fig. 12.6. In Fig. 12.6a and b, compared with the result obtained by active disturbance rejection control approach, there exists an obvious phase shift of the angle signal obtained by the PID method, which is the same as the first experiment that PID control leads an overshoot of the amplitude. The overshoot of the amplitude is large as the increase of the load mass m. The voltage curves u_1 and u_2 are also shown in Fig. 12.6c, d.

Through the analysis of experimental results, the PID control strategy leads to that the output of system (11.9) has the phenomenon of slow responses and serious overshoots. There exists fast responses and is not any overshoots for changing the load mass m using the active disturbance rejection control in this chapter. Therefore, the effectiveness of active disturbance rejection control to deal with varying load conditions is confirmed in the pneumatic motion simulation platform.

12.4 Conclusion

In this chapter, ADRC approach for a three-DoF pneumatic motion simulation system has been studied. In order to guarantee angles tracking accurately of the PMA mechanism system, a linear ESO has been introduced to estimate the total disturbances which include varying loads, internal disturbances, and other external disturbances. A nonlinear error feedback controller has been designed based on the state and disturbance estimation. Moreover, the corresponding convergence analysis of the linear ESO and the stabilization of the closed-loop system have also been presented. Finally, some compared experimental results indicate that the linear ESO based on the ADRC has advantage in dealing with the varying disturbances for the nonlinear system of PMA mechanism.

References

1. Hildebrandt A, Sawodny O, Neumann R, Hartmann A (2005) Cascaded control concept of a robot with two degrees of freedom driven by four artificial pneumatic muscle actuators. Am Control Conf 1:680–685
2. Tsagarakis N, Iqbal J, Khan H, Caldwell D (2014) A novel exoskeleton robotic system for hand rehabilitation-conceptualization to prototyping. Biocybern Biomed Eng 34(2):79–89
3. Aschemann H, Schindele D (2008) Sliding-mode control of a high-speed linear axis driven by pneumatic muscle actuators. IEEE Trans Ind Electron 55(11):3855–3864
4. Aschemann H, Schindele D (2014) Comparison of model-based approaches to the compensation of hysteresis in the force characteristic of pneumatic muscles. IEEE Trans Ind Electron 61(7):3620–3629
5. Jamwal PK, Xie SQ, Hussain S, Parsons JG (2014) An adaptive wearable parallel robot for the treatment of ankle injuries. IEEE-ASME Trans Mechatron 19(1):64–75
6. Iqbal S, Bhatti AI (2011) Load varying polytopic based robust controller design in LMI framework for a 2DoF stabilized platform. Arab J Forence Eng 36(2):311–327
7. Lee W, Lee CY, Jeong YH, Min BK (2015) Friction compensation controller for load varying machine tool feed drive. Int J Mach Tools Manuf 96:47–54
8. Xia Y, Shi P, Liu G-P, Rees D, Han J (2007) Active disturbance rejection control for uncertain multivariable systems with time-delay. IET Control Theory Appl 1(1):75–81
9. Liang G, Li W, Li Z (2013) Control of superheated steam temperature in large-capacity generation units based on active disturbance rejection method and distributed control system. Control Eng Pract 21(3):268–285
10. Huang C-E, Li D, Xue Y (2013) Active disturbance rejection control for the ALSTOM gasifier benchmark problem. Control Eng Pract 21(4):556–564
11. Zhao L, Yang Y, Xia Y, Liu Z (2015) Active disturbance rejection position control for a magnetic rodless pneumatic cylinder. IEEE Trans Ind Electron 62(9):5835–5846
12. Castañeda LA, Luviano-Juárez A, Chairez I (2015) Robust trajectory tracking of a delta robot through adaptive active disturbance rejection control. IEEE Trans Control Syst Technol 23(4):1387–1398
13. Zhu Z, Xu D, Liu J, Xia Y (2013) Missile guidance law based on extended state observer. IEEE Trans Ind Electron 60(12):5882–5891
14. Yang H, You X, Xia Y, Li H (2014) Adaptive control for attitude synchronisation of spacecraft formation via extended state observer. IET Control Theory Appl 8(18):2171–2185

15. Han J (2008) Active Disturbance Rejection Control Technique: The Technique for Estimating and Compensating the Uncertainties. National Defense Industry, Beijing
16. Trumpf J, Trentelman HL, Willems JC (2014) Internal model principles for observers. IEEE Trans Autom Control 59(7):1737–1749

Chapter 13
Linear Feedback Control

13.1 Introduction

As a novel type of pneumatic actuators, PAMs have attracted increasing attention in study and application. However, there exists strong nonlinearity and robustness in PAMs because of inflatable deformation, air elasticity, and compressibility [1]. Therefore, tracking control for nonlinear or uncertain systems is an important and practical problem which has attracted a lot of attentions [2, 3]. It is an important issue that how to design efficient and ordinary controllers for the PAM systems [4]. Some sliding mode control approaches are used for the nonlinear PAM systems [5]. In order to improve control performances and robustness on accurate trajectory tracking of the nonlinear PAM systems, a sliding mode controller has been designed based on a nonlinear disturbance observer in [6]. Although fruitful results on the nonlinear PAM systems have been found in recent publications, there is still a lot of space for further investigation yet. ADRC has been successfully applied to many engineering systems with well-controlled performances [7–9]. The ADRC is not only less dependent on accurate mathematical models but also has strong anti-interference abilities [10]. The ADRC consists of a tracking differentiator, an extended state observer, and an error feedback controller. The tracking differentiator can provide a desired transient and differential trajectory for a set value [11]. The extended state observer estimates total disturbances and uncertainties for nonlinear control systems [12–14]. The error feedback controller includes estimations to compensate the total disturbances and uncertainties in the nonlinear control systems [15]. To the best of our knowledge, very few results are available on the nonlinear pneumatic motion simulation systems using the ADRC, which motivated us to carry on this research work.

In this chapter, the ADRC is proposed for tracking control of the pneumatic motion simulate platform including a mechanism of PAMs. Not only convergence of the extended state observer is shown but also stabilization of the closed-loop pneumatic system is analyzed. Experiment results indicate that there exist stronger advantages in dealing with nonlinear and uncertain for the PAM system by the ADRC than PID control. The main contributions of this chapter are summarized as follows:

L. Zhao et al., *Pneumatic Servo Systems Analysis*, Advances in Industrial Control,
https://doi.org/10.1007/978-981-16-9515-5_13

i. An active disturbance rejection tracking control is presented for a nonlinear pneumatic muscle PAM system.
ii. Unmodeled dynamics, varying parameters, and external disturbances are merged into generalized disturbances in the PAM system.
iii. An error feedback controller is designed to guarantee tracking precision based on an extended state observer.

13.2 Main Results

13.2.1 Schematic Diagram of Control Method

For the nonlinear PAM system (11.9), the ADRC technique is developed in this chapter. The schematic diagram of ADRC is shown in Fig. 13.1. In Fig. 13.1, TD and ESO express the tracking differentiator and extended state observer, respectively, LSEFC is the linear state error feedback controller, i.e., $u(t) = \beta_1 r_1(t) + \beta_2 r_2(t)$, in which β_1 and β_2 are two adjustable parameters.

A transient process is arranged by the tracking differentiator. Then an abrupt input is changed to a continuous smooth signal in the transient process. The differential signal of the continuous smooth signal is also obtained by the tracking differentiator. For system (11.9), a second-order tracking differentiator is designed as

$$\begin{cases} f_h(t) = f_{han}(v_1(t) - v_0(t), v_2(t), r_0, h_0), \\ \dot{v}_1(t) = v_2(t), \\ \dot{v}_2(t) = f_h(t), \end{cases} \tag{13.1}$$

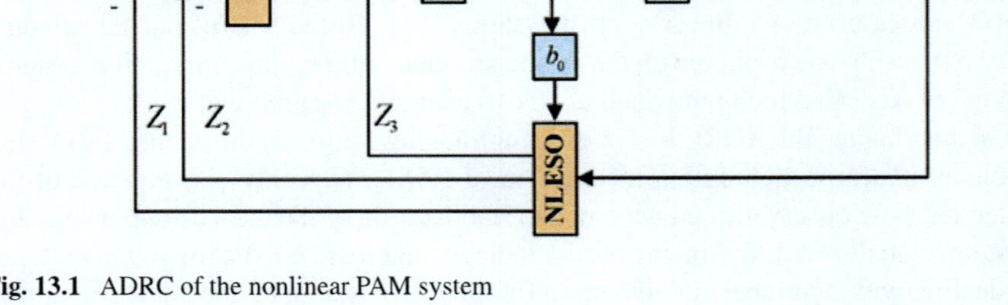

Fig. 13.1 ADRC of the nonlinear PAM system

$v_0(t)$ is a given input, $v_1(t)$ is an output tracking $v_0(t)$, and $v_2(t)$ is a differential signal of $v_1(t)$.

13.2.2 Nonlinear Extended State Observer

In this section, an extended state observer is introduced to deal with the unknown nonlinearity of system (11.9). Different from general observers, the extended state observer is not only used to observe state variables, but also to estimate total unknown nonlinear term. The third-order nonlinear extended state observer is designed as follows:

$$\begin{cases} e_1(t) = z_1(t) - y(t), \\ \dot{z}_1(t) = z_2(t) - \beta_{01}e_1(t), \\ \dot{z}_2(t) = z_3(t) - \beta_{02}fal(e_1(t), \alpha_1, \delta) + b_0u(t), \\ \dot{z}_3(t) = -\beta_{03}fal(e_1(t), \alpha_2, \delta), \end{cases} \tag{13.2}$$

where $z_1(t)$, $z_2(t)$, and $z_3(t)$ are the estimation states; $\beta_1 > 0$, $\beta_2 > 0$, and $\beta_3 > 0$ are the adjustable parameters; and $fal(e_1(t), \alpha_1, \delta)$ and $fal(e_1(t), \alpha_2, \delta)$ are the nonlinear functions. According to (12.2) and (13.2), it is shown that

$$\begin{cases} e_1(t) = z_1(t) - y(t), \\ \dot{e}_1(t) = e_2(t) - \beta_{01}e_1(t), \\ \dot{e}_2(t) = e_3(t) - \beta_{02}fal(e_1(t), \alpha_1, \delta), \\ \dot{e}_3(t) = -h(t) - \beta_{03}fal(e_1(t), \alpha_2, \delta). \end{cases} \tag{13.3}$$

Note that the balance point of the estimation error system (13.3) is zero with $h(t) = 0$.

For the estimation error system (13.3), the following two regions are given:

$$R_2 = \{(e_1(t), e_2(t), e_3(t)) : |h_2(e_1(t), e_2(t))| \leq g_1(e_1(t))\},$$
$$R_3 = \{(e_1(t), e_2(t), e_3(t)) : |h_3(e_1(t), e_2(t), e_3(t))| \leq g_2(e_1(t), e_2(t))\},$$

with

$$h_2(e_1(t), e_2(t)) = e_2(t) - \beta_{01}e_1(t) + k_1g_1(e_1(t))\text{sign}(e_1(t)), \tag{13.4}$$

$$\begin{aligned} h_3(e_1(t), e_2(t), e_3(t)) = &-\beta_{02}fal(e_1(t)) - \beta_{01}(e_2(t) - \beta_{01}e_1(t)) \\ &+k_2g_2(e_1(t), e_2(t))\text{sat}(h_2(t)/g_1(t)) + e_3(t), \end{aligned} \tag{13.5}$$

where $k_1 > 1, k_2 > 1, g_1(e_1(t))$, and $g_2(e_1(t), e_2(t))$ are arbitrary continuous positive definite functions with $g_1(0) = 0$ and $g_2(0, 0) = 0$, respectively.

For notational simplicity, let $g_1(t) = g_1(e_1(t)), g_2(t) = g_2(e_1(t), e_2(t)), g_1(t)s(t) = g_1(e_1(t))\text{sign}(e_1(t)), g_2(t)s(t) = g_2(e_1(t), e_2(t))\text{sat}(h_2(t)/g_1(t))$, and $h_2(t) = h_2(e_1(t), e_2(t)), h_3(t) = h_3(e_1(t), e_2(t), e_3(t))$ in the following.

Remark 13.1 In Fig. 2.2, there is a linear segment in the neighborhood of the origin where the error variable is small relatively. When $|e(t)| > \delta$, $fal(\cdot)$ is a concave function. Moreover, there exist two points which aren't differentiable for function $fal(e(t), \alpha, \delta)$. Since it isn't effective on the following theoretical analysis, let $\dot{fal}_1(t) = \frac{dfal(e_1(t), \alpha_1, \delta)}{de_1(t)} > 0$ in this paper.

The effectiveness of the extended state observer (13.2) is shown by the following three theorems.

Theorem 13.2 *Considering the error system (13.3), the trajectories of $e_1(t)$ and $e_2(t)$ converge to the origin after a certain time if $(e_1(t), e_2(t)) \in R_2$. That is, the observed states $z_1(t)$ and $z_2(t)$ converge to the actual states $x_1(t)$ and $x_2(t)$, respectively.*

Proof Let $V_{h_2(t)g_1(t)} = (h_2^2(t) - g_1^2(t))/2$. For $(e_1(t), e_2(t)) \in R_2$, one has that $h_2(t) - g_1(t) < 0$ and

$$- g_1(t) - k_1 g_1(t)s(t) < e_2(t) - \beta_{01}e_1(t) < g_1(t) - k_1 g_1(t)s(t). \tag{13.6}$$

Concerning the estimation error system (13.3), a candidate Lyapunov function is chosen as

$$V_0(t) = e_1^2(t)/2.$$

For (13.6), the derivative of V_0 is given as

$$\dot{V}_0(t) = e_1(t)\dot{e}_1(t) = e_1(t)(e_2(t) - \beta_{01}e_1(t)) < e_1(t)(g_1(t) - k_1 g_1(t)s(t)). \tag{13.7}$$

Based on the relationship between $e_1(t)$ and 0, two cases are shown in the following. Case 1. If $e_1(t) > 0$, one has that

$$\dot{V}_0(t) < e_1(t)(1 - k_1)g_1(t). \tag{13.8}$$

For $k_1 > 1$, (13.8) is rewritten as

$$\dot{V}_0(t) < -e_1(t)(k_1 - 1)g_1(t) < 0. \tag{13.9}$$

Case 2. If $e_1(t) < 0$, the inequality is given as

$$\dot{V}_0(t) < -|e_1(t)|(k_1 + 1)g_1(t) < 0. \tag{13.10}$$

It follows (13.9) and (13.10) that $\dot{V}_0(t) < 0$ holds for any estimation error $e_1(t) \neq 0$, which implies that the estimation error $e_1(t) \to 0$ after a certain time. Moreover, from (13.6), we have $e_2(t) \to 0$ for $e_1(t) \to 0$.

Theorem 13.3 *Considering the estimation error system (13.3), the trajectories of $e_1(t)$, $e_2(t)$, and $e_3(t)$ included in R_3 and outside R_2 converge to the origin after a certain time, if the following inequality*

$$g_2(t) > \frac{(k_1+1)^2}{k_2-1}\left|\frac{dg_1(t)}{de_1(t)}\right| |h_2(t)| \tag{13.11}$$

holds. That is, the observed states $z_1(t)$, $z_2(t)$, and $z_3(t)$ converge to the actual states $x_1(t)$, $x_2(t)$, and $x_3(t)$, respectively.

Proof Let

$$V_{h_3(t)g_2(t)} = (h_3^2(t) - g_2^2(t))/2.$$

For $(e_1(t), e_2(t), e_3(t)) \in R_3$, one has that $h_3(t) < g_2(t)$. The derivative of $V_{h_2(t)g_1(t)}$ is given as

$$\dot{V}_{h_2(t)g_1(t)} = h_2(t)\dot{h}_2(t) - g_1(t)\dot{g}_1(t). \tag{13.12}$$

From expression (13.4), the derivative of $h_2(t)$ is obtained as given below:

$$\begin{aligned}\dot{h}_2(t) &= \dot{e}_2(t) - \beta_{01}\dot{e}_1(t) + k_1\frac{dg_1(t)s(t)}{de_1(t)}\dot{e}_1(t)\\ &= -\beta_{02}fal(e_1(t)) - \beta_{01}(e_2(t) - \beta_{01}e_1(t))\\ &\quad +k_1\frac{dg_1(t)s(t)}{de_1(t)}(e_2(t) - \beta_{01}e_1(t)) + e_3(t).\end{aligned} \tag{13.13}$$

Substituting (13.4), (13.5), and (13.13) into (13.12), the following relation is obtained:

$$\begin{aligned}\dot{V}_{h_2(t)g_1(t)} &= h_2(t)[h_3(t) + k_1\frac{dg_1(t)s(t)}{de_1(t)}(h_2(t) - k_1g_1(t)s(t))\\ &\quad -k_2g_2(t)s(t)] - g_1(t)\frac{dg_1(t)}{de_1(t)}(h_2(t) - k_1g_1(t)s(t)).\end{aligned}$$

For $V_{h_3(t)g_2(t)} < 0$ and $V_{h_2(t)g_1(t)} \geq 0$, i.e., $h_3(t) - k_2g_2(t) \leq -(k_2-1)g_2(t)$, the following relation is obtained:

$$\begin{aligned}\dot{V}_{h_2(t)g_1(t)}(t) &\leq k_1\left|h_2(t)\frac{dg_1(t)s(t)}{de_1(t)}(h_2(t) - k_1g_1(t)s(t))\right| - |h_2(t)|\\ &\quad (k_2-1)g_2(t) + g_1(t)\left|\frac{dg_1(t)}{de_1(t)}(h_2(t) - k_1g_1(t)s(t))\right|\\ &\leq -|h_2(t)|(k_2-1)g_2(t) + h_2^2(t)(k_1+1)^2\left|\frac{dg_1(t)}{de_1(t)}\right||h_2(t)|\\ &< 0.\end{aligned}$$

Therefore, the trajectories of $e_1(t)$ and $e_2(t)$ converge into the region R_2. By Theorem 13.2, the trajectories converge to the origin ultimately. If $e_1(t) \to 0$ and $e_2(t) \to 0$, then we have $e_3(t) \to 0$ easily when $\omega(t) = 0$, i.e., $z_1(t) \to x_1(t)$, $z_2(t) \to x_2(t)$, and $z_3(t) \to x_3(t)$ after a certain time for the estimation error system (13.3).

The case of that the state error trajectories of system (13.3) are in the region outside R_3 is shown in the following theorem. Let

$$g_2(t) = \begin{cases} k_3|h_2(t)|, & |h_2(t)| \geq g_1(t), \\ k_3|g_1(t)|, & |h_2(t)| < g_1(t), \end{cases}$$

where k_3 is a constant and $k_3 > \frac{(k_1+1)^2}{k_2-1}\left|\frac{dg_1(t)}{de_1(t)}\right|$.

Theorem 13.4 *If the following inequality*

$$\begin{aligned}\beta_{01} > k_2k_3 &+ \frac{1}{k_3}\left|\beta_{03}\frac{|fal_2(t)|}{g_1(t)} - \beta_{02}k_1\dot{fal}_1(t) + k_1^2k_2k_3\frac{dg_1(t)s(t)}{de_1(t)}\right| \\ &+ \left|\frac{\beta_{02}\dot{fal}_1(t)}{k_3} - \beta_{01}k_2 - k_1k_2\frac{dg_1(t)s(t)}{de_1(t)} + k_2^2k_3\right| + \left|\frac{\partial g_2(t)}{\partial e_2(t)}\right| \\ &+ \frac{1}{k_3}\left|\frac{\partial g_2(t)}{\partial e_1(t)} + (\beta_{01} - k_2k_3)\frac{\partial g_2(t)}{\partial e_2(t)}\right| + \frac{k_1}{k_3}\left|\frac{\partial g_2(t)}{\partial e_1(t)} + \beta_{01}\frac{\partial g_2(t)}{\partial e_2(t)}\right| \end{aligned} \tag{13.14}$$

holds, then the trajectories outside region R_3 of the estimation error system (13.3) converge into R_3.

Proof The derivative of $V_{h_3(t)g_2(t)}$ is given as follows:

$$\begin{aligned}\dot{V}_{h_3(t)g_2(t)} &= h_3(t)\dot{h}_3(t) - g_2(t)\dot{g}_2(t) \\ &= h_3(t)\big((-\beta_{01} + k_2k_3)h_3(t) + (\beta_{01}k_2k_3 - \beta_{02}\dot{fal}_1(t) \\ &\quad + k_1k_2k_3\dot{g}_1(t)s(t) - k_2^2k_3^2\big)h_2(t) + (-\beta_{03}\frac{|fal_2(t)|}{g_1(t)} \\ &\quad + \beta_{02}\dot{fal}_1(t)k_1 - k_1^2k_2k_3\dot{g}_1(t)s(t))g_1(t)s(t)\big) \\ &\quad - g_2(t)\left(\frac{\partial g_2(t)}{\partial e_2(t)}h_3(t) + (\beta_{01}\frac{\partial g_2(t)}{\partial e_2(t)} - k_2k_3\frac{\partial g_2(t)}{\partial e_2(t)})h_2(t)\right. \\ &\quad \left. + \frac{\partial g_2(t)}{\partial e_1(t)} - k_1(\frac{\partial g_2(t)}{\partial e_1(t)} + \beta_{01}\frac{\partial g_2(t)}{\partial e_2(t)})g_1(t)s(t)\right). \end{aligned} \tag{13.15}$$

From $V_{h_3(t)g_2(t)} \geq 0$, one has that $|h_3(t)| \geq g_2(t) \geq \max\{k_3|h_2(t)|, k_3g_1(t)\}$. Moreover, (13.15) is rewritten as follows:

$$
\begin{aligned}
\dot{V}_{h_3(t)g_2(t)} \leq\ & |h_3(t)| \Big((-\beta_{01} + k_2k_3)|h_3(t)| + |\beta_{01}k_2k_3 - \beta_{02}\dot{fal}_1(t) - k_2^2k_3^2 \\
& + k_1k_2k_3\dot{g}_1(t)s(t)|\frac{|h_3(t)|}{k_3} + \Big|\beta_{02}\dot{fal}_1(t)k_1 - k_1^2k_2k_3\dot{g}_1(t)s(t) \\
& - \beta_{03}\frac{|fal_2(t)|}{g_1(t)}\Big|\frac{|h_3(t)|}{k_3}\Big) + |h_3(t)| \Bigg(\Bigg(\Big| -k_2k_3\frac{\partial g_2(t)}{\partial e_2(t)} + \beta_{01}\frac{\partial g_2(t)}{\partial e_2(t)} \\
& + \frac{\partial g_2(t)}{\partial e_1(t)}\Big| + k_1\Big|\frac{\partial g_2(t)}{\partial e_1(t)} + \beta_{01}\frac{\partial g_2(t)}{\partial e_2(t)}\Big|\frac{|h_3(t)|}{k_3}\Bigg) + \Big|\frac{\partial g_2(t)}{\partial e_2(t)}\Big||h_3|\Bigg).
\end{aligned}
$$

Then $\dot{V}_{h_3(t)g_2(t)} < 0$ holds for any β_{01}, β_{02}, and β_{03} in condition (13.14). Therefore, all the trajectories of the estimation error system (13.3) converge into the region R_3.

By Theorems 13.2–13.4, $\dot{V}_{h_3(t)g_2(t)} < 0$ with appropriate parameters β_{01}, β_{02}, and β_{03} when $(e_1(t), e_2(t), e_3(t)) \neq (0, 0, 0)$ holds. That is, $(e_1(t), e_2(t), e_3(t))$ converges to the origin after a certain time.

13.2.3 Linear Error Feedback Controller

An error feedback controller is designed such that an output of the mechanism angle $\theta(t)$ tracks the desired reference input. Error signals between the tracking differentiator (13.1) and extended state observer (13.2) are shown as

$$
\begin{aligned}
r_1(t) &= v_1(t) - z_1(t), \\
r_2(t) &= v_2(t) - z_2(t),
\end{aligned}
\tag{13.16}
$$

where $z_1(t)$ and $z_2(t)$ are the observations of $v_1(t)$ and $v_2(t)$ by observer (13.2), respectively. The state error feedback controller $u(t)$ is given as

$$
u(t) = \frac{\beta_1 r_1(t) + \beta_2 r_2(t) - z_3(t)}{b_0}, \tag{13.17}
$$

where β_1 and β_2 are controller gains, and $z_3(t)$ is the compensation for system (11.9).

The trajectory tracking error between the given signal and output signal is presented as follows:

$$
s_1(t) = v_1(t) - x_1(t). \tag{13.18}
$$

Then the derivative of $s_1(t)$ is given as follows:

$$
s_2(t) = v_2(t) - x_2(t). \tag{13.19}
$$

Then, the trajectory tracking errors are shown as

$$\begin{aligned} \dot{s}_1(t) &= s_2(t), \\ \dot{s}_2(t) &= v_3(t) - x_3(t) - b_0 u(t), \end{aligned} \tag{13.20}$$

where $v_3(t)$ is the derivative of $v_2(t)$, and it is continuous and bounded. From (13.3), (13.18), and (13.19), one has that

$$\begin{aligned} r_1(t) &= s_1(t) - e_1(t), \\ r_2(t) &= s_2(t) - e_2(t). \end{aligned} \tag{13.21}$$

Moreover, there exists

$$\dot{s}_2(t) = v_3(t) - \beta_1(s_1(t) - e_1(t)) - \beta_2(s_2(t) - e_2(t)) + e_3(t), \tag{13.22}$$

where $e_3(t) = z_3(t) - b_1 x_1(t) - b_2 x_2(t)$.

Theorem 13.5 *Consider the closed-loop system (11.9) with the error feedback controller (13.17). There exist appropriate positive variables β_1 and β_2 in controller (13.17) to ensure the stabilization of the closed-loop system (11.9).*

Proof For the closed-loop system (11.9), a candidate Lyapunov function is chosen as

$$V(t) = \frac{1}{2}(s_1(t) - e_1(t))^2 + \frac{1}{2}(s_2(t) - e_2(t))^2. \tag{13.23}$$

The first derivative of $V(t)$ along (13.3) and (13.22) yields

$$\begin{aligned} \dot{V}(t) &= (s_1(t) - e_1(t))(\dot{s}_1(t) - \dot{e}_1(t)) + (s_2(t) - e_2(t))(\dot{s}_2(t) - \dot{e}_2(t)) \\ &= -\beta_2\left((s_2(t) - e_2(t))^2 - (s_2(t) - e_2(t))\frac{N(t)}{\beta_2}\right), \end{aligned} \tag{13.24}$$

where

$$N(t) = (1 - \beta_1)(s_1(t) - e_1(t)) + v_3(t) + \beta_{02} fal(e_1(t), \alpha_1, \delta).$$

It follows from (13.24) that

$$\dot{V}(t) = -\beta_2\left((s_2(t) - e_2(t)) - \frac{N(t)}{2\beta_2}\right)^2 + \frac{N^2(t)}{4\beta_2}. \tag{13.25}$$

If $\beta_1 = 1$, then $N(t) = v_3(t) + \beta_{02} fal(e_1(t), \alpha_1, \delta)$ in which both $v_3(t)$ and $e_1(t)$ are bounded. Thereby, $s_1(t) - e_1(t)$ is restricted by the output in the process of experiment. That is, there exist appropriate parameters β_1 and β_2 to guarantee $\dot{V}(t) < 0$. Therefore, the closed-loop system (11.9) is stable when the appropriate positive parameters β_1 and β_2 of the controller (13.17) are given.

13.3 Experiments and Results

In this experiment, we study tracking control of a PMA mechanism which is a part of pneumatic motion simulate platform. Through the studying of static characteristics for PAMs and measuring of the experimental setup, all constant parameters of the PMA mechanism model are given in Table 13.1.

In this chapter, the experiment of the PMA mechanism includes four parts mainly. Firstly, a sinusoidal signal at 0.5Hz, 10° amplitude is given. Based on the analysis of Theorems 13.2–13.4, the parameters of observer (13.2) are set $\beta_{01} = 50, \beta_{02} = 7$, and $\beta_{03} = 315$. It follows from Theorem 13.5 that the parameters of controller (13.17) are set $\beta_1 = 5$ and $\beta_2 = 8$. The results of experiment are shown in Fig. 13.2. From Fig. 13.2a, one has that the designed tracking differentiator (13.1), extended state observer (13.2), and controller (13.17) are effective. That is, the output $v_1(t)$ of tracking differentiator tracks the given signal $v_0(t)$ quickly and steadily, the output $x_1(t)$ of system (11.9) also tracks the given signal $v_0(t)$ accurately, and the observed value $z_1(t)$ is almost coincident with $x_1(t)$ too. Under the condition of 2V preload voltage, the control voltages $u_1(t)$ and $u_2(t)$ are up and down around with 2V as shown in Fig. 13.2b, respectively. The estimation $z_3(t)$ of the uncertain nonlinear terms is shown in Fig. 13.2c.

The experimental results on tracking a step signal at 10° are shown in Figs. 13.3, 13.5. In Fig. 13.3a, the fact output $x_1(t)$ gets to the given angle $v_0(t)$ at 1.2s and the steady-state error is less than 0.05°. The output of tracking differentiator $v_1(t)$ tracks the given $v_0(t)$ quickly and accurately. The control voltages $u_1(t)$ and $u_2(t)$ of two PAMs are shown in Fig. 13.3b. The estimation of nonlinearity for system (11.9) is shown in Fig. 13.3c. Moreover, the experimental results based on PID controller are shown in Figs. 13.4 and 13.5.

Figure 13.4 shows the experimental results for tracking the same sinusoidal signal. Phase difference appears in the result as shown in Fig. 13.4a which implies that the

Table 13.1 The parameters in a PMA mechanism model

Parameters		
$k_0 = 0.09$	$\varepsilon_0 = 0.111$	$\mu_1 = 0.0501$
$\mu_2 = -0.0472$	$\mu_3 = 0.0106$	$r = 200\,\text{mm}$
$L_0 = 450\,\text{mm}$	$u_0 = 2\,\text{V}$	$k_u = 0.45$

Fig. 13.2 Experimental results for tracking a sinusoidal signal based on the ADRC approach

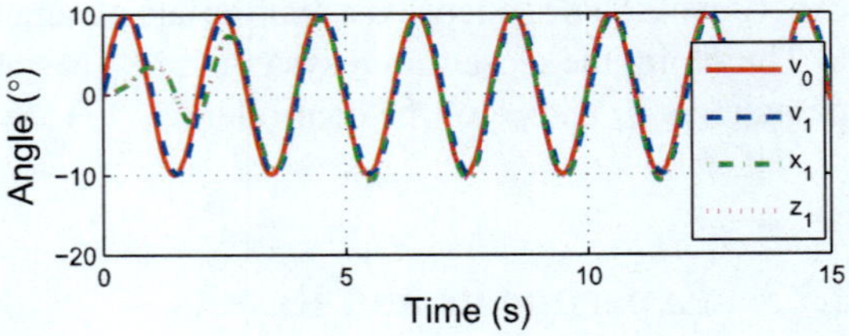

(a) Angle.

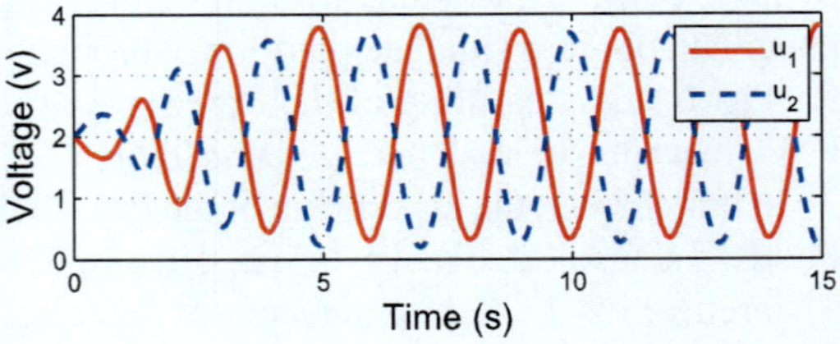

(b) Control input.

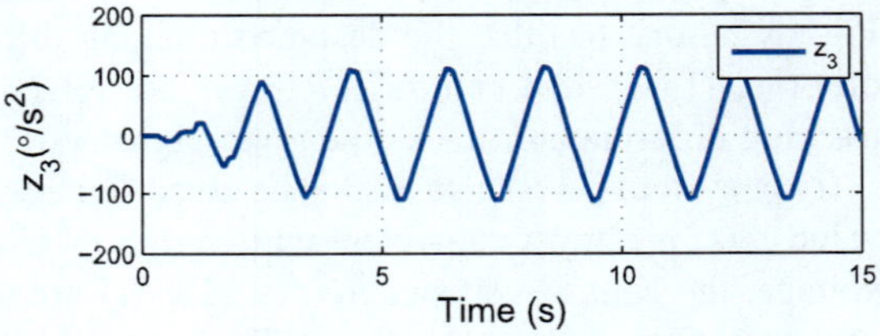

(c) Estimate of nonlinearity.

output signal $x_1(t)$ cannot track the given sinusoidal signal $v_0(t)$. The control voltages $u_1(t)$ and $u_2(t)$ are shown in Fig. 13.4b.

The experimental results for tracking the same step signal are shown in Fig. 13.5. There exists obvious overshoot phenomenon in Fig. 13.5a. Note that the output $x_1(t)$ reaches to the desired angle at about 2.5 s and the steady-state error is about 0.05°. Moreover, the control voltages $u_1(t)$ and $u_2(t)$ are shown in Fig. 13.5b. The comparison of two approaches indicates that the control speed and tracking effect are better based on the ADRC than PID control.

13.4 Conclusion

In this chapter, the advantage of ADRC has been fully used for a nonlinear PAM system in the pneumatic motion simulation platform. To deal with disturbances or uncertain terms, estimations of the extended state observer have been compensated to the feedback controller in real time. On convergence of the extended state observer, a

Fig. 13.3 Experimental results for tracking a step signal based on the ADRC approach

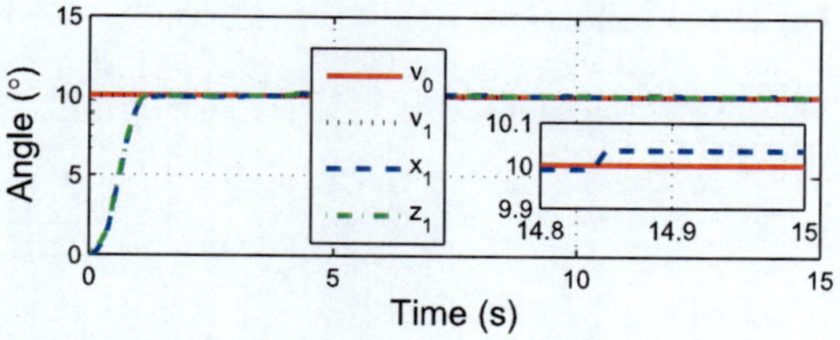

(a) Angle.

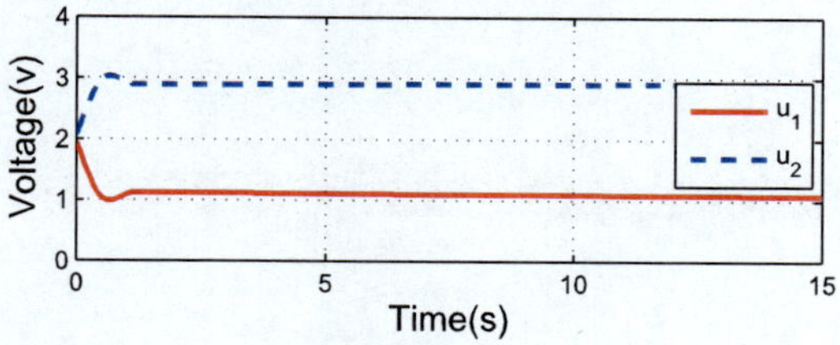

(b) Control input.

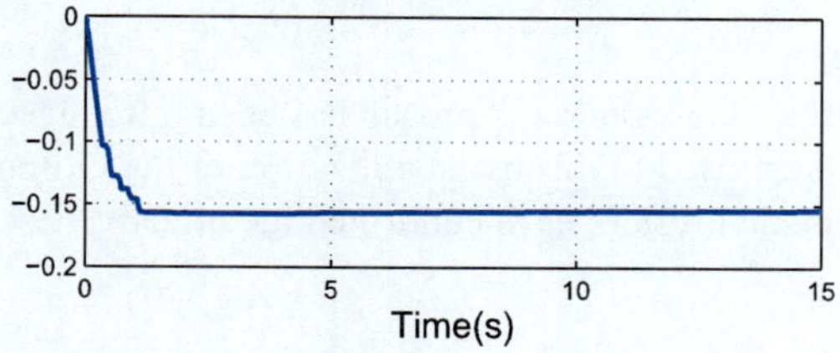

(c) Estimate of nonlinearity.

Fig. 13.4 Experimental results for tracking a sinusoidal signal based on the PID approach

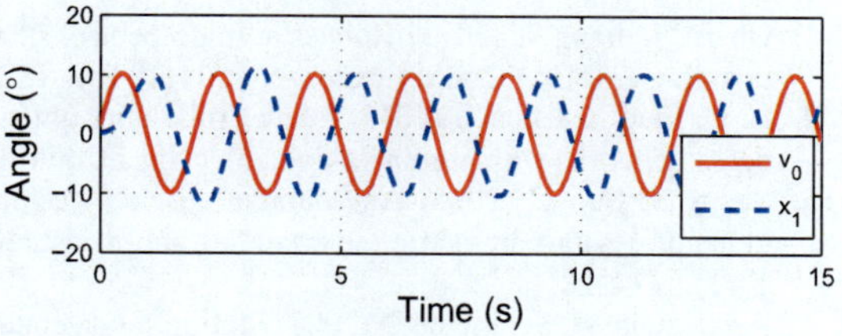

(a) Angle.

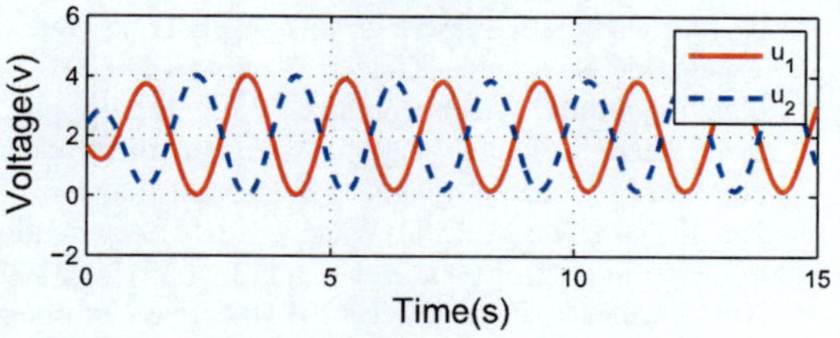

(b) Control input.

Fig. 13.5 Experimental results for tracking a step signal based on the PID approach

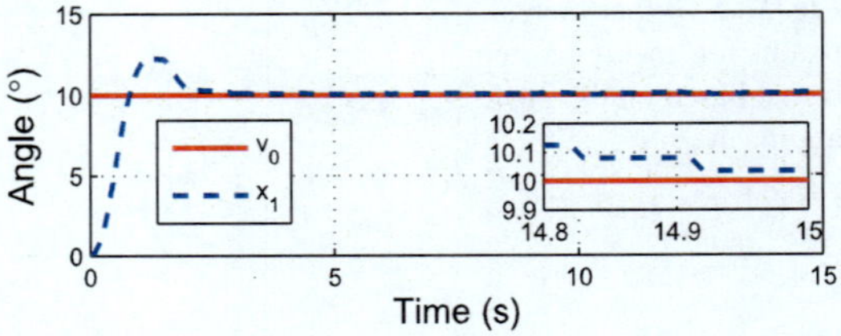

(a) Angle.

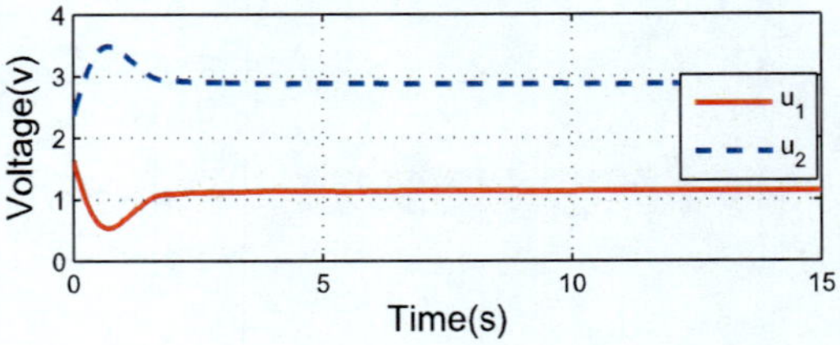

(b) Control input.

self-stable region approach has been introduced. Moreover, a Lyapunov method has been used to discuss stabilization of the closed-loop system. Finally, some experimental results have confirmed the effectiveness of the proposed method.

References

1. Chen M, Jiang B (2013) Robust attitude control of near space vehicles with time-varying disturbances. Int J Control Autom Syst 11(1):182–187
2. Li H, Liao X, Chen G (2013) Leader-following finite-time consensus in second-order multi-agent networks with nonlinear dynamics. Int J Control Autom Syst 11(2):422–426
3. Sheng S, Sun C (2016) An adaptive attitude tracking control approach for an unmanned helicopter with parametric uncertainties and measurement noises. Int J Control Autom Syst 14(1):217–228
4. Aschemann H, Schindele D (2008) Sliding-mode control of a high-speed linear axis driven by pneumatic muscle actuators. IEEE Trans Ind Electron 55(11):3855–3864
5. Lilly JH, Quesada PM (2004) A two-input sliding-mode controller for a planar arm actuated by four pneumatic muscle groups. IEEE Trans Neural Syst Rehabil Eng 12(3):349–359
6. Tsagarakis N, Iqbal J, Khan H, Caldwell D (2014) A novel exoskeleton robotic system for hand rehabilitation-conceptualization to prototyping. Biocybern Biomed Eng 34(2):79–89
7. Su Y, Zheng C, Duan B (2005) Automatic disturbances rejection controller for precise motion control of permanent-magnet synchronous motors. IEEE Trans Ind Electron 52(3):814–823
8. Xia Y, Dai L, Fu M, Li C, Wang C (2014) Application of active disturbance rejection control in tank gun control system. J Frankl Inst 351(4):2299–2314
9. Xia Y, Chen R, Pu F, Dai L (2014) Active disturbance rejection control for drag tracking in mars entry guidance. Adv Space Res 53(5):853–861
10. Xia Y, Liu B, Fu M (2012) Active disturbance rejection control for power plant with a single loop. Asian J Control 14(1):239–250
11. Sun G, Ren X, Li D (2015) Neural active disturbance rejection output control of multimotor servomechanism. IEEE Trans Control Syst Technol 23(2):746–753

12. Qiu Z, Wang B, Zhang X, Han J (2013) Direct adaptive fuzzy control of a translating piezoelectric flexible manipulator driven by a pneumatic rodless cylinder. Mech Syst Signal Process 36(2):290–316
13. Godbole AA, Kolhe JP, Talole SE (2013) Performance analysis of generalized extended state observer in tackling sinusoidal disturbances. IEEE Trans Control Syst Technol 21(6):2212–2223
14. Yang H, You X, Xia Y, Li H (2014) Adaptive control for attitude synchronisation of spacecraft formation via extended state observer. IET Control Theory Appl 8(18):2171–2185
15. Tang Z, Ge SS, Lee KP, He W (2016) Adaptive neural control for an uncertain robotic manipulator with joint space constraints. Int J Control 89(7):1428–1446

Chapter 14
Backstepping Control

14.1 Introduction

With the development of PAMs, control method for mechanisms of PAMs has been studied in more depth. A nonlinear PID-based antagonistic controller for positioning problems on PAMs has been designed in [1]. In order to improve performances of a compliant parallel ankle rehabilitation robot system, an impedance control strategy has been designed in [2]. However, these algorithms require accurate mathematical models which are obtained difficultly due to elastic forces of rubber, forces of friction, the end interface of PAMs, and other uncertain factors of PAM mechanisms. Therefore, it is necessary to improve the performances of PAMs via an algorithm that is less dependent on mathematical models. Active has been successfully applied in many engineering systems, such as tank gun control [3] and aviation [4]. As the kernel of ADRC, the ESO is designed to estimate nonlinear terms, external disturbances, and unmodeled dynamics of nonlinear systems [5]. A class of adaptive-ESO (AESO) has been proposed in [6]. The AESO combines both advantages of theoretical completeness in conventional LESO and good performances in conventional NESO. Hence, the AESO is an appropriate method to estimate nonlinear parts of pneumatic systems.

In this chapter, a backstepping strategy based on AESO is presented to improve the control precise and response rapidity of the pneumatic motion simulation platform. The TD is designed to arrange a transient process for getting a continuous smooth signal and its differential signal of abrupt inputs. In addition, the TD is also designed as a filter for time-varying multiplier in the PD-eigenvalues. The AESO is designed to estimate nonlinear terms, external disturbances, and unmodeled dynamics of the studied system. The backstepping-based controller is proposed to obtain good performances for the closed-loop pneumatic system. Both convergence of the AESO and stabilization of the closed-loop pneumatic system are analyzed. Finally, experimental results have shown that the control strategy has strong advantages in dealing with various uncertainties for the mechanism of PAMs.

L. Zhao et al., *Pneumatic Servo Systems Analysis*, Advances in Industrial Control,
https://doi.org/10.1007/978-981-16-9515-5_14

The main contributions of this chapter are summarized as follows:

i An adaptive backstepping controller based on an AESO is designed to improve the control precision and response rapidity for a mechanism of PAMs.
ii Three TDs are designed to obtain n-order derivatives of time-varying multipliers in PD-eigenvalues that are associated with gains of the AESO.
iii Unknown parameters and total disturbances for the mechanism of PAMs are handled by two adaptive laws and the AESO, respectively.

14.2 Main Results

14.2.1 Schematic Diagram of Control Method

In this chapter, the backstepping control strategy based on the AESO is introduced for the mechanism of PAMs. The schematic diagram of the proposed controller for the mechanism of PAMs is shown in Fig. 14.1.

For getting a continuous smooth signal of the given input, a TD is designed to arrange a transient process. Moreover, the differential signal of the given input is also obtained.

Considering system (11.9), the second-order TD is designed as follows:

$$\begin{cases} \dot{v}_1(t) = v_2(t) \\ \dot{v}_2(t) = f_{han}(v_1(t) - v_0(t), v_2(t), r_0, h_0), \end{cases}$$

where $v_0(t)$ is a given input, $v_1(t)$ is an output tracking signal of $v_0(t)$, and $v_2(t)$ is a differential signal of $v_1(t)$.

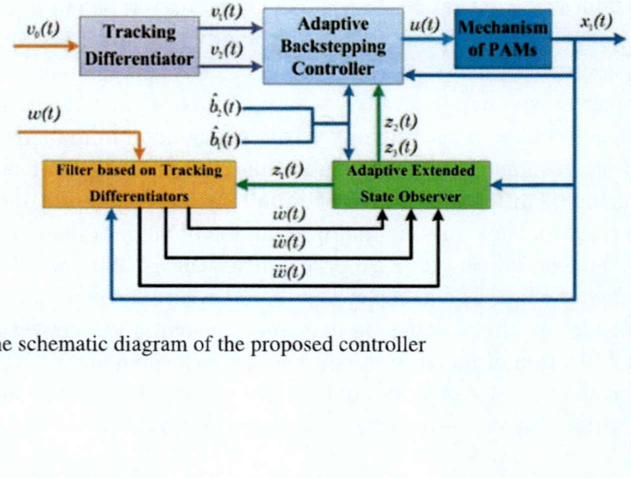

Fig. 14.1 The schematic diagram of the proposed controller

14.2.2 Adaptive Extended State Observer

In this chapter, an AESO is designed by modifying the traditional ESO as follows:

$$\begin{cases} e_1(t) = z_1(t) - y(t) \\ \dot{z}_1(t) = z_2(t) - g_1(t)e_1(t) \\ \dot{z}_2(t) = z_3(t) - g_2(t)e_1(t) + b_0u(t) \\ \dot{z}_3(t) = -g_3(t)e_1(t), \end{cases} \tag{14.1}$$

where $z_1(t)$, $z_2(t)$, and $z_3(t)$ are estimation values of $x_1(t)$, $x_2(t)$, and $x_3(t)$, and $e_1(t)$ is an estimation error of $z_1(t)$. Moreover, $g_1(t) > 0$, $g_2(t) > 0$, and $g_3(t) > 0$ are adjustable gains.

According to the extended system (11.9) and AESO (14.1), the estimation error system is obtained as follows:

$$\begin{cases} e_1(t) = z_1(t) - y(t) \\ \dot{e}_1(t) = e_2(t) - g_1(t)e_1(t) \\ \dot{e}_2(t) = e_3(t) - g_2(t)e_1(t) \\ \dot{e}_3(t) = -g_3(t)e_1(t) - h(t, X), \end{cases} \tag{14.2}$$

where $e_2(t)$ and $e_3(t)$ are estimation errors of $z_2(t)$ and $z_3(t)$, respectively. According to the estimation error system (14.2), a vector differential equation is written as follows:

$$\dot{e}(t) = A(t)e(t) + b(-h(t, X)), \tag{14.3}$$

where

$$e(t) = \begin{bmatrix} e_1(t) \\ e_2(t) \\ e_3(t) \end{bmatrix}, \; A(t) = \begin{bmatrix} -g_1(t) \; 1 \; 0 \\ -g_2(t) \; 0 \; 1 \\ -g_3(t) \; 0 \; 0 \end{bmatrix}, \; b = \begin{bmatrix} 0 \\ 0 \\ 1 \end{bmatrix}.$$

An equivalent canonical form is obtained by transforming the estimation error system (14.3) as follows:

$$\dot{l}(t) = A_c(t)l(t) + b_c(-h(t, X)), \tag{14.4}$$

where

$$A_c(t) = \begin{bmatrix} 0 & 1 & 0 \\ 0 & 0 & 1 \\ -a_1(t) & -a_2(t) & -a_3(t) \end{bmatrix}, \; l(t) = \begin{bmatrix} l_1(t) \\ l_2(t) \\ l_3(t) \end{bmatrix}, \; b_c = \begin{bmatrix} 0 \\ 0 \\ 1 \end{bmatrix}.$$

Moreover, $a_i(t)$ with $i = 1, 2, 3$ satisfy that

$$\left|\frac{d^n a_i(t)}{dt^n}\right| < c_n,$$

where c_n with $n = 0, 1, \ldots$ are positive constants. That is, $a_i(t)$ and its derivatives are smooth and bounded. In order to transform the estimation error system (14.3) into system (14.4), a definition of controllability matrix and a lemma are introduced in the following.

Definition 14.1 For a system $\dot{x}(t) = A(t)x(t) + b(t)u(t)$, where $x(t)$ is a n-dimensional state vector, $u(t)$ is a scalar input function, and $A(t)$ and $b(t)$ are time-varying matrices for appropriate order, a controllability matrix is defined as follows:

$$M(t) = \left[\, p_0(t), p_1(t), \ldots, p_{n-1}(t) \,\right],\ p_0(t) = b(t),$$
$$p_{k+1}(t) = -A(t)p_k(t) + \frac{d}{dt}p_k(t),\ k = 1, \ldots, n.$$

Lemma 14.2 ([7]) *The estimation error system (14.3) is uniformly controllable if its controllability matrix $M(t)$ has rank n everywhere. The estimation error system (14.3) is equivalent to system (14.4) if and only if the estimation error system (14.3) is uniformly controllable.*

According to Definition 14.1, the controllability matrix of the estimation error system (14.3) is obtained as follows:

$$M(t) = \begin{bmatrix} 0 & 0 & 1 \\ 0 & -1 & 0 \\ 1 & 0 & 0 \end{bmatrix}. \tag{14.5}$$

From the form of matrix $M(t)$, it is obvious that the estimation error system (14.3) is uniformly controllable. Therefore, the estimation error system (14.3) is equivalent to system (14.4). Let $l(t) = T(t)e(t)$, where $T(t)$ is a transformation matrix. According to [7], $T(t)$ is obtained as follows:

$$T(t) = M_c(t)M^{-1}(t), \tag{14.6}$$

where $M_c(t)$ is the controllability matrix of system (14.4). By Definition 14.1, $T(t)$ is calculated as follows:

$$T(t) = \begin{bmatrix} 1 & 0 & 0 \\ -a_3(t) & 1 & 0 \\ -a_2(t) + a_3^2(t) + \dot{a}_3(t) & -a_3(t) & 1 \end{bmatrix}. \tag{14.7}$$

The estimation error system (14.3) is equivalent to system (14.4) and $a_i(t)$ is smooth and bounded. $T(t)$ is continuous, nonsingular, bounded, and its derivative is continuous. From [8], it is easy to obtain that $T(t)$ is a Lyapunov transformation matrix by expressions (14.5)–(14.7). According to [6], the relationships between $g_i(t)$ and $a_i(t)$ are obtained as follows:

$$\begin{cases} g_1(t) = a_3(t) \\ g_2(t) = a_2(t) - 2\dot{a}_3(t) \\ g_3(t) = a_1(t) + \ddot{a}_3(t) - \dot{a}_2(t). \end{cases} \tag{14.8}$$

The estimation error system (14.3) has been transformed into a canonical form as in (14.4). The observer gain $g_i(t)$ has been expressed by $a_i(t)$ and its derivatives. Subsequently, the time-varying parameter $a_i(t)$ is designed to ensure convergence of the AESO (14.1) in the following.

Note that item $b_c(-h(t, X))$ in system (14.4) is regarded as a disturbance. A homogeneous form of system (14.4) is obtained without the disturbance as follows:

$$\dot{l}(t) = A_c(t)l(t). \tag{14.9}$$

Considering system (14.9), a scalar linear differential system is given as follows:

$$\xi^{(3)}(t) + a_1(t)\xi(t) + a_2(t)\dot{\xi}(t) + a_3(t)\ddot{\xi}(t) = 0. \tag{14.10}$$

Before moving on, the following definition and lemma are introduced as in [9, 10].

Definition 14.3 $\Re_a$ is expressed as a scalar polynomial differential operator (SPDO) of the scalar differential system (14.10). Then the scalar function $\lambda_i(t)$ with $i = 1, 2, 3$ in Eq. (14.12) is called series D-eigenvalues (SD-eigenvalues) of $\Re_a$. Note that $\rho(t) = \lambda_1(t)$ is called a parallel D-eigenvalue (PD-eigenvalue) of $\Re_a$ when $\xi(t) = \exp(\int \rho(t)dt)$ constitutes a solution of $\Re_a\{\xi\} = 0$. A multiset $\Gamma_a = \{\lambda_1(t), \lambda_2(t), \lambda_3(t)\}$ is called a series D-spectrum (SD-spectrum) of $\Re_a$, where $\lambda_i(t)$ with $i = 1, 2, 3$ are SD-eigenvalues of $\Re_a$. A multiset $\Upsilon_a = \{\rho_3(t), \rho_3(t), \rho_3(t)\}$ is called a parallel D-spectrum (PD-spectrum) of $\Re_a$, where $\rho_i(t)$ with $i = 1, 2, 3$ are PD-eigenvalues of $\Re_a$. Moreover, $\xi_i(t) = \exp(\int \rho_i(t)dt)$ constitute a fundamental set of solutions to Eq. (14.10).

Lemma 14.4 ([10]) *$\Re_a$ is expressed as the third-order SPDO with the PD-spectrum $\{\rho_1(t), \rho_2(t), \rho_3(t)\}$. The fundamental set of solutions for $\Re_a\{\xi\} = 0$ is given as $\{\xi_1(t), \xi_2(t), \xi_3(t)\}$, where $\xi_i(t) = \exp(\int \rho_i(t)dt)$. A diagonal matrix is denoted by $D(t) = diag\,\{\xi_1(t), \xi_2(t), \xi_3(t)\}$. Then an expression $W(t)D(t)^{-1} = V(t)$ is obtained, where a matrix $W(t)$ is known as a Wronskian matrix associated with $\{\xi_i(t)\}_{i=1}^3$. A canonical coordinate transformation matrix $V(t)$ is called a modal canonical matrix of $\Re_a$ associated with the PD-spectrum $\{\rho_1(t), \rho_2(t), \rho_3(t)\}$.*

Since system (14.9) and system (14.10) have equal properties, the DAST is applied to analyze system (14.10). From [9], a derivative operator is defined as $\delta = d/dt$, then system (14.10) is represented as $\Re_a\{\xi\} = 0$. For system (14.10), an expression of the SPDO is given as follows:

$$\Re_a = \delta^3 + a_3(t)\delta^2 + a_2(t)\delta + a_1(t). \tag{14.11}$$

Equation (14.11) is rewritten by using a Floquet factorization as

$$\Re_a = (\delta - \lambda_3(t))(\delta - \lambda_2(t))(\delta - \lambda_1(t)). \tag{14.12}$$

The transformation matrix $V(t)$ is obtained by using Definition 14.3 and Lemma 14.4. By the relationship between $\xi_i(t)$ and $\rho_i(t)$ with $i = 1, 2, 3$, $V(t)$ is expressed by $\rho_i(t)$ as follows:

$$\begin{aligned} V(t) &= W(t)D(t)^{-1} \\ &= \begin{bmatrix} \xi_1(t) & \xi_2(t) & \xi_3(t) \\ \dot{\xi}_1(t) & \dot{\xi}_2(t) & \dot{\xi}_3(t) \\ \ddot{\xi}_1(t) & \ddot{\xi}_2(t) & \ddot{\xi}_3(t) \end{bmatrix} \begin{bmatrix} \xi_1(t) & 0 & 0 \\ 0 & \xi_2(t) & 0 \\ 0 & 0 & \xi_3(t) \end{bmatrix}^{-1} \\ &= \begin{bmatrix} 1 & 1 & 1 \\ \frac{\dot{\xi}_1(t)}{\xi_1(t)} & \frac{\dot{\xi}_2(t)}{\xi_2(t)} & \frac{\dot{\xi}_3(t)}{\xi_3(t)} \\ \frac{\ddot{\xi}_1(t)}{\xi_1(t)} & \frac{\ddot{\xi}_2(t)}{\xi_2(t)} & \frac{\ddot{\xi}_3(t)}{\xi_3(t)} \end{bmatrix} \\ &= \begin{bmatrix} 1 & 1 & 1 \\ \rho_1(t) & \rho_2(t) & \rho_3(t) \\ \rho_1^2(t) + \dot{\rho}_1(t) & \rho_2^2(t) + \dot{\rho}_2(t) & \rho_3^2(t) + \dot{\rho}_3(t) \end{bmatrix}. \end{aligned} \tag{14.13}$$

To get the relationship between $a_i(t)$ and $\lambda_i(t)$ and the relationship between $\rho_i(t)$ and $\lambda_i(t)$, the following two lemmas [10] are recalled.

Lemma 14.5 *Let $\{\lambda_i(t)\}_{i=1}^p$ be the SD-spectrum for the pth-order SPDO $\Re_a$. Define a coefficient p for $\Re_a$ as $a_{p,j}(t)$, with $a_{k,0}(t) = 0$ and $a_{k,k+1}(t) = 1$. Then $a_{p,j}(t)$ is recursively calculated as*

$$a_{p,j}(t) = \dot{a}_{p-1,j}(t) - \lambda_p(t)a_{p-1,j}(t) + a_{p-1,j-1}(t), \tag{14.14}$$

where $j = 1, \dots, p$ and $k = 1, \dots, p-1$.

Lemma 14.6 *Let $\{\rho_i(t)\}_{i=1}^p$ be the PD-spectrum for the pth-order SPDO $\Re_a$. Let $V_k(t)$ be the determinant of kth-order modal canonical matrix for $\Re_a$. The third-order expression of $V_k(t)$ is shown in (14.13). The SD-spectrum $\{\lambda_i(t)\}_{i=1}^p$ for $\Re_a$ is given by*

$$\lambda_1(t) = \rho_1(t),\ \lambda_k(t) = \rho_k(t) + \frac{\dot{V}_k}{V_k} - \frac{\dot{V}_{k-1}}{V_{k-1}}, \tag{14.15}$$

where $k = 2, 3, \dots, p$.

From Lemma 14.5, the relationships between $a_i(t)$ and $\lambda_i(t)$ with $i = 1, 2, 3$ are shown as follows:

$$\begin{cases} a_1(t) = -\ddot{\lambda}_1(t) + \dot{\lambda}_1(t)\lambda_2(t) + \dot{\lambda}_2(t)\lambda_1(t) + \dot{\lambda}_1(t)\lambda_3(t) - \lambda_1(t)\lambda_2(t)\lambda_3(t) \\ a_2(t) = -2\dot{\lambda}_1(t) - \dot{\lambda}_2(t) + \lambda_2(t)\lambda_3(t) + \lambda_1(t)\lambda_3(t) + \lambda_1(t)\lambda_2(t) \\ a_3(t) = -\lambda_1(t) - \lambda_2(t) - \lambda_3(t). \end{cases} \tag{14.16}$$

Moreover, it follows from Lemma 14.6 that the relationships between $\rho_i(t)$ and $\lambda_i(t)$ with $i = 1, 2, 3$ are shown as follows:

$$\begin{cases} \lambda_1(t) = \rho_1(t) \\ \lambda_2(t) = \rho_2(t) + \frac{\dot{\rho}_2(t)-\dot{\rho}_1(t)}{\rho_2(t)-\rho_1(t)} \\ \lambda_3(t) = \rho_3(t) + \frac{\dot{J}}{J} - \frac{\dot{\rho}_2(t)-\dot{\rho}_1(t)}{\rho_2(t)-\rho_1(t)}, \end{cases} \tag{14.17}$$

where $J = (\rho_2(t) - \rho_1(t))(\rho_3^2(t) + \dot{\rho}_3(t) - \rho_1^2(t) - \dot{\rho}_1(t)) - (\rho_3(t) - \rho_1(t))(\rho_2^2(t) + \dot{\rho}_2(t) - \rho_1^2(t) - \dot{\rho}_1(t))$.

For convergence analysis of AESO (14.1), the PD-eigenvalues are designed in the following:

$$\rho_i(t) = \tilde{\rho}_i\omega(t),\ i = 1, 2, 3, \tag{14.18}$$

where $\omega(t)$ is a time-varying multiplier and $\tilde{\rho}_i$ represents nominal eigenvalues. Moreover, $\tilde{\rho}_1 = -\omega_0$, $\tilde{\rho}_{2,3} = (-0.5 \pm 0.866j)\omega_0$, where ω_0 is an adjustable parameter. According to the relational expressions (14.16), (14.17), and (14.8), the relationships between $g_i(t)$ and $\rho_i(t)$ with $i = 1, 2, 3$ are obtained as follows:

$$\begin{cases} g_1(t) = -(\tilde{\rho}_1 + \tilde{\rho}_2 + \tilde{\rho}_3)\omega(t) - 3\frac{\dot{\omega}(t)}{\omega(t)} \\ g_2(t) = (\tilde{\rho}_1\tilde{\rho}_3 + \tilde{\rho}_2\tilde{\rho}_3 + \tilde{\rho}_1\tilde{\rho}_2)\omega^2(t) - 3\frac{\dot{\omega}^2(t)}{\omega(t)} \\ \quad +(\tilde{\rho}_1 + \tilde{\rho}_2 + \tilde{\rho}_3)\dot{\omega}(t) - 5\frac{\ddot{\omega}(t)}{\omega(t)} \\ g_3(t) = -\tilde{\rho}_1\tilde{\rho}_2\tilde{\rho}_3\omega^3(t) - 2(\tilde{\rho}_1 + \tilde{\rho}_2 + \tilde{\rho}_3)\ddot{\omega}(t) - 2(\tilde{\rho}_1\tilde{\rho}_3 \\ \quad +\tilde{\rho}_2\tilde{\rho}_3 + \tilde{\rho}_1\tilde{\rho}_2)\omega(t)\dot{\omega}(t) - 2\frac{\dddot{\omega}(t)\omega(t)-\dot{\omega}(t)\ddot{\omega}(t)}{\omega^2(t)}. \end{cases} \tag{14.19}$$

Definition 14.7 An interval function is introduced as follows:

$$\text{fsg}(x, a, b) = \frac{\text{sign}(x - a) - \text{sign}(x - b)}{2} = \begin{cases} 1, & x \in [a, b] \\ 0, & \text{others}. \end{cases}$$

Then $\omega(t)$ is designed as follows:

$$\omega(t) = \text{fsg}(t, 0, T_w) + 10\text{fsg}(t, T_w, \infty) + c(z_1 - x_1)^2, \tag{14.20}$$

where T_w is a small time constant. The advantage of this design is that the gains of AESO (14.1) are maintained as small values in the beginning. Therefore, a peaking phenomenon is effectively weakened. Moreover, there exists a phenomenon of jump for $\omega(t)$ at T_w. Because TDs have advantage in processing step signal, $\dot{\omega}(t), \ddot{\omega}(t), \dddot{\omega}(t)$ in expression (14.19) are obtained by the TDs as filters. Then three TDs are designed in the following. The first one is

$$\begin{cases} \dot{\omega}_1(t) = \omega_2(t) \\ \dot{\omega}_2(t) = f_{han}(\omega_1(t) - \omega_0(t), \omega_2(t), r_{01}, h_{01}), \end{cases}$$

where $\omega_0(t) = \omega(t)$ is the given input, $\omega_1(t)$ is the output tracking signal of $\omega_0(t)$, and $\omega_2(t)$ is the differential signal of $\omega_1(t)$, i.e., $\omega_2(t) = \dot{\omega}(t)$. The second one is

$$\begin{cases} \dot{\omega}_3(t) = \omega_4(t) \\ \dot{\omega}_4(t) = f_{han}(\omega_3(t) - \omega_2(t), \omega_4(t), r_{02}, h_{02}), \end{cases}$$

where $\omega_2(t) = \dot{\omega}(t)$ is the given input, $\omega_3(t)$ is the output tracking signal of $\omega_2(t)$, and $\omega_4(t)$ is the differential signal of $\omega_3(t)$, i.e., $\omega_4(t) = \ddot{\omega}(t)$. The last one is

$$\begin{cases} \dot{\omega}_5(t) = \omega_6(t) \\ \dot{\omega}_6(t) = f_{han}(\omega_5(t) - \omega_4(t), \omega_6(t), r_{03}, h_{03}), \end{cases}$$

where $\omega_4(t) = \ddot{\omega}(t)$ is the given input, $\omega_5(t)$ is the output tracking signal of $\omega_4(t)$, and $\omega_6(t)$ is the differential signal of $\omega_5(t)$, i.e., $\omega_6(t) = \dddot{\omega}(t)$.

For analysis convergence of AESO (14.1), a definition and a theorem are given in the following.

Definition 14.8 Let $\sigma(t)$: $I \to R$ be a locally integrable function on $I = [t_0, \infty)$. An extended mean of $\sigma(t)$ over I is defined by

$$\mathrm{em}(\sigma(t)) = \limsup_{T\to+\infty} \frac{1}{T} \int_{t_0}^{t_0+T} \sigma(t)dt.$$

Lemma 14.9 ([6]) *For the following LTV system*

$$\dot{\zeta}(t) = A_\zeta(t)\zeta(t) + b(-h(x, \omega)) \tag{14.21}$$

define a Lyapunov transformation $z(t) = T_\zeta(t)\zeta(t)$ *that transforms (14.21) into the canonical form as follows:*

$$\dot{z}(t) = A_z(t)z(t) + b(-h(x, \omega)). \tag{14.22}$$

The homogeneous form for (14.22) is shown as

$$\dot{z}(t) = A_z(t)z(t). \tag{14.23}$$

Assume that the following conditions are satisfied:

(a) The PD-spectrum $\Upsilon_a = \{\rho_i(t)\}_{i=1}^{n+1}$ *of the homogeneous system (14.23), where* $\rho_i(t)$ *is smooth and bounded. Moreover,* $\rho_i(t)$ *is locally integrable on* $I = [t_0, \infty)$ *and satisfies* $\mathrm{em}(Re\rho_i(t)) = -m_i < 0,\ t \in I$ *and* $0 < m_i < \infty$.

(b) There exist $k_i > 0$ *and* $0 < \gamma_i < -d_i$ *such that*

$$\| v_i(t)u_i^T(t) \| \leq k_i \zeta^{-\gamma_i(t-\iota)}(t)$$

for all $t \geq \iota \geq t_0$, where $v_i(t)$ and $u_i^T(t)$ are defined as column PD-eigenvectors and row PD-eigenvectors of $\Re_a$.

(c) The disturbance item $b(-h(x, \omega))$ of (14.21) satisfies

$$\|b(-h(x, \omega))\| \leq \Delta_\varsigma < \sqrt{c_1/c_2}\eta_\varsigma r_\varsigma c_3/c_4, \tag{14.24}$$

where c_1, c_2, c_3, and c_4 are positive constants, $0 < \eta_\varsigma < 1$, $\forall t \geq 0$, $\varsigma(t) \in D$, $D = \{\varsigma(t) \in R^n | \|\varsigma(t)\| < r_\varsigma\}$. Then for all $\|\varsigma(t_0)\| < \sqrt{c_1/c_2}r$, the solution of (14.21) satisfies

$$\begin{cases} \|\varsigma(t)\| \leq k_\varsigma \varsigma^{-\gamma_\varsigma(t-t_0)} \|\varsigma(t_0)\|, & t_0 \leq t < t_0 + t_1 \\ \|\varsigma(t)\| \leq b_\varsigma, & t \geq t_0 + t_1 \end{cases}$$

for some finite t_1, where

$$k_\varsigma = \sqrt{\frac{c_2}{c_1}},\ \gamma_\varsigma = \frac{(1-\eta_\varsigma)c_3}{2c_2},\ b_\varsigma = \frac{c_4}{c_3}\sqrt{\frac{c_2}{c_1}}\frac{\Delta_\varsigma}{\eta_\varsigma}.$$

Theorem 14.10 *For the estimation error system (14.3), a Lyapunov transformation $l(t) = T(t)e(t)$ is given which transforms system (14.3) into system (14.4). Assume that the PD-spectrum $\Upsilon_a = \{\rho_1(t), \rho_2(t), \rho_3(t)\}$ of the homogeneous system (14.9) is satisfied with conditions in Lemma 14.9. The disturbance item $b(-h(t, X))$ of the estimation error system (14.3) satisfies the inequality as follows:*

$$\|b(-h(t, X))\| \leq \Delta_e < \sqrt{n_1/n_2}\eta_e r_e n_3/n_4. \tag{14.25}$$

Therefore, the solution of the estimation error system (14.3) is bounded for all $\|e(t_0)\| < \sqrt{n_1/n_2}r_e$ with finite time T.

Proof The PD-spectrum of the homogeneous system (14.9) and the disturbance item $b(-h(t, X))$ of the estimation error system (14.3) are satisfied with conditions in Lemma 14.9. Therefore, for all $\|e(t_0)\| < \sqrt{n_1/n_2}r_e$, solution $e(t)$ of the estimation error system (14.3) is bounded for finite time T, i.e.,

$$\begin{cases} \|e(t)\| \leq k_e e^{-\gamma_e(t-t_0)} \|e(t_0)\|, & t_0 \leq t < t_0 + T \\ \|e(t)\| \leq b_e, & t \geq t_0 + T, \end{cases}$$

where

$$k_e = \sqrt{\frac{n_2}{n_1}},\ \gamma_e = \frac{(1-\eta_e)n_3}{2n_2},\ b_e = \frac{n_4}{n_3}\sqrt{\frac{n_2}{n_1}}\frac{\Delta_e}{\eta_e}.$$

Remark 14.11 The constants γ, κ, and χ in Theorem 14.10 are related to PD-spectrum of the homogeneous system (14.9). According to [6, 11], the constants γ, κ, and χ satisfy $\|A(t)\| \leq \chi$ and $\|\Phi(t, t_0)\| \leq \kappa e^{-\gamma(t-t_0)}$, where $\|\Phi(t, t_0)\|$ is a state transition matrix of the homogeneous system (14.9), $\gamma = \min\{d_i\}$.

14.2.3 Nonlinear Backstepping Controller

In this section, the following nonlinear controller $u(t)$ is designed using the strategy of backstepping as

$$u(t) = \frac{-z_3(t) + \dot{\varpi}(t) - s_1(t) - k_2 fal(s_2(t), \alpha, \delta)}{b_0}, \tag{14.26}$$

where $\varpi(t) = v_2(t) - k_1 s_1(t)$ is a virtual control variable, $s_1(t)$ and $s_2(t)$ are error variables, and $fal(s_2(t), \alpha, \delta)$ is a nonlinear function to obtain good performance. The expression of $fal(s_2(t), \alpha, \delta)$ is given in expression (2.4).

Proof The error variables are as follows:

$$\begin{cases} s_1(t) = x_1(t) - v_1(t) \\ s_2(t) = z_2(t) - \varpi(t). \end{cases} \tag{14.27}$$

Step 1: Design a Lyapunov function $V_1(t)$ as follows:

$$V_1(t) = \frac{1}{2} s_1^2(t). \tag{14.28}$$

According to convergence analysis of AESO (14.1), equation $x_2(t) = z_2(t) - e_2(t)$ is given. From the estimation error system (14.2) and expression (14.27), equations $e_2(t) = \dot{e}_1(t) + g_1(t)e_1(t)$ and $z_2(t) = s_2(t) + \varpi(t)$ are obtained. Therefore, $x_2(t)$ is written as $x_2(t) = s_2(t) + \varpi(t) - (\dot{e}_1(t) + g_1(t)e_1(t))$. Then the derivative of $V_1(t)$ is obtained as follows:

$$\begin{aligned} \dot{V}_1(t) &= s_1(t)x_2(t) - s_1(t)v_2(t) \\ &= s_1(t)s_2(t) + s_1(t)\varpi(t) - s_1(t)(\dot{e}_1(t) + g_1(t)e_1(t)) - s_1(t)v_2(t) \\ &= s_1(t)s_2(t) - ks_1^2(t) - s_1(t)(\dot{e}_1(t) + g_1(t)e_1(t)), \end{aligned} \tag{14.29}$$

where $s_1^2(t) \geq 0$ and $s_1^2(t) = 0$ only when $s_1(t) = 0$. According to convergence analysis of AESO, an inequality $|\dot{e}_1(t) + g_1(t)e_1(t)| < \Delta_1$ is obtained, where Δ_1 is a bounded positive number. Therefore, expression (14.29) is rewritten as follows:

$$\dot{V}_1(t) < s_1(t)\Delta_1 + s_1(t)s_2(t) - ks_1^2(t).$$

If $s_2 = 0$, the first subsystem is stable by choosing proper k_1.

Step 2: The second Lyapunov function $V_2(t)$ is designed as follows:

$$V_2(t) = V_1(t) + \frac{1}{2}s_2^2(t). \tag{14.30}$$

The derivative of $V_2(t)$, it follows that

$$\begin{aligned}\dot{V}_2(t) = &-k_1 s_1^2(t) - s_1(t)(\dot{e}_1(t) + g_1(t)e_1(t)) \\ &+ s_2(t)(\dot{z}_2(t) - \dot{\varpi}(t)) + s_1(t)s_2(t).\end{aligned} \tag{14.31}$$

According to convergence analysis of AESO (14.1), equations $z_2(t) = e_2(t) + x_2(t)$, $\dot{z}_2(t) = \dot{e}_2(t) + \dot{x}_2(t)$. From system (11.9) and the estimation error system (14.2), expressions $\dot{x}_2(t) = x_3(t) + b_0 u(t)$ and $\dot{e}_2(t) = e_3(t) - g_2(t)e_1(t)$ are obtained. Then $\dot{V}_2(t)$ is rewritten as follows:

$$\begin{aligned}\dot{V}_2(t) = &-k_1 s_1^2(t) + s_1(t)s_2(t) - s_1(t)(\dot{e}_1(t) + g_1(t)e_1(t)) \\ &+ s_2(t)(x_3(t) + b_0 u(t) + e_3(t) - g_2(t)e_1(t) - \dot{\varpi}(t)).\end{aligned}$$

Then the nonlinear controller is designed as follows:

$$u(t) = \frac{-z_3(t) + \dot{\varpi}(t) - s_1(t) - k_2 fal(s_2(t), \alpha, \delta)}{b_0}.$$

Note that $fal(s_2(t), \alpha, \delta)$ is formulated with $fal(s_2(t))$ for simplicity. Substituting (14.26) into (14.30), $\dot{V}_2(t)$ is rewritten as follows:

$$\begin{aligned}\dot{V}_2(t) &= s_1(t)s_2(t) - s_1(t)(\dot{e}_1(t) + g_1(t)e_1(t)) \\ &\quad - k_1 s_1^2(t) + s_2(t)(x_3(t) - z_3(t) - s_1(t) \\ &\quad - k_2 fal(s_2(t)) + e_3(t) - g_2(t)e_1(t)) \\ &= -k_1 s_1^2(t) - s_1(t)(\dot{e}_1(t) + g_1(t)e_1(t)) \\ &\quad - k_2 s_2(t) fal(s_2(t)) - s_2(t)g_2(t)e_1(t).\end{aligned} \tag{14.32}$$

Similarly, according to convergence analysis of AESO, $|g_2(t)e_1(t)| < \Delta_2$ is obtained, where Δ_2 is a bounded positive number. Therefore, expression (14.31) is rewritten as follows:

$$\dot{V}_2(t) < s_1(t)\Delta_1 + s_2(t)\Delta_2 - k_1 s_1^2(t) - k_2 s_2(t) fal(s_2(t)).$$

Then the closed-loop system (11.9) is stable by choosing appropriate parameters k_1 and k_2. That is, the output of system (11.9) $x_1(t)$ can track to the desired input $v_0(t)$.

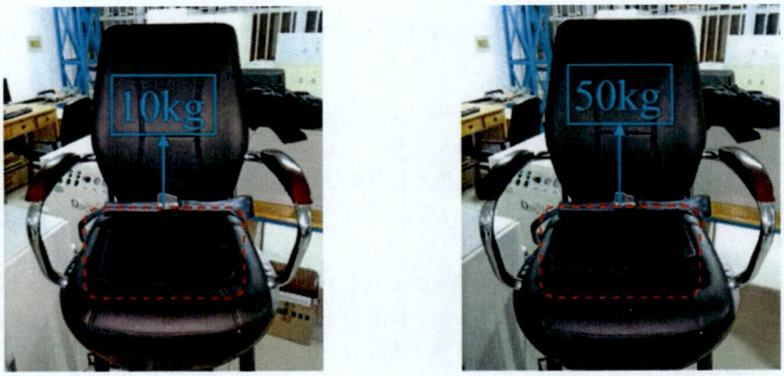

(a) Load mass $m = 10$kg. (b) Load mass $m = 50$kg.

Fig. 14.2 Experiment platform with variable loads

Table 14.1 The parameters PAM mechanism model

Parameters		
$k_0 = 0.09$	$k_1 = 0.0501$	$k_2 = -0.0472$
$k_3 = 0.0106$	$k_u = 0.45$	$r = 200\,\text{mm}$
$L_0 = 450\,\text{mm}$	$u_0 = 2\,\text{V}$	$\varepsilon_0 = 0.111$

Table 14.2 Main parameters of controller and PD-eigenvalues

Sinusoidal signal	Parameters		
Controller	$k_1 = 1$	$k_2 = 1$	$b_0 = 260$
PD-eigenvalues $\rho_i(t)$	$\omega_0 = 1$	$c = 1$	$T_w = 0.5\,\text{s}$
Step signal	Parameters		
Controller	$k_1 = 1$	$k_2 = 0.1$	$b_0 = 800$
PD-eigenvalues $\rho_i(t)$	$\omega_0 = 0.5$	$c = 1$	$T_w = 0.5\,\text{s}$

14.3 Experiments and Results

In this chapter, tracking control for one mechanism of PAMs is studied. The variable load experiments are introduced to indicate the effectiveness of the proposed control method as shown in Fig. 14.2. Moreover, comparison experiments between the proposed control method and the control method in [12] are carried out in this experiment. Note that masses of two loads are $m = 10\,\text{kg}$ and $m = 50\,\text{kg}$ in Fig. 14.2a and b, respectively. The other parameters of the experiment platform are given in Table 14.1. The main adjustable parameters in controller (14.26) and PD-eigenvalues (14.18) of the tracking experiment for a sinusoidal signal and a step signal are listed in Table 14.2.

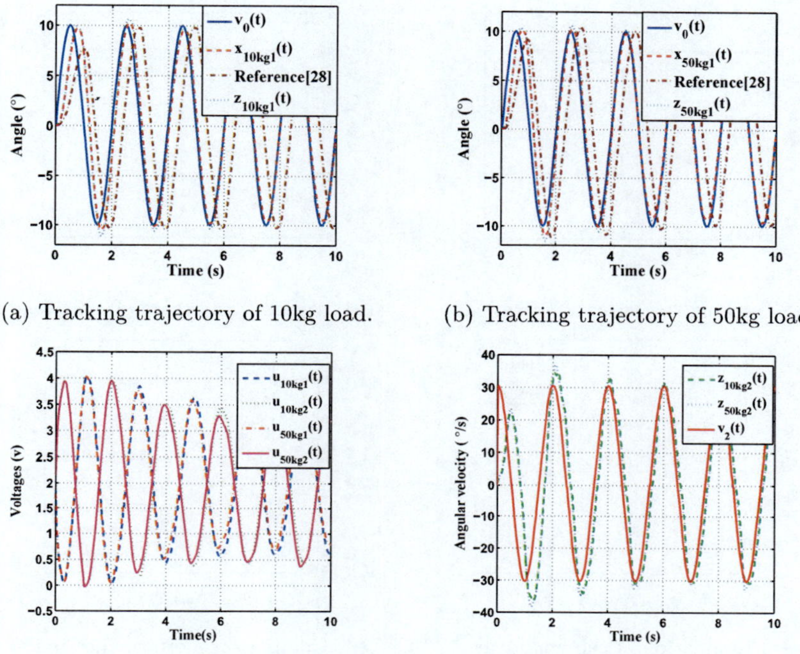

(a) Tracking trajectory of 10kg load.

(b) Tracking trajectory of 50kg load.

(c) Control voltages.

(d) Angular velocity.

Fig. 14.3 Experimental results of the tracking experiment for sinusoidal signal with loads of 10 and 50 kg

In the first experiment, the sinusoidal signal is given with 0.5 Hz and 10° amplitude as a desired input signal. The results of experiment are shown in Fig. 14.3. In Fig. 14.3a, the angle output $x_{10kg1}(t)$ tracks the given input single $v_0(t)$ accurately with the load of 10 kg. The tracking curve follows the given curve at about 1.8 s, and the motion error is less than 0.5°. Note that the delay time is reduced about 0.5 s compared with [12]. The observed value $z_1(t)$ is almost coincident with $x_1(t)$. The similar results are obtained with the load of 50 kg as shown in Fig. 14.3b. The control voltages $u_1(t)$ and $u_2(t)$ of two PAMs for $m = 10$ kg and $m = 50$ kg are shown in Fig. 14.3c. Moreover, the observed values $z_2(t)$ with loads of 10 and 50 kg are presented in Fig. 14.3d.

In the second experiment, a step signal at 10° is given as the desired input signal $v_0(t)$. The experimental results are shown in Fig. 14.4. In Fig. 14.4a, the actual position tracking $x_{10kg1}(t)$ for the mechanism of PAMs responds quickly and accurately with a load with 10 kg. The response time is 1.5 s and the steady-state error is less than 0.2°. Moreover, the output signal $v_1(t)$ of the tracking differentiator tracks the given input signal $v_0(t)$ quickly and steadily. Similar results are also obtained with a load with 50 kg as shown in Fig. 14.4b. A better performance of the mechanism for PAMs based on the proposed control method is obtained than the one in [12]. In

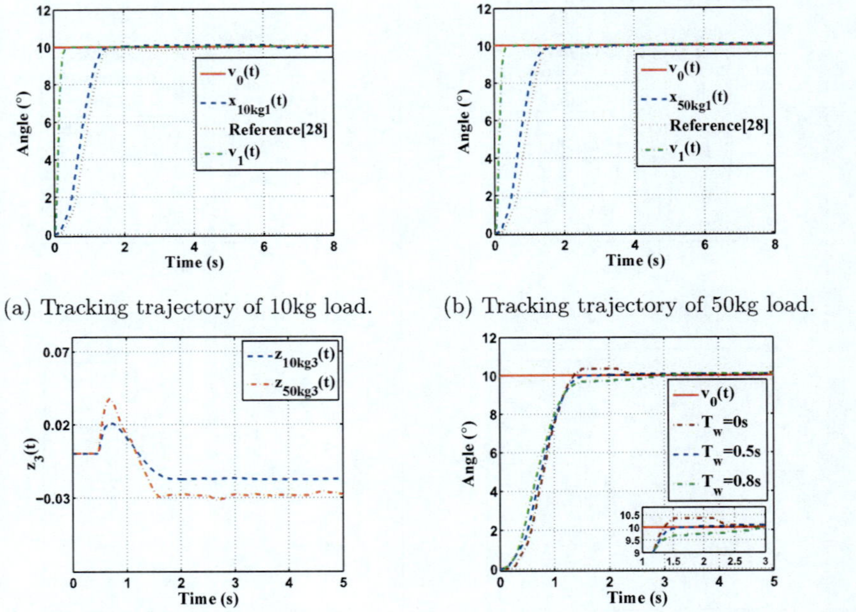

(a) Tracking trajectory of 10kg load.

(b) Tracking trajectory of 50kg load.

(c) Estimation of the nonlinear.

(d) Comparison performance under different T_w in $\rho_i(t)$.

Fig. 14.4 Experimental results of the tracking experiment for step signal with loads of 10 and 50 kg

addition, the estimated results of nonlinear terms, external disturbances, and unmodeled dynamics for nonlinear systems are shown in Fig. 14.4c. It indicates different total disturbances under different loads. Because the given input signal is a step signal, $z_3(t)$ tends to a stable value when the tracking signal follows the given signal. The estimated value $z_3(t)$ has slight tremor due to external interferences. In order to illustrate the advantages of $\omega(t)$ in (14.20), comparison results under different values of T_w in $\omega(t)$ are shown in Fig. 14.4d. A good performance of the mechanism for PAMs is also obtained under $T_w = 0.5$ s. Moreover, smooth experimental curves are obtained due to advantages of the proposed filter based on the TD for processing a step signal. The effectiveness of the backstepping control strategy based on the AESO is shown by experiments. That is, experimental results indicate that the control strategy has strong advantages in dealing with various uncertainties for the mechanism of PAMs.

14.4 Conclusion

In this chapter, the backstepping control strategy based on an AESO for mechanism of PAMs has been introduced. The transient process has been arranged by a TD. In addition, the time-varying multiplier in the PD-eigenvalues has been designed as a novel form. Because TDs have an advantage in processing step signals, the filters have been designed via TDs. The backstepping control strategy has been applied to design a nonlinear controller. Convergence of the AESO has been analyzed by DAST. Moreover, stabilization of closed-loop system has been discussed via Lyapunov theory. The experimental results indicate that the control strategy has strong advantages in dealing with various uncertainties for the mechanism of PAMs in the pneumatic motion simulation platform.

References

1. Andrikopoulos G, Nikolakopoulos G, Manesis S (2014) Advanced nonlinear PID-based antagonistic control for pneumatic muscle actuators. IEEE Trans Ind Electron 61(12):6926–6937
2. Jamwal PK, Hussain S, Ghayesh MH, Rogozina SV (2016) Impedance control of an intrinsically compliant parallel ankle rehabilitation robot. IEEE Trans Ind Electron 63(6):3638–3647
3. Xia Y, Dai L, Fu M, Li C, Wang C (2014) Application of active disturbance rejection control in tank gun control system. J Frank Inst 351(4):2299–2314
4. Xia Y, Chen R, Pu F, Dai L (2014) Active disturbance rejection control for drag tracking in mars entry guidance. Adv Space Res 53(5):853–861
5. Yang H, You X, Xia Y, Li H (2014) Adaptive control for attitude synchronisation of spacecraft formation via extended state observer. IET Control Theory Appl 8(18):2171–2185
6. Pu Z, Yuan R, Yi J, Tan X (2015) A class of adaptive extended state observers for nonlinear disturbed systems. IEEE Trans Ind Electron 62(9):5858–5869
7. Silverman L (1966) Transformation of time-variable systems to canonical (phase-variable) form. IEEE Trans Autom Control 11(2):300–303
8. Zadeh LA, Desoer CA (1963) Linear system theory: the state space approach. McGraw-Hill, New York
9. Choi J, Lee HC, Zhu J (2001) Decoupling and tracking control using eigenstructure assignment for linear time-varying systems. Int J Control 74(5):453–464
10. Zhu X, Tao G, Yao B, Cao J (2008) Adaptive robust posture control of a parallel manipulator driven by pneumatic artificial muscles. Automatica 44(9):2248–2257
11. Liu Y, Zhu JJ (2007) Regular perturbation analysis for trajectory linearization control. In: American control conference, pp 3053–3058
12. Yang H, Yu Y, Zhang J (2017) Angle tracking of a pneumatic muscle actuator mechanism under varying load conditions. Control Eng Pract 61:1–10

References

Part IV
Pneumatic Rod Cylinder Servo System

Chapter 15
Platform Introduction

15.1 Application Background

Pneumatic rod cylinder servo systems are always applied in industrial automation, medical instrument, national defense, and other fields, due to their unique advantages such as simple mechanism, zero pollution, low cost, high power weight ratio, and so on. In recent years, with the further development of pneumatic technology, pneumatic rod cylinder servo systems are also used in intelligent manufacturing and medical technology. In Fig. 15.1a, a pneumatic actuator invented by Wuxi Huatong Pneumatic Manufacture Co., Ltd. It is used in position control of air door in the power plant and the pneumatic rod cylinder is the key component. In Fig. 15.1b, a pneumatic surgical robot driven by pneumatic cylinders has been designed by a field research laboratory for Tokyo Institute of Technology in Japan [1]. Moreover, along with the development of control technology, pneumatic rod cylinder servo systems will have more application value than before.

15.2 Platform Structure

15.2.1 Platform Components

The experiment platform of pneumatic rod cylinder servo system is shown in Fig. 15.2, which is called pneumatic rod cylinder servo platform. The working stroke of the rod cylinder is 200 mm, range of motion is 0–200 mm. The experiment platform contains six main parts: a pneumatic rod cylinder (SMC, MBBQ80-200B), a five-way proportional valve (Festo, MPYE-5-3/8-010-B), a displacement sensor with class 0.1 (Fuxin, LVDT-PB41MS150X), an acquisition card (Advantech, PCL-812), an output card (Advantech, PCI-1727), and an industrial control computer (Advantech, IPC-610H).

L. Zhao et al., *Pneumatic Servo Systems Analysis*, Advances in Industrial Control,
https://doi.org/10.1007/978-981-16-9515-5_15

(a) Huatong pneumatic actuator. (b) Pneumatic surgical robot

Fig. 15.1 Manipulator system driven by pneumatic cylinders

Fig. 15.2 Platform of pneumatic servo system

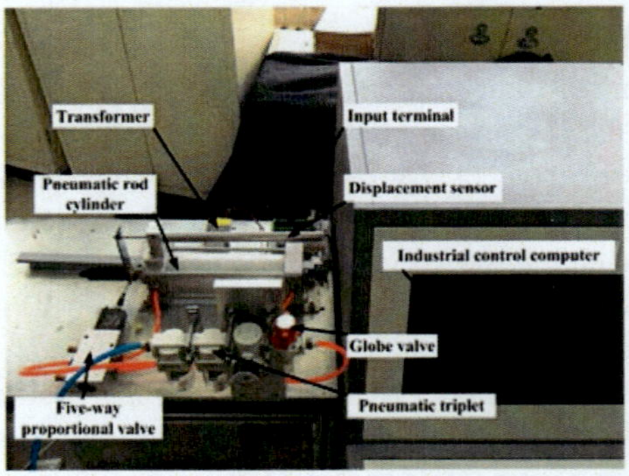

The displacement sensor is an important element in the pneumatic rod cylinder servo system platform, and it can measure the displacement of the piston rod in real time. The displacement sensor used in the pneumatic circuit is shown in Fig. 15.3. Its range is 200 mm, accuracy class is 0.1, rated voltage is +12 V and ambient temperature is 0–60 °C. The reliability of the displacement sensor is further improved by the use of the linear variable differential transformer (LVDT) principle and the latest SMD technology. And the secondary voltage of LVDT is demodulated and filtered into a DC slow change signal which is linearly related to the displacement.

In this chapter, main components of the experimental platform are shown in Table 15.1.

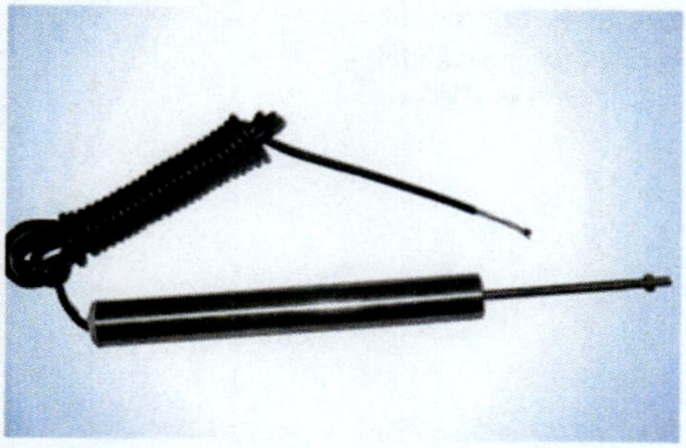

Fig. 15.3 Displacement sensor

Table 15.1 Main components of pneumatic rod cylinder servo platform

Component name	Component model	Main performance indicators
Cylinder	MBBQ80-200B (SMC)	Working stroke: 200 mm
		Working pressure: 0–1.0 MPa
Proportional directional control valve	MPYE-5-3/8-010-B (Festo)	Output voltage: 0–10VDC
		Working pressure: 0–1.0 MPa
		Rate of flow: 100–200 L/min
Acquisition card	PCL-812	16 path A/D input
		Input range: 0–10 V
Output card	PCI-1727	12 path D/A output
		Output range: 0–10 V

15.2.2 Control Circuit

In the pneumatic rod cylinder servo system platform, the rod cylinder is a controlled plant. The displacement sensor is installed on the top of the cylinder to measure displacement of the rod cylinder continuously. The proportional directional control valve is used to control flow of gas into the cylinder. The proportional directional control valve is the same as the one in pneumatic motion simulation system. The industrial control computer with cards is used to collect sensor data, algorithm operation, and output control signal.

A control process for the pneumatic rod cylinder servo system platform is shown in Fig. 15.4. In Fig. 15.4a, the left chamber is filled with gas; the voltage signal is 4 V; thus the rod cylinder stays at initial position. In Fig. 15.4b, gas enters the right chamber and the pressure difference of two chambers pushes the rod to move. The voltage signal changes in real time. In Fig. 15.4c, the rod cylinder reaches the ideal position and the voltage signal stays at a fixed value.

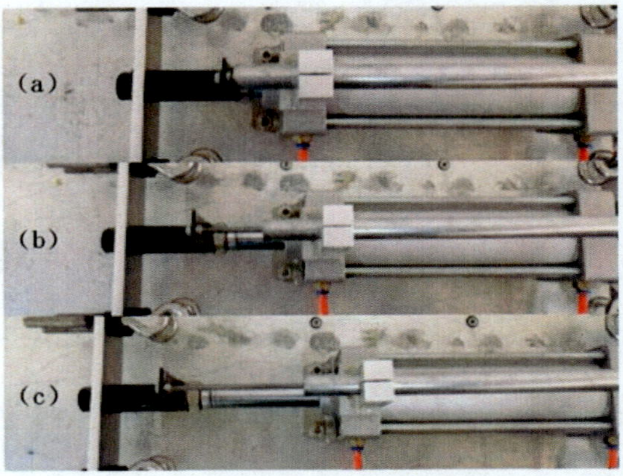

Fig. 15.4 The control process for the parallel mechanism of PMAs

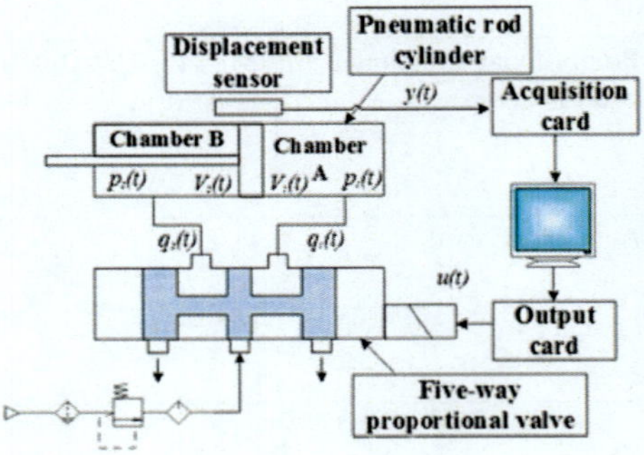

Fig. 15.5 The control structure for the experimental platform

According to the control process for the pneumatic rod cylinder servo platform, a control circuit for pneumatic rod cylinder servo platform is shown in Fig. 15.5. In Fig. 15.5, $u(t)$ denotes the control voltage of the proportional directional control valve, $p_1(t)$ and $p_2(t)$ are pressures in Chamber A and Chamber B, respectively, $q_1(t)$ and $q_2(t)$ are mass flow rate in Chamber A and Chamber B, respectively, $V_1(t)$ and $V_2(t)$ are volumes in Chamber A and Chamber B, respectively.

15.3 System Model

Considering [2] and related mechanics theories, a mathematical model of the pneumatic servo system with rod cylinder is given as follows:

$$m\ddot{y}(t) = -F(t) + (A_1 p_1(t) - A_2 p_2(t)), \tag{15.1}$$

where m describes the mass of the piston, $y(t)$ is the displacement of the piston, $\dot{y}(t)$ is the velocity, $\ddot{y}(t)$ represents the acceleration, $F(t)$ stands for the friction of the pneumatic rod cylinder, A_1 and A_2 denote the areas of chamber A and chamber B, respectively, $p_1(t)$ and $p_2(t)$ are the pressures of chamber A and chamber B, respectively.

According to expression of the friction, Eq. (15.1) is rewritten to an incremental expression as

$$m\Delta\ddot{y}(t) = -\Delta F(t) + (A_1\Delta p_1(t) - A_2\Delta p_2(t)), \tag{15.2}$$

where $\Delta\ddot{y}(t)$, $\Delta F(t)$, $\Delta p_1(t)$, and $\Delta p_2(t)$ are the increments of $\ddot{y}(t)$, $F(t)$, $p_1(t)$, and $p_2(t)$, respectively. According to [3], pressures of the pneumatic rod cylinder in chamber A and chamber B are obtained as follows:

$$q_{(1,2)}(t) = \frac{1}{QT}\left(\frac{V_{(1,2)}(t)}{K}\dot{p}_{(1,2)}(t) + p_{(1,2)}(t)\dot{V}_{(1,2)}(t)\right), \tag{15.3}$$

where subscripts 1 and 2 represent attributes of chamber A and chamber B, respectively, $q_{(1,2)}(t)$ is the mass flow rate, $p_{(1,2)}(t)$ is the pressure, $V_{(1,2)}(t)$ is the volume, K is the radio of specific heats of air, Q is the universal gas constant, and T is the system temperature. Note that the position of piston is set in the middle of the pneumatic rod cylinder. Thereby, the following expression is obtained as

$$\begin{cases} q_1(t) = 0 + \Delta q_1(t) \\ p_1(t) = p_{10} + \Delta p_1(t) \\ V_1(t) = V_{10} + \Delta V_1(t) \end{cases} \tag{15.4}$$

where p_{10} and V_{10} represent the pressure and volume of chamber A in the initial state, respectively, $\Delta q_1(t)$ and $\Delta V_1(t)$ are the increments of mass flow rate and volume of chamber A, respectively. In order to simplify, $\Delta p_1(t)$ and $\Delta V_1(t)$ would be neglected if the piston is located in a suitable displacement. Considering (15.3) and (15.4), the following expression is desired as

$$\Delta q_1(t) = \frac{1}{QT}\left(p_{10}A_1\Delta\dot{y}(t) + \frac{V_{10}}{K}\Delta\dot{p}_1(t)\right), \tag{15.5}$$

where $\Delta\dot{y}(t)$ is the increment of $\dot{y}(t)$. In the same way, the equation of working chamber B is obtained as follows:

$$-\Delta q_2(t) = \frac{1}{QT}\left(-p_{20}A_2\Delta\dot{y}(t) + \frac{V_{20}}{K}\Delta\dot{p}_2(t)\right), \tag{15.6}$$

where p_{20} and V_{20} represent the pressure and volume of chamber B in the initial state, respectively, $\Delta q_2(t)$ is the increment of mass flow rate of chamber B. According to [4], the five-way proportional valve is modeled as follows:

$$q_i(t) = 0.0404A(x(t))C_d f\left(\frac{p_d(t)}{p_u(t)}\right) \tag{15.7}$$

with

$$f\left(\frac{p_d(t)}{p_u(t)}\right) = \frac{p_u(t)}{\sqrt{T}} \begin{cases} 1, & \dfrac{p_a}{p_u(t)} < \dfrac{p_d(t)}{p_u(t)} < 0.528 \\ l_1 l_2, & 0.528 < \dfrac{p_d(t)}{p_u(t)} < 1 \end{cases}$$

where $i = 1, 2$, $A(x(t))$ represents the effective valve orifice area, C_d is the discharge coefficient, $p_d(t)$ is the upstream pressure, $p_u(t)$ is the downstream pressure, p_a is the atmospheric pressure and

$$l_1 = \sqrt{\frac{2}{k-1}\left(\frac{k+1}{2}\right)^{\frac{k+1}{k-1}}},$$

$$l_2 = \sqrt{\left(\frac{p_d(t)}{p_u(t)}\right)^{\frac{2}{k}} - \left(\frac{p_d(t)}{p_u(t)}\right)^{\frac{k+1}{k}}},$$

in which k is the ratio of specific heats. Using linearization for the expression (15.7), a relationship between the pressure and the flow of the proportional valve near zero position is rewritten as follows:

$$\begin{cases} \Delta q_1(t) = k_q \Delta x(t) - k_p \Delta p_1(t) \\ -\Delta q_2(t) = k_q \Delta x(t) + k_p \Delta p_2(t), \end{cases} \tag{15.8}$$

where $\Delta x(t)$ represents the effective valve opening displacement. Let

$$\begin{cases} k_q = \left.\dfrac{\partial q_1(t)}{\partial x(t)}\right|_0 = \left.\dfrac{\partial q_2(t)}{\partial x(t)}\right|_0 \\ k_p = -\left.\dfrac{\partial q_1(t)}{\partial p_1(t)}\right|_0 = \left.\dfrac{\partial q_2(t)}{\partial p_2(t)}\right|_0. \end{cases} \tag{15.9}$$

Considering [4, 5] and regardless of dead zone of the proportional valve, the relationship between $u(t)$ and $\Delta x(t)$ is linearized as follows:

$$\Delta x(t) = k_u u(t), \tag{15.10}$$

where k_u represents the proportional coefficient and $u(t)$ is the voltage signal of proportional valve. According to (15.5), (15.6), (15.8), and (15.10), the following equation is obtained as

$$\begin{cases} \dfrac{1}{QT}\left(p_{10} A_1 \Delta \dot{y}(t) + \dfrac{V_{10}}{K} \Delta \dot{p}_1(t)\right) = k_q k_u u(t) - k_p \Delta p_1(t) \\ \dfrac{1}{QT}\left(-p_{20} A_2 \Delta \dot{y}(t) + \dfrac{V_{20}}{K} \Delta \dot{p}_2(t)\right) = k_q k_u u(t) + k_p \Delta p_2(t) \end{cases} \tag{15.11}$$

Remark 15.1 According to (15.2) and (15.11), it is shown that the voltage signal $u(t)$ does not control directly the pneumatic rod cylinder servo system (15.2). However, it is easy to obtain a relationship between the voltage signal $u(t)$ and the time derivative of pressures from (15.11). Therefore, a direct relationship between the voltage signal $u(t)$ and the pneumatic rod cylinder servo system is gotten by differentiating (15.2).

Differentiating (15.2), there exists

$$m\Delta\dddot{y}(t) = -\Delta\dot{F}(t) + (A_1\Delta\dot{p}_1(t) - A_2\Delta\dot{p}_2(t)), \tag{15.12}$$

where $\Delta\dddot{y}(t)$ is the time derivative of the acceleration $\Delta\ddot{y}(t)$, and it is also commonly called as jerk [6]. Considering (15.11) and (15.12), it follows that

$$\begin{aligned} m\Delta\dddot{y}(t) = &-\left(\frac{KA_1^2p_{10}}{V_{10}} + \frac{KA_2^2p_{20}}{V_{20}}\right)\Delta\dot{y}(t) \\ &+\left(\frac{A_1}{V_{10}} + \frac{A_2}{V_{20}}\right)KQTk_qk_uu(t) \\ &+\left(-\frac{A_1\Delta p_1(t)}{V_{10}} + \frac{A_2\Delta p_2(t)}{V_{20}}\right)KQTk_2 - \Delta\dot{F}(t), \end{aligned} \tag{15.13}$$

where $k_p = -\left.\dfrac{\partial q_1(t)}{\partial p_1(t)}\right|_0 = 0$ is obtained from (15.7) and (15.9). The pneumatic rod cylinder servo system (15.13) is rewritten as follows:

$$\begin{cases} \dot{x}_1(t) = x_2(t) \\ \dot{x}_2(t) = x_3(t) \\ \dot{x}_3(t) = -\dfrac{\Delta\dot{F}(t)}{m} - \left(\dfrac{KA_1^2p_{10}}{mV_{10}} + \dfrac{KA_2^2p_{20}}{mV_{20}}\right)x_2(t) \\ \qquad\quad + \left(\dfrac{A_1}{mV_{10}} + \dfrac{A_2}{mV_{20}}\right)KQTk_qk_uu(t), \end{cases} \tag{15.14}$$

where $\Delta y(t)$ is the increment of $y(t)$. $x_1(t)$, $x_2(t)$, and $x_3(t)$ are set as $\Delta y(t)$, $\Delta\dot{y}(t)$, and $\Delta\ddot{y}(t)$, respectively.

In system (15.14), the unknown nonlinear term

$$-\frac{\Delta\dot{F}(t)}{m} - \left(\frac{KA_1^2p_{10}}{mV_{10}} + \frac{KA_2^2p_{20}}{mV_{20}}\right)x_2(t)$$

is continuously differentiable and bounded, and it is extended to a new state $x_4(t)$. Meanwhile, the following section

$$\left(\frac{A_1}{mV_{10}} + \frac{A_2}{mV_{20}}\right)KRTk_qk_u$$

is a gain b_0 of the input control voltage signal $u(t)$. Moreover, the pneumatic rod cylinder servo system (15.14) is rewritten as follows:

$$\begin{cases} \dot{x}_1(t) = x_2(t) \\ \dot{x}_2(t) = x_3(t) \\ \dot{x}_3(t) = x_4(t) + b_0 u(t) \\ \dot{x}_4(t) = w(t) \end{cases} \tag{15.15}$$

where $w(t)$ is the derivative of $x_4(t)$. Note that $w(t)$ is bounded in practice.

Considering [2, 7] and related mechanics theories, the following mathematical model of pneumatic rod cylinder servo system with varying loads is also given as

$$\ddot{y}(t) = d(t) + \frac{p_1(t)A_1 - p_2(t)A_2}{m}, \tag{15.16}$$

where $d(t)$ denotes effect of various friction forces. According to Eq. (15.3) of Chap. 17 and (15.16), the pneumatic rod cylinder servo system will become complicated if the relationship between mass flow rates and pressures is taken into account. For simplicity, the pneumatic rod cylinder servo system is taken as a second-order system [7, 8]. The Eq. (15.16) is processed as follows:

$$\begin{cases} \dfrac{p_1(t)A_1 - p_2(t)A_2}{m} = b_0 u(t) + \Delta u(t) \\ , \\ \Delta u(t) = f(t, X(t)) \end{cases} \tag{15.17}$$

where b_0 is the control gain, $u(t)$ is the control input signal, $\Delta u(t)$ is the nonlinear input signal, $X(t) = [x_1(t)\ x_2(t)]^T$ is the state vector of pneumatic servo system, $f(t, X(t))$ is various uncertain factors of the pneumatic rod cylinder servo system. Then, the Eq. (15.16) is rewritten as follows:

$$\begin{cases} \dot{x}_1(t) = x_2(t) \\ \dot{x}_2(t) = f(t, X(t)) + d(t) + b_0 u(t) \\ y(t) = x_1(t), \end{cases} \tag{15.18}$$

where $x_1(t)$ and $x_2(t)$ are set as $y(t)$ and $\dot{y}(t)$, respectively. Therefore, the pneumatic rod cylinder servo system is also able to rewrite as a second-order system.

Remark 15.2 At present, many control strategies have been applied to improve the performances of pneumatic servo systems. However, due to various complex factors of pneumatic servo systems, there exist some problems for these control strategies. Such as in [9], it is difficult to obtain the ability of strong anti-interference by using a PID controller with simple structure. The satisfied response rapidity has not been easy to obtain by using a traditional linear active disturbance rejection controller [10]. Therefore, an optimized controller is expected to design for obtaining good anti-interference ability and response rapidity. Then, the optimized controller is designed in the following.

15.4 Simulation and Results

In this section, a simulation is given to shown the effectiveness of the dynamic system (15.15) for the pneumatic rod cylinder servo system platform. The parameters of the pneumatic rod cylinder servo system (15.15) are given in Table 15.2, please refer to [11, 12].

In simulation experiments, two tracking signals, which contains a step signal at 0.06 m and a sinusoidal signal at 0.2 Hz, 0.06 m amplitude, are given to verify the effectiveness of optimized linear active disturbance rejection controller for pneumatic rod cylinder servo system (15.15). At the beginning, the supply absolute pressure P_s is set as 0.6 MPa and the atmospheric absolute pressure P_a is set as 0.1 MPa. The simulation results are shown in Figs. 15.6 and 15.7. Note that r is the given tracking signal, N_1, N_2, and PID are the tracking trajectories of pneumatic rod cylinder servo system based on a linear active disturbance rejection controller, an optimized linear active disturbance rejection controller based on least squares support vector machine and a PID controller, respectively.

According to the simulation results shown in Figs. 15.6 and 15.7, the pneumatic rod cylinder servo system (15.15) is able to represent the pneumatic rod cylinder servo system platform.

Table 15.2 Various parameters in simulation

Parameters		
$P_{20} = 4.65 \times 10^5$ Pa	$T = 300$ K	$A_1 = 5.024 \times 10^{-3}$ m^2
$V_{20} = 3.064 \times 10^{-4}$ m^3	$P_{10} = 4 \times 10^5$ Pa	$A_2 = 3.044 \times 10^{-3}$ m^2
$V_{10} = 5.024 \times 10^{-4}$ m^3	$m = 1$ kg	$F_s = 29$ N
$K = 1.4$	$b_0 = 3501$	$F_c = 20$ N
$v_t = 0.126$ m/s	$v_{sp} = 0.252$ m/s	$C = 64.5$ N · s/m
$Q = 287$ N · m/kg · K		

Fig. 15.6 Control results of three controllers for tracking the step signal

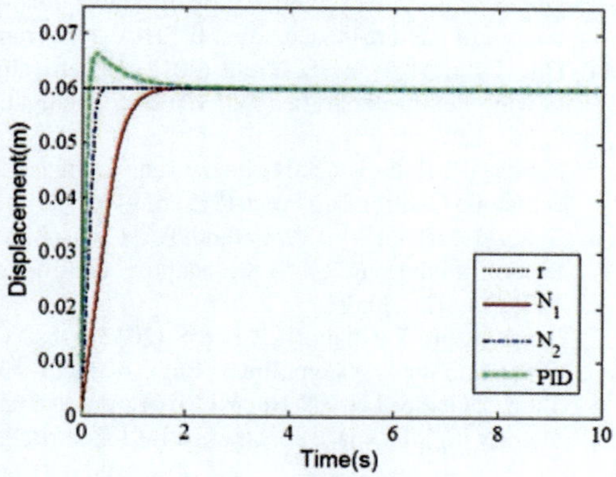

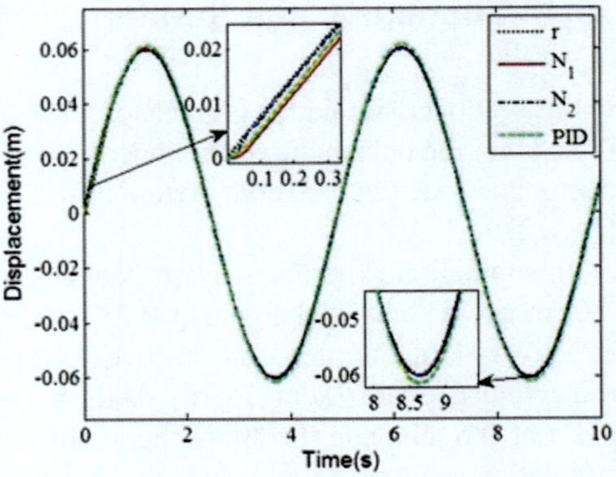

Fig. 15.7 Control results of three controllers for tracking the sinusoidal signal

15.5 Conclusion

In this chapter, a pneumatic rod cylinder servo platform has been shown in detail. Platform components of the pneumatic rod cylinder servo platform have been shown and listed in the chapter. Pneumatic circuit and control circuit have been given to show structure of the pneumatic rod cylinder servo platform. Moreover, a dynamic nonlinear system has been derived to express the pneumatic rod cylinder servo platform. Finally, effectiveness of the dynamic nonlinear system has been shown by simulation.

References

1. http://www.kk.pi.titech.ac.jp/archives/research
2. Rao Z, Bone GM (2008) Nonlinear modeling and control of servo pneumatic actuators. IEEE Trans Control Syst Technol 16(3):562–569
3. Gulati N, Barth EJ (2009) A globally stable, load-independent pressure observer for the servo control of pneumatic actuators. IEEE-ASME Trans Mechatron 14(3):295–306
4. Qiu Z, Shi M, Wang B, Xie Z (2012) Genetic algorithm based active vibration control for a moving flexible smart beam driven by a pneumatic rod cylinder. J Sound Vib 331(10):2233–2256
5. Schulte H, Hahn H (2004) Fuzzy state feedback gain scheduling control of servo-pneumatic actuators. Control Eng Pract 12(5):639–650
6. Sobczyk MR, Gervini VI, Perondi EA, Cunha MAB (2016) A continuous version of the LuGre friction model applied to the adaptive control of a pneumatic servo system. J Frankl Inst 353(13):3021–3039
7. Liu Y, Kung T, Chang K, Chen S (2013) Observer-based adaptive sliding mode control for pneumatic servo system. Precis Eng 37(3):522–530
8. Tang Z, Ge SS, Lee KP, He W (2016) Adaptive neural control for an uncertain robotic manipulator with joint space constraints. Int J Control 89(7):1428–1446

9. Situm Z, Pavkovic D, Novakovic B (2004) Servo pneumatic position control using fuzzy PID gain scheduling. J Dyn Syst Meas Control 126(2):376–387
10. Shi S, Zhao S, Li J (2014) On tracking ability analysis of linear extended state observer for uncertain system. Comput Modell New Technol 18(2):175–179
11. Specker T, Buchholz M, Dietmayer K (2014) A new approach of dynamic friction modelling for simulation and observation. In: International federation of automatic control, pp 4523–4528
12. Meng D, Tao G, Liu H, Chen J (2012) Analysis of friction characteristics of pneumatic cylinders based on LuGre model. J Zhejiang Univ 46(6):1027–1033

Chapter 16
Finite-Time Control

16.1 Introduction

Achieving high-precision and fast-response for the control of pneumatic rod cylinders is a challenging engineering task as a reason of high nonlinearity induced by variable mechanical friction and air compressibility [1, 2]. During the last decades, many research works have been devoted to deal with this challenge [3]. A proportional adaptive neuro-fuzzy controller is proposed to achieve position control for a pneumatic cylinder which is modified in [4]. An adaptive fuzzy-PD controller is presented to deal with some typical time-varying and nonlinearity dynamics in [5]. A sliding mode control method is applied to reduce dry friction of a pneumatic system in [6]. An adaptive sliding mode controller is designed to settle unknown friction, input dead-zone and external disturbances for a nonlinear pneumatic servo system in [7]. However, most of the methods mentioned above need precise mathematical models which are tough to be obtained because of the uncertainty induced by the time-varying friction and loads [8]. Therefore, it is necessary to design an effective control strategy, which does not rely on precise mathematical models and improves control performance for pneumatic rod cylinder servo systems.

In this paper, a STESO and a finite-time controller are applied on a pneumatic rod cylinder servo system with friction and varying loads. The target is to investigate the effectiveness of the proposed control strategy for a pneumatic rod cylinder servo system to handle friction and varying loads. According to Lyapunov approaches, both the effectiveness of the STESO and the finite-time convergence of the closed-loop system are analyzed by Lyapunov approaches. Furthermore, the proposed controller is tested and verified on the pneumatic rod cylinder servo platform. Experimental results show that the performances of the proposed control strategy are better than the ones of ISM controller associated with a GNESO in the pneumatic rod cylinder servo system to handle friction and varying loads.

L. Zhao et al., *Pneumatic Servo Systems Analysis*, Advances in Industrial Control,
https://doi.org/10.1007/978-981-16-9515-5_16

The main contributions of this paper are summarized as follows.

i A STESO is designed to estimate total disturbances in a pneumatic rod cylinder servo system such that estimated errors of disturbances converge to zero in finite time.
ii A finite-time controller based on a power integrator technology is proposed to stabilize the pneumatic cylinder servo system and make system errors reach finite-time convergence.
iii Experimental results, which are better than the ones in [9], show that an stable tracking error is 0.15 mm within 1.65 s for a step signal and an amplitude error is 4.6 mm with delay time 0.2 s for a sinusoidal signal.

16.2 Main Results

16.2.1 Schematic Diagram of Control Method

In this section, a STESO is designed to estimate the total dynamics of the friction and the varying loads for the pneumatic rod cylinder servo system platform. Then the estimation is used for the controller to compensate the total disturbance. A control schematic diagram of the proposed control strategy is shown in Fig. 16.1. In order to prove the stability of the pneumatic rod cylinder servo system (15.18), some lemmas are introduced in the following.

Lemma 16.1 ([1]) *If a positive definite function $V(\chi(t))$ is found for a dynamics model $\dot{\chi}(t) = g(\chi(t))$ with $\chi \in R^n$ and $g(0) = 0$, the following inequality holds as*

$$\dot{V}(\chi(t)) \leq -\rho V^{\alpha}(\chi(t)) - \nu V(\chi(t)),$$

where $\rho > 0$, $0 < \alpha < 1$ and $\nu > 0$, then finite-time stability is achieved for the dynamics model. Moreover, convergence time is given as follows

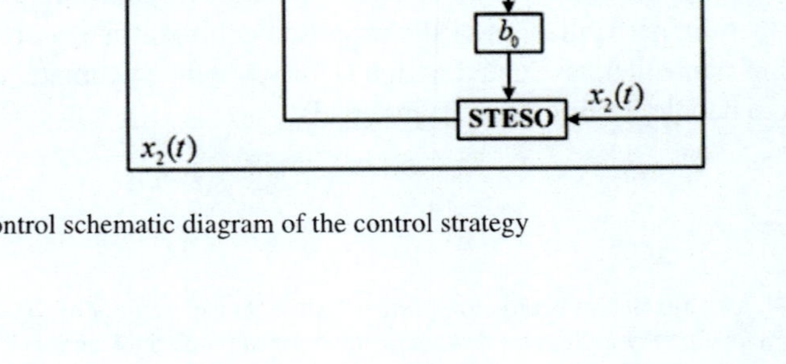

Fig. 16.1 Control schematic diagram of the control strategy

$$t \leq \frac{1}{\nu(1-\alpha)} ln \frac{\nu V^{1-\alpha}(\chi(0)) + \rho}{\rho}.$$

If $\rho > 0$, $0 < \alpha < 1$ and $\nu = 0$, then finite-time stability is also achieved. Convergence time is given as follows:

$$t \leq \frac{1}{\alpha\rho} V^{\alpha}(\chi(0)),$$

where $\chi(0)$ is an initial state.

Lemma 16.2 ([10]) *If d and f are two positive numbers, j and k are two constants, a function $\eta(j,k) > 0$ holds, then there exists the following inequality:*

$$|j|^f |k|^d \leq \frac{f\eta(j,k)|j|^{f+d}}{f+d} + \frac{d\eta^{-\frac{f}{d}}(j,k)|k|^{f+d}}{f+d},$$

Lemma 16.3 ([10]) *If $0 < a \leq 1$ and d_i is any real number with $i = 1, 2, \ldots, n$, then there exists the following inequality:*

$$(|d_1| + \cdots + |d_n|)^a \leq |d_1|^a + \cdots + |d_n|^a.$$

If $j > 0$ and $k > 0$ are odd integers and $a = j/k \leq 1$, then there exists

$$|c^a - s^a| \leq 2^{1-a}|c - s|^a,$$

where c and s are real numbers.

16.2.2 Super-Twisting Extended State Observer

In this section, a STESO is proposed to estimate the total disturbance which contains the friction and the varying loads in the pneumatic rod cylinder servo system (15.18). Since a velocity signal is obtained by the nonlinear tracking differentiator (TD), then the pneumatic rod cylinder servo system (15.18) is constructed as a first-order system. Thus, the total disturbance $L(t) = f(t, x(t)) + d(t)$ is redefined as state $x_3(t)$. Considering the pneumatic rod cylinder servo system (15.18), one has that

$$\begin{cases} \dot{x}_2(t) = x_3(t) + b_0 u(t) \\ \dot{x}_3(t) = -J(t), \end{cases} \tag{16.1}$$

where $-J(t)$ is the derivative of $x_3(t)$.

Based on system (16.1), an STESO is proposed as

$$\begin{cases} \dot{z}_1(t) = z_2(t) + b_0 u(t) - \beta_1 \phi_1(w_1(t)) \\ \dot{z}_2(t) = -\beta_2 \phi_2(w_1(t)), \end{cases} \tag{16.2}$$

where

$$\phi_1(w_1(t)) = |w_1(t)|^{\frac{1}{2}} \operatorname{sign}(w_1(t)) + w_1(t)$$
$$\phi_2(w_1(t)) = \frac{1}{2} \operatorname{sign}(w_1(t)) + w_1(t) + \frac{3}{2} |w_1(t)|^{\frac{1}{2}} \operatorname{sign}(w_1(t))$$

Note that $z_1(t)$ and $z_2(t)$ are the observations of $x_2(t)$ and $x_3(t)$, respectively, β_1 and β_2 are two tuning parameters. According to system (16.1) and the STESO (16.2), an observation error system of the STESO (16.2) is obtained as follows:

$$\begin{cases} \dot{w}_1(t) = w_2(t) - \beta_1 \phi_1(w_1(t)) \\ \dot{w}_2(t) = -\beta_2 \phi_2(w_1(t)) + J(t), \end{cases} \tag{16.3}$$

where $w_i(t) = z_i(t) - x_{i+1}(t)$ with $i = 1, 2$.

In this section, a proof of the finite-time convergence for the STESO (16.2) is presented based on observation error system (16.3). First, according to Lemma 16.1, a Lyapunov function is given as follows:

$$V_0(\xi(t)) = \xi(t)^T P \xi(t) \tag{16.4}$$

where $\xi(t)^T = [\phi_1(w_1(t)),\ w_2(t)]$. $\dot{\xi}(t)$ is obtained as

$$\begin{aligned} \dot{\xi}(t) &= \phi_1'(w_1(t)) \begin{bmatrix} -\beta_1 \phi_1(w_1(t)) + w_2(t) \\ -\beta_2 \phi_1(w_1(t)) + \frac{J(t)}{\phi_1'(w_1(t))} \end{bmatrix} \\ &= \phi_1'(w_1(t))(A\xi(t) + B\varsigma(t)) \end{aligned}$$

where

$$\varsigma(t) = J(t)/\phi_1'(w_1(t)),\ A = \begin{bmatrix} -\beta_1 & 1 \\ -\beta_2 & 0 \end{bmatrix}$$
$$\phi_1'(w_1(t)) = \left(\frac{|w_1(t)|^{-\frac{1}{2}}}{2} + 1 \right),\ B = \begin{bmatrix} 0 \\ 1 \end{bmatrix}$$

Assumption 16.1 ([11]) The perturbation term $\varsigma(t)$ is satisfied with a sector condition for all $\xi \in R^2$ at any time as follows:

$$\Delta(\xi(t), \varsigma(t)) = \begin{bmatrix} \xi(t) \\ \varsigma(t) \end{bmatrix}^T \begin{bmatrix} R & S^T \\ S & -\theta \end{bmatrix} \begin{bmatrix} \xi(t) \\ \varsigma(t) \end{bmatrix} \geq 0 \tag{16.5}$$

where $R \geq 0$, $\theta \geq 0$, $S = [c\ h]$, c and h are two constants.

Subsequently, the finite-time convergence of the observation error system (16.3) is formally presented as the following theorem.

Theorem 16.4 *If there is the following matrix inequality*

$$\begin{bmatrix} A^T P + PA + \epsilon P + R & PB + S^T \\ B^T P + S & -\theta \end{bmatrix} \leq 0 \tag{16.6}$$

where θ and ϵ are two positive constants, P is a symmetric and positive definite matrix, then observation error system (16.3) achieves finite time stable. Furthermore, the convergence time of the trajectories is shorter than T_1 as follows:

$$T_1 = \frac{2}{\epsilon} \ln \left(\frac{2}{\lambda_{\min}^{\frac{1}{2}}\{P\}} V_0^{\frac{1}{2}}(\xi(0)) + 1 \right), \tag{16.7}$$

where $\xi(0)$ is an initial state.

Proof In the light of observation error system (16.3), the derivative of $V_0(\xi(t))$ is

$$\begin{aligned} \dot{V}_0(\xi(t)) &= \phi_1'(w_1(t)) \left[\xi^T(t)(A^T P + PA)\xi(t) \right. \\ &\quad \left. + \varsigma(t) B^T P \xi(t) + \xi^T(t) P B \varsigma(t) \right] \\ &= \phi_1'(w_1(t)) \begin{bmatrix} \xi(t) \\ \varsigma(t) \end{bmatrix}^T W \begin{bmatrix} \xi(t) \\ \varsigma(t) \end{bmatrix} \end{aligned} \tag{16.8}$$

where

$$W = \begin{bmatrix} A^T P + PA & PB \\ B^T P & 0 \end{bmatrix}$$

Then the following inequality is obtained as

$$\lambda_{\min}\{P\} \|\xi(t)\|_2^2 \leq \xi^T(t) P \xi(t) \leq \lambda_{\max}\{P\} \|\xi(t)\|_2^2, \tag{16.9}$$

where

$$\begin{aligned} \|\xi(t)\|_2^2 &= \phi_1^2(w_1(t)) + w_2^2(t) \\ &= |w_1(t)| + 2|w_1(t)|^{\frac{3}{2}} + w_1^2(t) + w_2^2(t), \end{aligned}$$

which is the Euclidean norm of $\xi(t)$. According to the inequality (16.9), the following inequality

$$|w_1(t)|^{\frac{1}{2}} \leq \|\xi(t)\|_2 \leq \frac{V_0^{\frac{1}{2}}}{\lambda_{\min}^{\frac{1}{2}}\{P\}} \tag{16.10}$$

is also satisfied.

Combining the inequalities (16.5), (16.6), (16.10), and the equality (16.8), the following inequality as

$$\begin{aligned}
\dot{V}_0(\xi(t)) &\leq \phi_1^{'}(w_1(t))\left(\begin{bmatrix} \xi(t) \\ \varsigma(t) \end{bmatrix}^T W \begin{bmatrix} \xi(t) \\ \varsigma(t) \end{bmatrix} + \Delta(\xi(t), \varsigma(t)\right) \\
&= \phi_1^{'}(w_1(t)) \begin{bmatrix} \xi(t) \\ \varsigma(t) \end{bmatrix}^T Q \begin{bmatrix} \xi(t) \\ \varsigma(t) \end{bmatrix} \\
&\leq \phi_1^{'}(w_1(t)) \begin{bmatrix} \xi(t) \\ \varsigma(t) \end{bmatrix}^T \begin{bmatrix} -\epsilon P & 0 \\ 0 & 0 \end{bmatrix} \begin{bmatrix} \xi(t) \\ \varsigma(t) \end{bmatrix} \\
&\leq -\phi_1^{'}(w_1(t))\epsilon V_0(\xi(t)) \\
&= -\frac{1}{2|w_1(t)|^{\frac{1}{2}}}\epsilon V_0(\xi(t)) - \epsilon V_0(\xi(t)) \\
&\leq -\frac{\epsilon\lambda_{\min}^{\frac{1}{2}}\{P\}}{2} V_0^{\frac{1}{2}}(\xi(t)) - \epsilon V_0(\xi(t)) \\
&\leq 0
\end{aligned} \tag{16.11}$$

where

$$Q = \begin{bmatrix} A^T P + PA + R & PB + S^T \\ B^T P + S & -\theta \end{bmatrix}$$

According to the inequality (16.4) and the inequality (16.11), it is obvious that the STESO (16.2) is effective.

According to Lemma 16.1 and the inequality (16.11), convergence time for observation error system (16.3) is shorter than T_1 as follows:

$$T_1 = \frac{2}{\epsilon}\ln\left(\frac{2}{\lambda_{\min}^{\frac{1}{2}}\{P\}} V_0^{\frac{1}{2}}(\xi(0)) + 1\right).$$

That is, the observation error system (16.3) achieves finite-time stability.

16.2.3 Finite-Time Controller

In this section, a finite-time controller is designed for the pneumatic rod cylinder servo system (15.18) by using a power integrator technology. For a desire signal $y_0(t)$, system errors of the finite-time controller are obtain as

$$\begin{cases} e_1(t) = y_0(t) - x_1(t) \\ e_2(t) = \dot{y}_0(t) - x_2(t), \end{cases} \tag{16.12}$$

where $\dot{y}_0(t)$ is the derivative of $y_0(t)$, $e_1(t)$ is the displacement error, $e_2(t)$ is the speed error. Considering the pneumatic rod cylinder servo system (15.18) and the system errors in (16.12), the differentials of $e_1(t)$ and $e_2(t)$ are obtained as follows:

$$\begin{cases} \dot{e}_1(t) = e_2(t) \\ \dot{e}_2(t) = \ddot{y}_0(t) - f_d(t, X(t)) - b_0 u(t), \end{cases} \tag{16.13}$$

where $\ddot{y}_0(t)$ is the derivative of $\dot{y}_0(t)$. Then a finite-time controller is proposed as follows:

$$u(t) = \frac{1}{b_0}(-\Lambda(t) - z_2(t) + \ddot{y}_0(t)) \tag{16.14}$$

with

$$\Lambda(t) = -\left(2 - \frac{1}{\varrho}\right) 2^{1-\frac{1}{\varrho}} k_1^{1+\varrho} k_2 (e_2^{\varrho}(t) + k_1^{\varrho} e_1(t))^{\frac{2}{\varrho}-1},$$

where $1 < \varrho = r_1/r_2 < 2$, r_1 and r_2 are positive odd integers, k_1 and k_2 are two positive constants.

The convergence analysis for tracking error system (16.13) using STESO (16.2) and finite-time controller (16.14) is divided into two steps. Firstly, tracking error system (16.13) is bounded stable when STESO (16.2) doesn't reach finite-time stability within $t < T_1$. Secondly, tracking error system (16.13) converge to zero when STESO (16.2) achieves finite-time stability when $t \geq T_1$.

Proof **Step1**: When $t < T_1$, an energy storage function is constructed as

$$V(t) = \frac{1}{2}[x_1^2(t) + x_2^2(t) + z_1^2(t) + z_2^2(t) + e_1^2(t) + e_2^2(t)].$$

Then $\dot{V}(t)$ is obtained as

$$\begin{aligned} \dot{V}(t) = {} & x_1(t)\dot{x}_1(t) + x_2(t)\dot{x}_2(t) + z_1(t)\dot{z}_1(t) \\ & + z_2(t)\dot{z}_2(t) + e_1(t)\dot{e}_1(t) + e_2(t)\dot{e}_2(t). \end{aligned} \tag{16.15}$$

There exists an inequality $|\tau|^{\eta} < 1 + |\tau|$, where τ is a real number and $\eta \in (0, 1)$. The inequality can be applied in following calculation and analysis. Let $K_1 = k_1$ and $K_2 = (2 - 1/\varrho)2^{1-1/\varrho}k_1^{1+\varrho}k_2$. Then the equality (16.15) is divided into the following parts as

$$
\begin{aligned}
x_1(t)\dot{x}_1(t) &\le \frac{1}{2}(x_1^2(t) + x_2^2(t)) \\
x_2(t)\dot{x}_2(t) &= x_2(t)f_d(t, X(t)) - x_2(t)z_2(t) + x_2(t)\ddot{y}_0(t) + K_1^{\varrho}e_1(t))^{\frac{2}{\varrho}-1} \\
&\quad + x_2(t)K_2(e_2^{\varrho}(t) \\
&\le \frac{1}{2}K_2(x_2^2(t) + e_2^2(t)) + \frac{1}{2}(x_2^2(t) + w_2^2(t)) \\
&\quad + \frac{1}{2}[(K_2 + K_2K_1^{2-\varrho} + \ddot{y}_0(t))^2 + x_2^2(t)] \\
&\quad + \frac{1}{2}K_1^{2-\varrho}K_2(x_2^2(t) + e_1^2(t)) \\
z_1(t)\dot{z}_1(t) &= z_1(t)K_2(e_2^{\varrho}(t) + z_1(t)K_1^{\varrho}e_1(t))^{\frac{2}{\varrho}-1} + z_1(t)\ddot{y}_0(t) \\
&\quad - \beta_1 z_1(t)\phi_1(w_1(t)) \\
&\le |z_1(t)||(K_2 + K_1^{2-\varrho}K_2 + \beta_1 + \ddot{y}_0(t))| + K_2|z_1(t)||e_2(t)| \\
&\quad + K_1^{2-\varrho}K_2|z_1(t)||e_1(t)| + 2|\beta_1||z_1(t)||w_1(t)| \\
&\le \frac{1}{2}K_2(z_1^2(t) + e_2^2(t)) + |\beta_1|(z_1^2(t) + w_1^2(t)) \\
&\quad + \frac{1}{2}[(\beta_1 + K_2 + K_2K_1^{2-\varrho} + \ddot{y}_0(t))^2 + z_1^2(t)] \\
&\quad + \frac{1}{2}K_1^{2-\varrho}K_2(z_1^2(t) + e_1^2(t)) \\
z_2(t)\dot{z}_2(t) &= z_2(t)(-\beta_2\phi_2(w_1(t))) \\
&\le |\beta_2||z_2(t)|\left(2 + \frac{5}{2}|w_1(t)|\right) \\
&\le \frac{1}{2}|\beta_2|(z_2^2(t) + 4) + \frac{5}{4}|\beta_2|(z_2^2(t) + w_1^2(t)) \\
e_1(t)\dot{e}_1(t) &\le \frac{1}{2}(e_1^2(t) + e_2^2(t)) \\
e_2(t)\dot{e}_2(t) &= -e_2(t)[K_2(e_2(t)^{\varrho} + K_1^{\varrho}e_1(t))^{\frac{2}{\varrho}-1} + w_2(t)] \\
&\le |e_2(t)|(K_2 + K_2K_1^{2-\varrho}) + K_2e_2^2(t) + |e_2(t)||w_2(t)| \\
&\quad + K_2K_1^{2-\varrho}|e_1(t)||e_2(t)| \\
&\le \frac{1}{2}(K_2 + K_2K_1^{2-\varrho})^2 + (K_2 + 1)e_2^2(t) + \frac{1}{2}w_2^2(t) \\
&\quad + \frac{1}{2}K_2K_1^{2-\varrho}(e_1^2(t) + e_2^2(t))
\end{aligned}
$$

When $t < T_1$, it is known that the observation error system (16.3) is bounded. There exist $|w_1| \leq \wp_1$ and $|w_2| \leq \wp_2$, where $\wp_1$ and $\wp_2$ are two positive constants. Then the equality (16.15) is constructed as

$$\dot{V}(t) \leq \mathfrak{g}V(t) + \mathfrak{c}, \tag{16.16}$$

where

$$\begin{aligned}
\mathfrak{g} = 2 \times \max \bigg\{ &\frac{3}{2} + \frac{1}{2}K_2 + \frac{1}{2}K_2K_1^{2-\varrho},\ \frac{3}{2} + 2K_2 + \frac{1}{2}K_2K_1^{2-\varrho} \\
&\frac{1}{2} + |\beta_1| + \frac{1}{2}K_2 + \frac{1}{2}K_2K_1^{2-\varrho},\ \frac{1}{2} + \frac{3}{2}K_2K_1^{2-\varrho},\ \frac{1}{2},\ \frac{7}{4}|\beta_2| \bigg\} \\
\mathfrak{c} = &\frac{1}{2}(\dddot{y}_0(t) + K_2 + K_2K_1^{2-\varrho})^2 + |\beta_1|\wp_1^2 + \frac{5}{4}|\beta_2|\wp_1^2 + 2|\beta_2| + \wp_2^2 \\
&+ \frac{1}{2}(\beta_1 + K_2 + K_2K_1^{2-\varrho} + \ddot{y}_0(t))^2 + \frac{1}{2}(k_2 + K_2K_1^{2-\varrho})^2
\end{aligned}$$

A time-varying coefficient $e^{-\mathfrak{g}t}$, where e is the Euler number, is multiplied on both sides of the inequality (16.16), then a follow inequality is obtained as

$$\frac{dV(t)}{dt}e^{-\mathfrak{g}t} \leq \mathfrak{g}V(t)e^{-\mathfrak{g}t} + \mathfrak{c}e^{-\mathfrak{g}t} \tag{16.17}$$

The integral for the equality (16.17) is

$$V(t)e^{-\mathfrak{g}t} + \frac{1}{\mathfrak{g}}\mathfrak{c}e^{-\mathfrak{g}t} \leq V(0) + \frac{\mathfrak{c}}{\mathfrak{g}} \tag{16.18}$$

For $t < T_1$, the inequality (16.18) is rewritten as follows:

$$V(t) \leq \left(V(0) + \frac{\mathfrak{c}}{\mathfrak{g}}\right)e^{\mathfrak{g}t} - \frac{\mathfrak{c}}{\mathfrak{g}}.$$

Therefore, the tracking error system (16.13) is bounded stable within $t < T_1$ using finite-time controller (16.14).

Step2: When $t \geq T_1$, the tracking error system (16.13) is constructed as

$$\begin{cases} \dot{e}_1(t) = \dot{e}_2(t) \\ \dot{e}_2(t) = -\left(2 - \frac{1}{\varrho}\right)2^{1-\frac{1}{\varrho}}k_1^{1+\varrho}k_2(e_2^{\varrho}(t) + k_1^{\varrho}e_1(t))^{\frac{2}{\varrho}-1} \end{cases} \tag{16.19}$$

We choose a Lyapunov function as follows:

$$V_1(t) = \frac{1}{2}e_1^2(t).$$

Then $\dot{V}_1(t)$ is obtained as

$$\dot{V}_1(t) = e_1(t)(e_2(t) - e_2^*(t)) + e_1 e_2^*(t), \tag{16.20}$$

where $e_2^*(t)$ is a virtual control law which is defined as $e_2^*(t) = -k_1 e_1^{1/\varrho}(t)$. From the equality (16.20), a following inequality is obtained as

$$\dot{V}_1(t) \leq e_1(t)(e_2(t) - e_2^*(t)) - k_1 e_1^{1+\frac{1}{\varrho}}(t)$$

Set

$$\xi_2(t) = e_2^{\varrho}(t) - e_2^{*\varrho}(t) \tag{16.21}$$

and choose a Lyapunov function $V_2(t)$ as follows:

$$V_2(t) = V_1(t) + \frac{\int_{e_2^*(t)}^{e_2(t)} [s^{\varrho} - e_2^{*\varrho}(t)]^{2-\frac{1}{\varrho}} ds}{(2 - \frac{1}{\varrho}) 2^{1-\frac{1}{\varrho}} k_1^{1+\varrho}}.$$

Due to $\partial e_2^{*\varrho}(t)/\partial e_1(t) = -k_1^{\varrho}$, $\dot{V}_2(t)$ is constructed as

$$\begin{aligned}\dot{V}_2(t) \leq{}& \frac{e_2(t)}{2^{1-\frac{1}{\varrho}} k_1} \int_{e_2^*(t)}^{e_2(t)} [s^{\varrho} - e_2^{*\varrho}(t)]^{1-\frac{1}{\varrho}} ds - k_1 e_1^{1+\frac{1}{\varrho}}(t) \\ &+ \frac{\xi_2(t)^{2-\frac{1}{\varrho}} \Lambda(t)}{(2 - \frac{1}{\varrho}) 2^{1-\frac{1}{\varrho}} k_1^{1+\varrho}} + e_1(t)(e_2(t) - e_2^*(t)).\end{aligned} \tag{16.22}$$

According to Lemma 16.3, it is obtained that

$$|e_2(t) - e_2^*(t)| \leq 2^{1-1/\varrho} |\xi_2(t)|^{1/\varrho}.$$

Thus, the inequality (16.22) becomes the following inequality as

$$\begin{aligned}\dot{V}_2(t) \leq{}& -k_1 e_1^{1+\frac{1}{\varrho}}(t) + 2^{1-\frac{1}{\varrho}} |e_1(t)||\xi_2(t)|^{\frac{1}{\varrho}} \\ &+ \frac{|e_2(t)||\xi_2(t)|}{k_1} + \frac{\xi_2^{2-\frac{1}{\varrho}}(t) \Lambda(t)}{(2 - \frac{1}{\varrho}) 2^{1-\frac{1}{\varrho}} k_1^{1+\varrho}}.\end{aligned} \tag{16.23}$$

By Lemma 16.2, an inequality is obtained as follows:

$$2^{1-\frac{1}{\varrho}} |e_1(t)||\xi_2(t)|^{\frac{1}{\varrho}} \leq \frac{2^{1-\frac{1}{\varrho}} \varrho}{1+\varrho} \gamma e_1^{1+\frac{1}{\varrho}}(t) + \frac{2^{1-\frac{1}{\varrho}}}{1+\varrho} \gamma^{-\varrho} \xi_2^{1+\frac{1}{\varrho}}(t), \tag{16.24}$$

where γ is an arbitrary positive constant. Let $\gamma/(1+\varrho) = k_1/4$, then the inequality (16.24) is rewritten as follows:

$$2^{1-\frac{1}{\varrho}}|e_1(t)||\xi_2(t)|^{\frac{1}{\varrho}} \leq 2^{1-\frac{1}{\varrho}}\varrho\frac{k_1}{4}e_1^{1+\frac{1}{\varrho}}(t) + \frac{2^{1-\frac{1}{\varrho}}}{1+\varrho}\left[\frac{4}{(1+\varrho)k_1}\right]^{\varrho}\xi_2^{1+\frac{1}{\varrho}}(t). \quad (16.25)$$

Using Lemma 16.3, a following inequality

$$|e_2(t)| \leq |\xi_2(t)|^{\frac{1}{\varrho}} + |e_2^{*\varrho}(t)|^{\frac{1}{\varrho}} = |\xi_2(t)|^{\frac{1}{\varrho}} + k_1|e_1(t)|^{\frac{1}{\varrho}}$$

is acquired. Therefore, the following inequality

$$|e_2(t)\xi_2(t)|/k_1 \leq \xi_2(t)^{1+\frac{1}{\varrho}}/k_1 + |e_1(t)|^{\frac{1}{\varrho}}|\xi_2(t)|$$

is obtained. According to Lemma 16.2, one has that

$$|e_1(t)|^{\frac{1}{\varrho}}|\xi_2(t)| \leq \frac{\gamma}{1+\varrho}e_1^{1+\frac{1}{\varrho}}(t) + \frac{\varrho}{1+\varrho}\gamma^{-\frac{1}{\varrho}}\xi_2^{1+\frac{1}{\varrho}}(t).$$

Letting $\gamma/(1+\varrho) = k_1/4$, the following inequality is obtained as

$$\frac{|e_2(t)\xi_2(t)|}{k_1} \leq \frac{\xi_2^{1+\frac{1}{\varrho}}(t)}{k_1} + \frac{k_1}{4}e_1^{1+\frac{1}{\varrho}}(t) + \frac{\varrho}{1+\varrho}\left[\frac{4}{(1+\varrho)k_1}\right]^{\frac{1}{\varrho}}\xi_2^{1+\frac{1}{\varrho}}(t). \quad (16.26)$$

Substituting the inequalities (16.25) and (16.26) into the inequality (16.23) yields

$$\begin{aligned}\dot{V}_2(t) \leq & -\frac{3}{4}k_1 e_1^{1+\frac{1}{\varrho}}(t) + 2^{1-\frac{1}{\varrho}}\varrho\frac{k_1}{4}e_1^{1+\frac{1}{\varrho}}(t) + \delta\xi_2^{1+\frac{1}{\varrho}}(t) \\ & + \frac{\xi_2^{2-\frac{1}{\varrho}}(t)\Lambda(t)}{(2-\frac{1}{\varrho})2^{1-\frac{1}{\varrho}}k_1^{1+\varrho}}, \end{aligned} \quad (16.27)$$

where

$$\delta = \frac{2^{1-\frac{1}{\varrho}}}{1+\varrho}\left[\frac{4}{(1+\varrho)k_1}\right]^{\varrho} + \frac{1}{k_1} + \frac{\varrho}{1+\varrho}\left[\frac{4}{(1+\varrho)k_1}\right]^{\frac{1}{\varrho}}.$$

Then $\Lambda(t)$ is as follows:

$$\Lambda(t) = -\left(2-\frac{1}{\varrho}\right)2^{1-\frac{1}{\varrho}}k_1^{1+\varrho}k_2(e_2^{\varrho}(t) + k_1^{\varrho}e_1(t))^{\frac{2}{\varrho}-1}, \quad (16.28)$$

where $k_2 > \delta$. Substituting the equality (16.28) into the inequality (16.27), results in an inequality as follows:

$$\dot{V}_2(t) \leq -\frac{k_1}{4} e_1^{1+\frac{1}{\varrho}}(t) - k_3 \xi_2^{1+\frac{1}{\varrho}}(t), \tag{16.29}$$

where $k_3 = k_2 - \delta > 0$. Furthermore, the inequality (16.29) is transformed as the following form:

$$\dot{V}_2(t) + l V_2^{\frac{1+\varrho}{2\varrho}}(t) \leq -k \left(e_1^{1+\frac{1}{\varrho}}(t) + \xi_2^{1+\frac{1}{\varrho}}(t) \right) \leq 0, \tag{16.30}$$

where

$$k = \max \left\{ \frac{k_1}{4} - \frac{l}{2^{\frac{1+\varrho}{2\varrho}}},\ k_3 - l \left(\frac{\varrho}{2\varrho - 1} k_1^{1+\varrho} \right)^{\frac{1+\varrho}{2\varrho}} \right\}$$

$$l = \max \left\{ 2^{\frac{1-3\varrho}{2\varrho}} k_1;\ \frac{(2\varrho - 1)^{\frac{1+\varrho}{2\varrho}} k_3}{\varrho^{\frac{1+\varrho}{2\varrho}} k_1^{\frac{(1+\varrho)^2}{2\varrho}}} \right\}.$$

For the inequality (16.30), finite-time controller (16.14) is proven to be effective for pneumatic rod cylinder servo system (15.18).

According to Lemma 16.1 and the inequality (16.30), the convergence time for tracking error system (16.19) is shorter than $T_1 + T_2$, where

$$T_2 = \frac{2\varrho}{(1+\varrho)c} V_2^{\frac{1+\varrho}{2\varrho}}(T_1). \tag{16.31}$$

Therefore, there exists the following theorem as

Theorem 16.5 *Consider tracking error system (16.13) with STESO (16.2) and finite-time controller (16.14). If* $1 < \varrho = r_1/r_2 < 2$, r_1 *and* r_2 *are two positive odd integers,* k_1 *is a positive constant and* k_2 *satisfies the following inequality as*

$$k_2 > \frac{2^{1-\frac{1}{\varrho}}}{1+\varrho} \left[\frac{4}{(1+\varrho)k_1} \right]^{\varrho} + \frac{1}{k_1} + \frac{\varrho}{1+\varrho} \left[\frac{4}{(1+\varrho)k_1} \right]^{\frac{1}{\varrho}},$$

then tracking error system (16.13) achieves finite-time stability. Moreover, the stabilization time T_2 *is as follows:*

$$T_2 = \frac{2\varrho}{(1+\varrho)c} V_2^{\frac{1+\varrho}{2\varrho}}(T_1).$$

That is, stabilization time for the pneumatic rod cylinder servo system (15.18) is shorter than T_0, where

$$T_0 = \frac{2}{\epsilon}\ln\left(\frac{2}{\lambda_{\min}^{\frac{1}{2}}\{P\}}V_0^{\frac{1}{2}}(\xi(0))+1\right)+\frac{2\varrho}{(1+\varrho)c}V_2^{\frac{1+\varrho}{2\varrho}}(T_1).$$

Remark 16.6 The difference between $e_2(t)$ and $e_2^*(t)$ is not always greater than zero. However, it is sure that the Lyapunov function $V_2(t)$ is always positive. The difference between $e_2(t)$ and $e_2^*(t)$ has two cases, i.e., $e_2^*(t) - e_2(t) \leq 0$ and $e_2^*(t) - e_2(t) > 0$. For $e_2^*(t) - e_2(t) \leq 0$, there exists $e_2^*(t) \leq s \leq e_2(t)$. One has that $[s^\varrho - e_2^{*\varrho}(t)]^{2-1/\varrho} \geq 0$ holds. Therefore, it is obtained that

$$\int_{e_2^*(t)}^{e_2(t)} [s^\varrho - e_2^{*\varrho}(t)]^{2-\frac{1}{\varrho}} ds \geq 0.$$

By the same way, there exists $e_2(t) \leq s \leq e_2^*(t)$ for $e_2^*(t) - e_2(t) > 0$. Therefore, it is shown that $[s^\varrho - e_2^{*\varrho}(t)]^{2-1/\varrho} < 0$ by which the following inequality is obtained as

$$\int_{e_2^*(t)}^{e_2(t)} [s^\varrho - e_2^{*\varrho}(t)]^{2-\frac{1}{\varrho}} ds > 0.$$

Moreover, $(2 - 1/\varrho)2^{1-1/\varrho}k_1^{1+\varrho} \geq 0$ holds for the reason of that there exist $1 < \varrho < 2$ and $k_1 > 0$. It is obtained that

$$\frac{\int_{e_2^*(t)}^{e_2(t)} [s^\varrho - e_2^{*\varrho}(t)]^{2-\frac{1}{\varrho}} ds}{(2-\frac{1}{\varrho})2^{1-\frac{1}{\varrho}}k_1^{1+\varrho}} \geq 0.$$

Then there exists $V_2(t) \geq 0$ by $V_1(t) \geq 0$. That is, the Lyapunov function $V_2(t)$ is positive in this paper.

Remark 16.7 The stabilization time for tracking error system (16.13) is obtained after STESO (16.2) achieves finite-time stability when $t \geq T_1$. Therefore, T_1 is chosen as a new initial time according to Lemma 16.1. Comparing with ISMC via a GNESO in [9], the finite-time convergence of the closed-loop system is proven by Lyapunov approaches. Moreover, the stabilization time for the pneumatic rod cylinder servo system (15.18) is given in this paper.

16.3 Experiments and Results

Experiments for a step signal and a sinusoidal signal are conducted to show effectiveness of the proposed control strategy. Due to the lack of a speed sensor, a velocity signal of the pneumatic rod cylinder cannot be obtained directly. The nonlinear TD is applied to the pneumatic rod cylinder servo system (15.18) to obtain an estimated

Table 16.1 Parameters for a step signal

Parameters					
β_1	6000	β_2	800	b_0	100000
r_1	73	r_2	71	h_0	0.001
k_1	25.05	k_2	0.2295	r_0	3000000

velocity signal of the pneumatic rod cylinder. Moreover, varying loads are produced by four springs which are installed on the pneumatic rod cylinder. When the displacement of the pneumatic rod cylinder changes, both the lengths and spring forces of the springs also change. Note that the spring forces are the load of the pneumatic rod cylinder servo system. Therefore, the load varies during the displacement change of the pneumatic rod cylinder. The elasticity coefficients of the four springs are 0.4 N/mm. The absolute pressure of the supply gas is 0.6 MPa.

In step experiments, a step signal with magnitude of 80 mm is given as the tracking reference, and the sampling period h is 0.01 s. Parameters of both STESO (16.2) and finite-time controller (16.14) are listed in Table 16.1. Solving the matrix inequality (16.6), the symmetric and positive definite matrix as

$$P = \begin{bmatrix} 79728788.54519317 & -13288.35682222242 \\ -13288.35682222242 & 2.689093662869980 \end{bmatrix}$$

is obtained.

In order to show the effectiveness of the proposed control strategy, the ISMC via a GNESO in [9] is applied into the pneumatic rod cylinder servo system (15.18). In Fig. 16.2a, $x_1(t)$ is the displacement of the pneumatic rod cylinder and $v_1(t)$ is the tracking signal of $x_1(t)$. It is easy to know that nonlinear TD is effective and $v_2(t)$ can be used as a speed of the pneumatic rod cylinder for the proposed control strategy. In Fig. 16.2b, $z_1(t)$ is the observation of $v_2(t)$, $z_2(t)$ represents the estimation of the total disturbance, the effectiveness of STESO (16.2) is shown. In Fig. 16.2c, curve $x_1(t)$ represents displacement signal by the proposed control strategy in this paper, curve ISMC represents displacement signal by ISMC via a GNESO in [9]. In Fig. 16.2d, an estimated value of the total disturbance is shown. In Fig. 16.2e, it shows control inputs of the proposed control strategy and ISMC via a GNESO in [9]. In Fig. 16.2f, it shows that the proposed control strategy's stable tracking error is 0.15 mm and the stable tracking error for ISMC via a GNESO in [9] is 0.25 mm.

The detailed performance comparisons are shown in Table 16.2. For 1.65 s $<$ 1.75 s and 0.15 mm $<$ 0.25 mm, the proposed control strategy in this paper has achieved better control performance than the one in [9].

Moreover, a sinusoidal experiment is conducted to show effectiveness of the finite-time control strategy which also compares with ISMC via a GNESO in [9]. A frequency of the sinusoidal experiment is 0.4 Hz. An amplitude of the sinusoidal experiment is 60 mm. A sampling period h is also set as 0.01 s. Parameters of both the

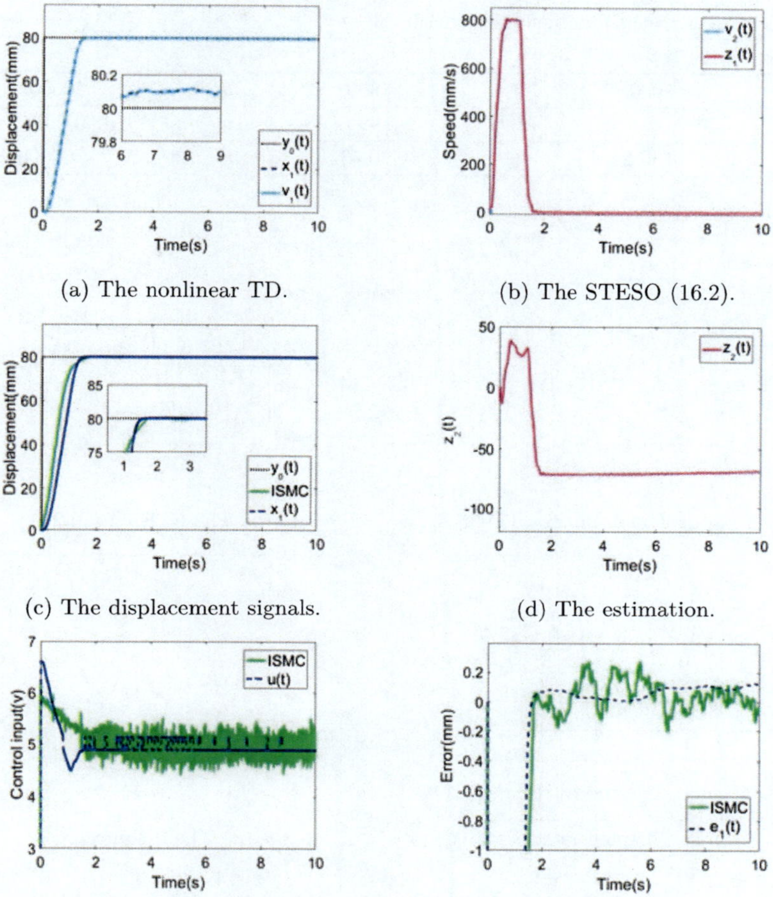

(a) The nonlinear TD.

(b) The STESO (16.2).

(c) The displacement signals.

(d) The estimation.

(e) The control input.

(f) The position error.

Fig. 16.2 Experimental data for a step signal

Table 16.2 Performances comparison

Controller	Convergence time (s)	Tracking error (mm)
ISMC in [9]	1.75	0.25
In this paper	1.65	0.15

STESO (16.2) and the finite-time controller (16.14) are listed in Table 16.3. Since the parameters β_1 and β_2 of the STESO (16.2) don't change, the symmetric and positive definite matrix P also does not change. In Fig. 16.3a, $x_1(t)$ is the displacement of the pneumatic rod cylinder and $v_1(t)$ stands for the tracking signal of $x_1(t)$. According to the closeness of $x_1(t)$ and $v_1(t)$, it is shown that the nonlinear TD is effective. In

Table 16.3 Parameters for a sinusoidal signal

Parameters					
β_1	6000	β_2	800	b_0	200000
r_1	147	r_2	99	h_0	0.001
k_1	130.5	k_2	0.43	r_0	3000000

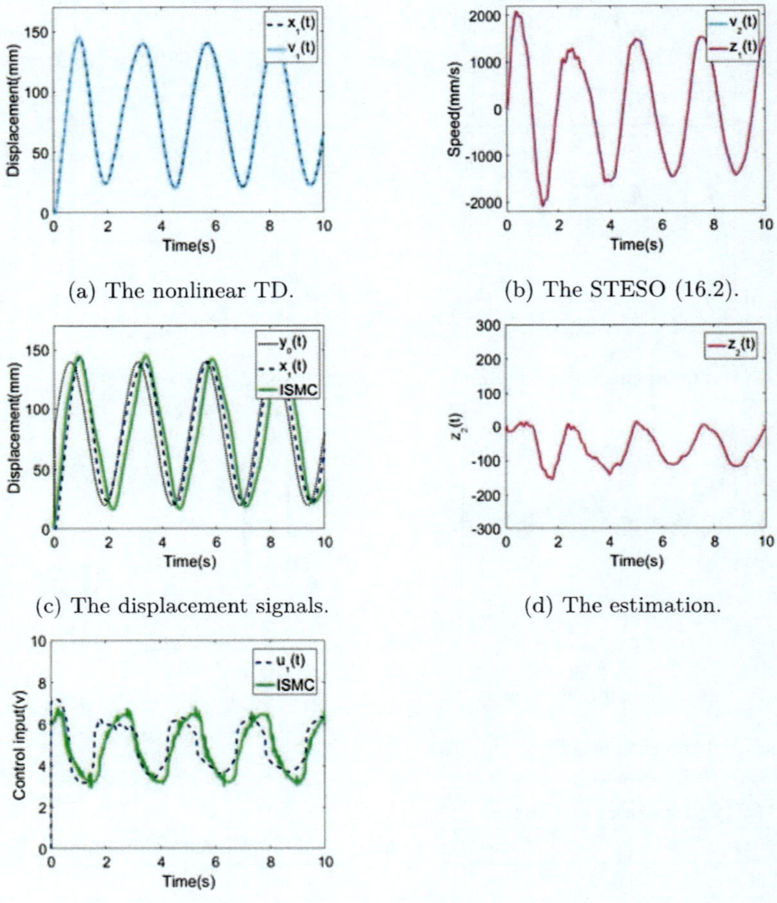

(a) The nonlinear TD.

(b) The STESO (16.2).

(c) The displacement signals.

(d) The estimation.

(e) The control input.

Fig. 16.3 Experimental data for a sinusoidal signal

Fig. 16.3b, $z_1(t)$ is the observation value of $v_2(t)$, $z_2(t)$ represents an estimate value of a total disturbance. The effectiveness of the STESO (16.2) is known on the basis of the closeness of $z_1(t)$ and $v_2(t)$ in Fig. 16.3b. For the two control algorithms, two displacement signals of the pneumatic rod cylinder are compared with a desire signal

Table 16.4 Performances comparison

Controller	Delay time (s)	Amplitude error (mm)
ISMC in [9]	0.36	5.3
In this paper	0.2	4.6

$y_0(t)$ in Fig. 16.3c. In Fig. 16.3d, an estimated value of the total disturbance is shown. Two control input signals of the two control algorithms are shown in Fig. 16.3e. The detailed performance comparisons are shown in Table 16.4. For $0.2\,\mathrm{s} < 0.36\,\mathrm{s}$ and $4.6\,\mathrm{mm} < 5.3\,\mathrm{mm}$, the proposed control strategy in this paper has also achieved better control performance than the one in [9].

Remark 16.8 Four springs are installed on both sides of the pneumatic rod cylinder to simulate varying loads in practice. The load changing in other forms can be dealt with by the algorithm in this paper for the reason of that the varying load is modelled as an unknown time-varying nonlinear function which is not dependent on the changing characteristics of the four springs. Note that the STESO (16.2) is applied to estimate the unknown time-varying nonlinear function in this paper.

16.4 Conclusion

This chapter has proposed an STESO based on a super-twisting algorithm to estimate the dynamics caused by mechanical friction and varying loads in a pneumatic rod cylinder servo system. By a power integrator technology, a finite-time controller was designed to achieve finite-time displacement tracking for the pneumatic rod cylinder servo system. Furthermore, both the effectiveness of the STESO and the finite-time convergence of the closed-loop system were analyzed in the sense of Lyapunov approaches. Finally, we implemented the proposed method in an experimental platform, and the results validated the effectiveness of the proposed control strategy. Specifically, comparing with ISMC via a GNESO in [9],the proposed control strategy exhibited an obvious improvement in both control precision and response rapidity.

References

1. Zhao L, Zhang B, Yang H, Wang Y (2017) Finite-time tracking control for pneumatic servo system via extended state observer. IET Control Theory Appl 11(16):2808–2816
2. Zhao L, Xia Y, Yang Y, Liu Z (2017) Multicontroller positioning strategy for a pneumatic servo system via pressure feedback. IEEE Trans Ind Electron 64(6):4800–4809
3. Yao Z, Ma G, Guo Y, Zeng T (2016) A multi power reaching law of sliding mode control design and analysis. Acta Autom Sinica 42(3):466–472

4. Azman MA, Faudzi AAM, Elnimair MO, Hikmat OF, Osman K, Kai CC (2013) P-adaptive neuro-fuzzy and PD-fuzzy controller design for position control of a modified single acting pneumatic cylinder. In: IEEE/ASME international conference on advanced intelligent mechatronics, pp 176–181
5. Gao X, Feng Z-J (2005) Design study of an adaptive Fuzzy-PD controller for pneumatic servo system. Control Eng Pract 13(1):55–65
6. Ameur O, Massioni P, Scorletti G, Brun X, Smaoui M (2016) Lyapunov stability analysis of switching controllers in presence of sliding modes and parametric uncertainties with application to pneumatic systems. IEEE Trans Control Syst Technol 24(6):1953–1964
7. Liu Y, Kung T, Chang K, Chen S (2013) Observer-based adaptive sliding mode control for pneumatic servo system. Precis Eng 37(3):522–530
8. Taghizadeh M, Najafi F, Ghaffari A (2010) Multimodel PD-control of a pneumatic actuator under variable loads. Int J Adv Manuf Technol 48(5–8):655–662
9. Zhao L, Zhang B, Yang H, Wang Y (2018) Observer-based integral sliding mode tracking control for a pneumatic cylinder with varying loads. IEEE Trans Syst Man Cybern: Syst 1–9
10. Sahabi ME, Li G, Wang X, Li S (2006) Disturbance compensation based finite-time tracking control of rigid manipulator. In: Mathematical problems in engineering, pp 1–12
11. Moreno JA (2009) A linear framework for the robust stability analysis of a generalized super-twisting algorithm. In: International conference on electrical engineering, computing science and automatic control, pp 12–17

Chapter 17
Nonsingular Fast Terminal Sliding Mode Control

17.1 Introduction

At present, many researchers have put forward a lot of control strategies to improve negative problems of pneumatic cylinder servo systems, which is caused by nonlinearity of servo valves, various friction forces, air compressibility, and so on [1, 2]. A neural network control scheme based on PID is introduced to ensure good tracking performance of a pneumatic X–Y table [3]. Methods in [4, 5] are presented to modulate pneumatic servo systems with variable payload. However, these control schemes require highly complete mathematical models which are hardly obtained because of strong nonlinearity, unmodeled dynamics and external disturbances . It is a key idea to estimate these uncertain factors effectively by using an ESO for a pneumatic rod cylinder servo system. In the pneumatic servo system, it is expected that stabilization of the closed-loop system is achieved in finite time. However, traditional linear SMC only guarantees asymptotic tracking performance, which means the tracking error converges to zero only when time goes to infinity [6]. Therefore, to obtain finite-time stabilization of a system, terminal SMC has been presented in [7]. The terminal SMC not only has strong robustness with respect to various uncertainties but also makes the tracking error converge to zero in finite-time [8]. However, there exists a singularity problem for the terminal SMC, please refer to [9]. To solve the singularity problem, a nonsingular terminal SMC method has been proposed in [10]. Then a nonsingular fast terminal sliding mode (NFTSM) control method has been presented to avoid the singularity problem, and it has a faster convergence rate than a nonsingular terminal SMC method [11]. To the best of our knowledge, very few results are available on the extended state observer based NFTSM finite-time tracking control for the pneumatic rod cylinder servo system, which motivates us to carry on this research work.

In this chapter, to deal with these uncertain factors effectively, the finite-time tracking control for a pneumatic rod cylinder servo system is investigated by combining an extended state observer with a NFTSM controller. The extended state observer is used to estimate strong nonlinearity, unknown modelling and external

L. Zhao et al., *Pneumatic Servo Systems Analysis*, Advances in Industrial Control,
https://doi.org/10.1007/978-981-16-9515-5_17

disturbances of the pneumatic rod cylinder servo system. Moreover, the NFTSM controller is applied to ensure satisfactory performances of the closed-loop system. Then, both sufficiently small observation error and stabilisation of the closed-loop system are analyzed in finite time. Finally, experimental results are given to illustrate the effectiveness of the proposed method. The main contributions of this chapter are summarized as below:

i An extended state observer-based NFTSM finite-time tracking control scheme is proposed to ensure response rapidity and control precision for pneumatic rod cylinder servo system.
ii An extended state observer is proposed to deal with incomplete mathematical models via estimating strong nonlinearity, unmodeled dynamics and external disturbances.
iii Both sufficiently small observation error and stabilization of the closed-loop system are proved in finite time by using appropriate Lyapunov functions.

17.2 Main Results

17.2.1 Schematic Diagram of Control Method

Considering strong nonlinearity, unknown modeling and external disturbances of the pneumatic rod cylinder servo system (15.18), an extended state observer-based NFTSM finite-time tracking control scheme is designed to study the positioning problem of the system. The block diagram of the proposed design is shown in Fig. 17.1. In addition, in order to make this chapter express clearly, three lemmas are introduced in the following.

Lemma 17.1 ([12]) *Consider the system of differential equations*

$$\dot{x}_a(t) = f(x_a(t)), \tag{17.1}$$

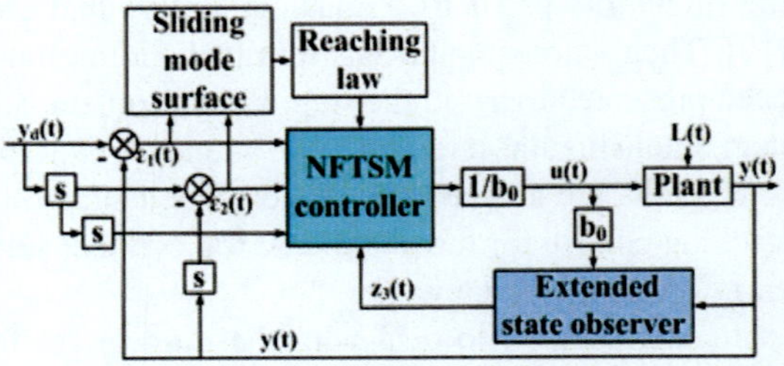

Fig. 17.1 Block diagram of the proposed design

where $f : \mathcal{D} \to R^n$ is continuous on an open neighborhood $\mathcal{D} \subseteq R^n$ of the origin and $f(0) = 0$. Suppose there exists a continuous function $V : \mathcal{D} \to R$ such that the following conditions hold:

(i) V is positive definite.

(ii) There exist $c > 0$, $\varrho \in (0, 1)$ and an open neighborhood $\varpi \subseteq \mathcal{D}$ of the origin such that

$$\dot{V}_a(x_a(t)) + c(V_a(x_a(t)))^{\varrho} \leq 0, \; x_a(t) \in \varpi \backslash \{0\}. \tag{17.2}$$

Then the origin is a finite-time stable equilibrium of (17.1). Furthermore, the setting-time function T_1 is shown as follows:

$$T_1 \leq \frac{1}{c(1-\varrho)}(V_a(x_a(t_0)))^{1-\varrho}, \tag{17.3}$$

where $V_a(x_a(t_0))$ is the initial value of $V_a(x_a(t))$ and T_1 is continuous. In addition, if $\mathcal{D} = R^n$, $V_a(x_a(t))$ is proper, and $\dot{V}_a(x_a(t))$ takes negative values on $R^n \backslash \{0\}$, then the origin is a globally finite-time stable equilibrium of (17.1).

Lemma 17.2 ([13]) *Consider the nonlinear system $\dot{x}_b(t) = f(x_b(t))$, $f(0) = 0$ and $x_b(t) \in R^n$. Suppose there exists a positive definite scalar function $V_b(x_b(t))$ such that*

$$\dot{V}_b(x_b(t)) \leq -\tau_1 V_b(x_b(t)) - \tau_2 V_b(x_b(t))^{\theta}, \tag{17.4}$$

where $\tau_1 > 0$, $\tau_2 > 0$, and $0 < \theta < 1$, then the system is finite-time stable. Furthermore, the setting-time T_2 is obtained as follows:

$$T_2 \leq \frac{1}{\tau_1(1-\theta)} \ln \frac{\tau_1 V_b^{1-\theta}(x_b(t_0)) + \tau_2}{\tau_2}, \tag{17.5}$$

where $V_b(x_b(t_0))$ is the initial value of $V_b(x_b(t))$.

Lemma 17.3 ([14]) *Suppose a NFTSM surface is chosen as follows:*

$$s_1(t) = \varepsilon(t) + \hat{k}_1 sign^{\varsigma_1}(\varepsilon(t)) + \hat{k}_2 sign^{\varsigma_2}(\dot{\varepsilon}(t)), \tag{17.6}$$

where $\hat{k}_1 > 0$, $\hat{k}_2 > 0$, $1 < \varsigma_2 < 2$, and $\varsigma_1 > \varsigma_2$. If $s_1 = 0$, the convergence time T_3 of $\varepsilon(t)$ is given as follows:

$$\begin{aligned} T_3 &= \int_0^{|\varepsilon(0)|} \frac{\hat{k}_2^{1/\varsigma_2}}{(\varepsilon(t) + \hat{k}_1 \varepsilon^{\varsigma_1}(t))^{1/\varsigma_2}} dx = \frac{\varsigma_2 |\varepsilon(0)|^{1-1/\varsigma_2}}{\hat{k}_2^{-1/\varsigma_2}(\varsigma_2 - 1)} \\ &\cdot F\left(\frac{1}{\varsigma_2}, \frac{\varsigma_2 - 1}{(\varsigma_1 - 1)\varsigma_2}; 1 + \frac{\varsigma_2 - 1}{(\varsigma_1 - 1)\varsigma_2}; -\hat{k}_1 |\varepsilon(0)|^{\varsigma_1 - 1}\right), \end{aligned} \tag{17.7}$$

where $\varepsilon(0)$ is the initial value of $\varepsilon(t)$, $F(\cdot)$ is the Gaussian Hypergeometric function.

Assumption 17.1 In the pneumatic rod cylinder servo system (15.18), $f(t, X(t))$ and $d(t)$ are regarded as a class of generalized disturbance $L(t)$, i.e., $L(t) = f(t, X(t)) + w(t)$. In addition, $L(t)$ is continuous differentiable and bounded, i.e., $|\dot{L}(t)| \le L_d$, where L_d is a positive constant.

Remark 17.4 According to [15, 16], the Gaussian Hypergeometric function is given as follows

$$F(a, b; c; z) = \sum_{n=0}^{\infty} \frac{(a)_n (b)_n}{(c)_n n!} z^n,$$

where z is the variable, real numbers a, b, c are the function parameters and $c \neq 0, -1, -2, \ldots$. The Pochhammer symbol $(x)_n$ is defined by $(x)_0 = 1$ and $(x)_n = x(x+1)\cdots(x+n-1)$. Since $1 < \varsigma_2 < 2$ and $\varsigma_1 > \varsigma_2$, function $F(\cdot)$ in equation (11) is the Gaussian Hypergeometric function. According to [33], the conditions of ς_1, ς_2 and $\hat{k}_1$ induce that function $F(\cdot)$ will keep convergent. Moreover, the exact form of function $F(\cdot)$ changes with the involved parameters. For example,

$$F(a, b; b; z) = (1 - z)^{-a}, \quad F\left(\frac{1}{2}, 1; \frac{3}{2}; -z^2\right) = z^{-1} \arctan z$$

17.2.2 Extended State Observer

In this section, various negative factors, which contain strong nonlinearity, unknown modeling and external disturbances of the pneumatic rod cylinder servo system (15.18), are estimated by using an ESO. In pneumatic rod cylinder servo system (15.18), the generalized disturbance $L(t) = f(t, X(t)) + d(t)$ is regarded as an extended state $x_3(t)$, i.e., $L(t) = x_3(t)$. Then Eq. (15.18) is rewritten as follows:

$$\begin{cases} \dot{x}_1(t) = x_2(t) \\ \dot{x}_2(t) = x_3(t) + b_0 u(t) \\ \dot{x}_3(t) = \dot{L}(t). \end{cases} \tag{17.8}$$

Based on Eq. (17.8), an ESO is presented as follows:

$$\begin{cases} \dot{z}_1(t) = z_2(t) - \beta_1 \text{sign}^{\frac{\alpha+1}{2}}(z_1(t) - x_1(t)) \\ \dot{z}_2(t) = z_3(t) - \beta_2 \text{sign}^{\frac{\alpha+1}{2}}(z_1(t) - x_1(t)) + b_0 u(t) \\ \dot{z}_3(t) = -\beta_3 \text{sign}^{\alpha}(z_1(t) - x_1(t)), \end{cases} \tag{17.9}$$

where α is the given parameter with $\alpha \in (0, 1)$, $z_i(t)$ is the observation value of $x_i(t)$, β_i is the observation gain with $i = 1, 2, 3$. Let $e_i(t) = z_i(t) - x_i(t)$, where $e_i(t)$ is the observation error. Considering Eqs. (17.8) and (17.9), the following observer error system is given as

$$\begin{cases} \dot{e}_1(t) = e_2(t) - \beta_1 \text{sign}^{\frac{\alpha+1}{2}}(e_1(t)) \\ \dot{e}_2(t) = e_3(t) - \beta_2 \text{sign}^{\frac{\alpha+1}{2}}(e_1(t)) \\ \dot{e}_3(t) = -\dot{L}(t) - \beta_3 \text{sign}^{\alpha}(e_1(t)). \end{cases} \tag{17.10}$$

Subsequently, finite-time sufficiently small observation errors are analyzed by Lyapunov theory in the following.

Theorem 17.5 *Considering the extended state observer (17.9) and Assumption 17.1, there exist gains* β_1, β_2, β_3, *and* α $(\beta_1 > 0, \beta_2 > 0, \beta_3 > 0, 1 > \alpha > 0)$ *such that the following inequation is obtained in finite-time* t_s $(t_s > 0)$ *as*

$$\|\eta(t)\|_2 \leq \frac{I L_d}{\sigma_{\min}\{A_2\}\sigma_{\min}\{P\}}, \tag{17.11}$$

where

$$\begin{aligned} \eta(t) &= [sign^{(\alpha+1)/2}(e_1(t)), e_2(t), e_3(t)]^T, \\ I &= \sqrt{\beta_3^2 + 4}, \\ A_2 &= \begin{bmatrix} \beta_1 & -1 & 0 \\ \beta_2 & 0 & -1 \\ \beta_3 & 0 & 0 \end{bmatrix}, \\ P &= \begin{bmatrix} 2\beta_1/(\alpha+1) + \beta_2^2 + \beta_3^2 & -\beta_2 & -\beta_3 \\ -\beta_2 & 2 & 0 \\ -\beta_3 & 0 & 2 \end{bmatrix}. \end{aligned}$$

Moreover, select the appropriate parameters β_1, β_2 *and* β_3 *to meet* $I L_d \ll \sigma_{\min}\{A_2\}\sigma_{\min}\{P\}$, *then* $\|\eta(t)\|_2$ *is limited to be small enough in finite time.*

Proof An appropriate Lyapunov function is constructed as follows

$$V(\eta(t)) = \eta^T(t) P \eta(t) \tag{17.12}$$

The following formula is confirmed as

$$\begin{aligned} V(\eta(t)) = &\frac{2\beta_1}{\alpha+1}|e_1(t)|^{\alpha+1} + e_2^2(t) + e_3^2(t) \\ &+ \left[e_2(t) - \beta_2 \text{sign}^{\frac{\alpha+1}{2}}(e_1(t))\right]^2 \\ &+ \left[e_3(t) - \beta_3 \text{sign}^{\frac{\alpha+1}{2}}(e_1(t))\right]^2 \\ &\geq 0. \end{aligned} \tag{17.13}$$

Remark 17.6 Obviously, in addition to $e_1 = 0$, $V(\eta(t))$ is continuous and differentiable everywhere. Moreover, before reaching the origin, the observer error system (17.10) is not possible to stay in $e_1(t) = 0$. Therefore, when $e_1(t) \neq 0$, $\dot{V}(\eta(t))$ is calculated in accordance with the conventional method.

The equation is calculated as follows:

$$\frac{d\left[\text{sign}^{(\alpha+1)/2}(e_1(t))\right]}{dt} = \frac{(\alpha+1)|e_1(t)|^{(\alpha-1)/2}\dot{e}_1(t)}{2}. \tag{17.14}$$

Hence, the following equation is obtained as

$$\begin{aligned}
\dot{\eta}(t) &= \begin{bmatrix} \frac{\alpha+1}{2}|e_1(t)|^{\frac{\alpha-1}{2}}\left(e_2(t) - \beta_1 \text{sign}^{\frac{\alpha+1}{2}}(e_1(t))\right) \\ e_3(t) - \beta_2 \text{sign}^{\frac{\alpha+1}{2}}(e_1(t)) \\ -\beta_3 \text{sign}^{\alpha}(e_1(t)) - \dot{L}(t) \end{bmatrix} \\
&= \begin{bmatrix} -r\mu\beta_1 & r\mu & 0 \\ -\beta_2 & 0 & 1 \\ -\mu\beta_3 & 0 & 0 \end{bmatrix} \eta - \begin{bmatrix} 0 \\ 0 \\ 1 \end{bmatrix} \dot{L}(t) \\
&= A\eta(t) - B\dot{L}(t),
\end{aligned} \tag{17.15}$$

where $1 > r = (\alpha+1)/2 > 1/2$, $\mu = |e_1(t)|^{(\alpha-1)/2} > 0$. The characteristic equation of the matrix A is written as follows:

$$\begin{aligned}
G(\tilde{s}) &= |\tilde{s}E - A| \\
&= \begin{vmatrix} \tilde{s} + r\mu\beta_1 & -r\mu & 0 \\ \beta_2 & \tilde{s} & -1 \\ \mu\beta_3 & 0 & \tilde{s} \end{vmatrix} \\
&= \tilde{s}^3 + r\mu\beta_1\tilde{s}^2 + r\mu\beta_2\tilde{s} + r\mu^2\beta_3,
\end{aligned} \tag{17.16}$$

where E is the identity matrix. If there exists $\beta_i > 0$ with $i = 1, 2, 3$, then all coefficients of $G(\tilde{s})$ is positive. Therefore, the matrix A is Hurwitz which means that A is stable. The time differential equation of $V(\eta(t))$ is expressed as follows:

$$\begin{aligned}
\dot{V}(\eta(t)) &= \dot{\eta}^T(t)P\eta(t) + \eta^T(t)P\dot{\eta}(t) \\
&= (A\eta(t) - B\dot{L}(t))^T P\eta(t) + \eta^T(t)P(A\eta(t) - B\dot{L}(t)) \\
&\quad + \eta^T(t)P(A\eta(t) - B\dot{L}(t)) \\
&= \eta^T(t)(A^T P + PA)\eta(t) + 2\dot{L}(t)\hat{B}^T\eta(t),
\end{aligned} \tag{17.17}$$

where $\hat{B}^T = -B^T P = [\beta_3 \ 0 \ -2]$, then $I = \|\hat{B}\|_2 = \sqrt{\beta_3^2 + 4}$. Since the matrix A is Hurwitz, there exists a corresponding symmetrical positive definite matrix $Q > 0$ such that P is the solution of Lyapunov equation. One has that

$$A^T P + PA = -Q \tag{17.18}$$

According to Eq. (17.12), the following inequality is obtained as

$$\lambda_{\min}\{P\}\|\eta(t)\|_2^2 \le V(\eta(t)) \le \lambda_{\max}\{P\}\|\eta(t)\|_2^2, \tag{17.19}$$

where $\|\eta(t)\|_2^2 = |e_1(t)|^{\alpha+1} + e_2^2(t) + e_3^2(t)$. Meanwhile, the following inequality is acquired as

$$\|\eta(t)\|_2 \ge |e_1(t)|^{\frac{\alpha+1}{2}}. \tag{17.20}$$

According to Eqs. (17.17) and (17.18), the inequality is obtained as follows:

$$\begin{aligned}\dot{V}(\eta(t)) &= -\eta^T(t)Q\eta(t) + 2\dot{L}(t)\hat{B}\eta(t)\\ &\le -\lambda_{\min}\{Q\}\|\eta(t)\|_2^2 + 2IL_d\|\eta(t)\|_2\\ &= -(\lambda_{\min}\{Q\}\|\eta(t)\|_2 - 2IL_d)\|\eta(t)\|_2.\end{aligned} \tag{17.21}$$

For the symmetrical positive definite matrix Q, the following equation is gotten as

$$\sigma_{\min}\{Q\} = \lambda_{\min}\{Q\}. \tag{17.22}$$

Considering Eq. (17.18), there exists

$$Q = (-A)^T P + P(-A). \tag{17.23}$$

Note that A and P are nonsingular matrices when $e_1 \ne 0$, it follows that

$$\sigma_{\min}\{Q\} = 2\sigma_{\min}\{-AP\} \ge 2\sigma_{\min}\{-A\}\sigma_{\min}\{P\}. \tag{17.24}$$

The nonsingular matrix A is rewritten as follows:

$$-A = \begin{bmatrix} r\mu & 0 & 0 \\ 0 & 1 & 0 \\ 0 & 0 & \mu \end{bmatrix} \begin{bmatrix} \beta_1 & -1 & 0 \\ \beta_2 & 0 & -1 \\ \beta_3 & 0 & 0 \end{bmatrix} = A_1 A_2 \tag{17.25}$$

One has that

$$\sigma_{\min}\{-A\} = \sigma_{\min}\{A_1 A_2\} \ge \sigma_{\min}\{A_1\}\sigma_{\min}\{A_2\}. \tag{17.26}$$

Since A_1 is a diagonal matrix and $r\mu < \mu$, there exists

$$\sigma_{min}\{A_1\} = \begin{cases} 1, & |e_1(t)| < \left(\dfrac{2}{\alpha+1}\right)^{\frac{2}{\alpha-1}} \\ r\mu, & |e_1(t)| \geq \left(\dfrac{2}{\alpha+1}\right)^{\frac{2}{\alpha-1}} \end{cases} \tag{17.27}$$

If $|e_1(t)| \geq \left(\dfrac{2}{\alpha+1}\right)^{\frac{2}{\alpha-1}}$, then $\|\eta(t)\|_2 \geq \left(\dfrac{2}{\alpha+1}\right)^{\frac{\alpha+1}{\alpha-1}}$. One has that

$$\lambda_{\min}\{Q\} = \sigma_{\min}\{Q\} \geq 2r\mu\sigma_{\min}\{A_2\}\sigma_{\min}\{P\}. \tag{17.28}$$

Considering formulas (17.20), (17.21) and (17.28), the following relationship is obtained as

$$\begin{aligned} q_1 &= \lambda_{\min}\{Q\}\|\eta(t)\|_2 - 2IL_d \\ &\geq 2r\mu\sigma_{\min}\{A_2\}\sigma_{\min}\{P\}|e_1(t)|^{\frac{\alpha+1}{2}} - 2IL_d \\ &= (\alpha+1)\sigma_{\min}\{A_2\}\sigma_{\min}\{P\}|e_1(t)|^{\alpha} - 2IL_d \\ &\geq (\alpha+1)\left(\frac{2}{\alpha+1}\right)^{\frac{2\alpha}{\alpha-1}}\sigma_{\min}\{A_2\}\sigma_{\min}\{P\} - 2IL_d. \end{aligned} \tag{17.29}$$

Since α is the given parameter with $\alpha \in (0, 1)$, there exists a minimum value $\hbar$ such that

$$(\alpha+1)\left(\frac{2}{\alpha+1}\right)^{\frac{2\alpha}{\alpha-1}} \geq \hbar. \tag{17.30}$$

Therefore, inequality (17.29) is rewritten as follows:

$$q_1 \geq \hbar\sigma_{\min}\{A_2\}\sigma_{\min}\{P\} - 2IL_d = q_{1,\min}. \tag{17.31}$$

Note that β_1, β_2, and β_3 are the adjustable parameters. Before β_1, β_2 and β_3 are given, it is necessary to take the adjustable parameters into $q_{1,\min} = \hbar\sigma_{\min}\{A_2\}\sigma_{\min}\{P\} - 2IL_d$. When $q_{1,\min} \leq 0$, the adjustable parameters β_1, β_2, and β_3 are reassigned such that $q_{1,\min} > 0$. When $q_{1,\min} > 0$, corresponding experiments are carried out using the adjustable parameters β_1, β_2, and β_3. That is, β_1, β_2, and β_3 are determined when both the condition $\hbar\sigma_{\min}\{A_2\}\sigma_{\min}\{P\} > 2IL_d$ and experimental effects are satisfied. Therefore, inequality $q_{1,\min} > 0$ is established by selecting the appropriate β_1, β_2 and β_3. Then it follows that

$$\dot{V}(\eta(t)) < 0. \tag{17.32}$$

Considering formulas (17.19), (17.21) and (17.31), one has that

$$\dot{V}(\eta(t)) \le -q_1\|\eta(t)\|_2 \le -\frac{q_{1,\min}}{\sqrt{\lambda_{\max}\{P\}}}V^{\frac{1}{2}}(\eta(t)) = -C_1V^{\frac{1}{2}}(\eta(t)). \quad (17.33)$$

According to Lemma 17.1, the observer error system (17.10) converges into the region $\|\eta(t)\|_2 < \left(\frac{2}{\alpha+1}\right)^{\frac{\alpha+1}{\alpha-1}}$ in finite-time time t_{s1}. Meanwhile, the finite-time time t_{s1} is expressed as follows:

$$t_{s1} \le \frac{1}{C_1/2}V^{\frac{1}{2}}(\eta(t)) \le \frac{1}{C_1/2}V^{\frac{1}{2}}(\eta(t_0)). \quad (17.34)$$

If $\|\eta(t)\|_2 < \left(\frac{2}{\alpha+1}\right)^{\frac{\alpha+1}{\alpha-1}}$, then $|e_1(t)| < \left(\frac{2}{\alpha+1}\right)^{\frac{2}{\alpha-1}}$. Therefore, there exists

$$\lambda_{\min}\{Q\} = \sigma_{\min}\{Q\} \ge 2\sigma_{\min}\{A_2\}\sigma_{\min}\{P\}. \quad (17.35)$$

According to formulas (17.20), (17.21) and (17.35), the following relationship is acquired as

$$\begin{aligned} q_2 &= \lambda_{\min}\{Q\}\|\eta(t)\|_2 - 2IL_d \\ &\ge 2\sigma_{\min}\{A_2\}\sigma_{\min}\{P\}\|\eta(t)\|_2 - 2IL_d. \end{aligned} \quad (17.36)$$

If the following relationship

$$\left(\frac{2}{\alpha+1}\right)^{\frac{\alpha+1}{\alpha-1}} > \|\eta(t)\|_2 > \frac{IL_d}{\sigma_{\min}\{A_2\}\sigma_{\min}\{P\}} = M_d \quad (17.37)$$

is satisfied, then there exist $q_2 > 2\sigma_{\min}\{A_2\}\sigma_{\min}\{P\}M_d - 2IL_d = 0$ and $\dot{V}(\eta(t)) < 0$. Similar to (17.33) and (17.34), it follows that

$$\dot{V}(\eta(t)) \le -q_2\|\eta(t)\|_2 \le -\frac{q_2}{\sqrt{\lambda_{\max}\{P\}}}V^{\frac{1}{2}}(\eta(t)) = -C_2V^{\frac{1}{2}}(\eta(t)), \quad (17.38)$$

$$t_{s2} \le \frac{1}{C_2/2}V^{\frac{1}{2}}(\eta(t)) \le \frac{1}{C_2/2}V^{\frac{1}{2}}(\eta(t_1)). \quad (17.39)$$

Therefore, $\|\eta\|_2$ is a decreasing function of time. Meanwhile, the following inequation is obtained in finite time $t_s = t_{s1} + t_{s2}$ as

$$\|\eta(t)\|_2 \le \frac{IL_d}{\sigma_{\min}\{A_2\}\sigma_{\min}\{P\}}. \quad (17.40)$$

Obviously, if the appropriate parameters β_1, β_2, and β_3 are selected to make $\sigma_{\min}\{A_2\}\sigma_{\min}\{P\}$ large enough, then the observation error $\|\eta(t)\|_2$ is limited to be sufficiently small in finite time. This completes the proof.

17.2.3 Nonsingular Fast Terminal Sliding Mode Controller

In this section, a NFTSM controller is designed to ensure good performances of the pneumatic rod cylinder servo system (15.18). The tracking error is presented as follows:

$$\begin{cases} \varepsilon_1(t) = y(t) - y_d(t) = x_1(t) - y_d(t) \\ \varepsilon_2(t) = \dot{x}_1(t) - \dot{y}_d(t) = x_2(t) - \dot{y}_d(t), \end{cases} \tag{17.41}$$

where $y_d(t)$ is the reference signal. According to Eqs. (15.18) and (17.41) as well as Assumption 17.1, the following tracking error system is given as

$$\begin{cases} \dot{\varepsilon}_1(t) = \varepsilon_2(t) \\ \dot{\varepsilon}_2(t) = L(t) + b_0 u(t) - \ddot{y}_d(t). \end{cases} \tag{17.42}$$

Considering Eq. (17.41), the sliding mode surface $s(t)$ is designed as follows

$$s(t) = \varepsilon_1(t) + k_1 \text{sign}^{\gamma_1} \varepsilon_1(t) + k_2 \text{sign}^{\gamma_2} \varepsilon_2(t), \tag{17.43}$$

where $k_1 > 0, k_2 > 0, 1 < \gamma_2 < 2$ and $\gamma_1 > \gamma_2$. The sliding mode reaching condition is expressed as follows:

$$s(t)\dot{s}(t) \leq 0. \tag{17.44}$$

Considering inequation (17.44), the following reaching law is selected as

$$\dot{s}(t) = -k_3 \text{sign}(s(t)) - k_4 s(t), \tag{17.45}$$

where $k_3 > 0$, $k_4 > 0$, $k_4 = k_l + k_d$, $k_l > 0$, and $k_d > 0$. In order to obtain satisfactory tracking position, the controller $u(t)$ is designed as follows:

$$\begin{aligned} u(t) = &-\frac{1}{b_0}[z_3(t) + \frac{1}{k_2\gamma_2}\text{sign}^{2-\gamma_2}\varepsilon_2(t) \times (1 + k_1\gamma_1|\varepsilon_1(t)|^{\gamma_1-1}) \\ &+ k_3\text{sign}(s(t)) + k_4 s(t) - \ddot{y}_d(t)]. \end{aligned} \tag{17.46}$$

The controller $u(t)$ is a voltage signal, which is used to determine the opening of proportional valve [17].

Remark 17.7 Total disturbances are estimated and system states are tracked in real time by using the ESO. Moreover, the estimation and tracking processes are synchronized with controlling the system. According to Theorem 17.5, the observation error $\|\eta(t)\|_2$ is limited to be sufficiently small in finite time. Meanwhile, total disturbances of the pneumatic rod cylinder servo system are bounded in practice. The disturbance estimation z_3 is also bounded. Therefore, system states do not diverge, please refer to [18–25]. That is, states are not divergent so seriously that the closed-loop system can not be controlled using the extended state observer.

Subsequently, finite-time stabilization of the closed-loop system (17.42) is analyzed by Lyapunov theory in the following.

Theorem 17.8 *Consider the closed-loop system (17.42) with the NFTSM controller (17.46). When the sliding mode surface is selected as (17.43), and by choosing appropriate parameters $k_4 = k_l + k_d$ such that $(L(t) - z_3(t))s(t) \le k_d s^2(t)$, $|L(t) - z_3(t)| \le k_4|s(t)|$, then tracking error of the closed-loop system (17.42) converges to zero in finite time.*

Proof Considering the closed-loop system (17.42), the following Lyapunov function is designed as

$$V_1(s(t)) = \frac{1}{2}s^2(t). \tag{17.47}$$

Differentiating Eq. (17.47), there exists

$$\begin{aligned}\dot{V}_1(s(t)) &= s(t)\dot{s}(t)\\ &= s(t)(\varepsilon_2(t) + k_1\gamma_1|\varepsilon_1(t)|^{\gamma_1-1}\varepsilon_2(t)\\ &\quad +k_2\gamma_2|\varepsilon_2(t)|^{\gamma_2-1}\dot{\varepsilon}_2(t)).\end{aligned} \tag{17.48}$$

According to Eq. (17.42), the Eq. (17.48) is rewritten as follows:

$$\begin{aligned}\dot{V}_1(s(t)) = s(t)[&\varepsilon_2(t) + k_1\gamma_1|\varepsilon_1(t)|^{\gamma_1-1}\varepsilon_2(t)\\ &+k_2\gamma_2|\varepsilon_2(t)|^{\gamma_2-1}(L(t) + b_0u(t) - \ddot{y}_d(t))].\end{aligned} \tag{17.49}$$

Substituting Eq. (17.46) into Eq. (17.49), it follows that

$$\begin{aligned}\dot{V}_1(s(t)) &= k_2\gamma_2|\varepsilon_2(t)|^{\gamma_2-1}(L(t) - z_3(t))s(t)\\ &\quad -k_2\gamma_2|\varepsilon_2(t)|^{\gamma_2-1}k_3|s(t)| - k_2\gamma_2|\varepsilon_2(t)|^{\gamma_2-1}k_4s^2(t)\\ &= k_2\gamma_2|\varepsilon_2(t)|^{\gamma_2-1}(L(t) - z_3(t))s(t)\\ &\quad -k_2\gamma_2|\varepsilon_2(t)|^{\gamma_2-1}k_3|s(t)| - k_2\gamma_2|\varepsilon_2(t)|^{\gamma_2-1}k_ds^2(t)\\ &\quad -k_2\gamma_2|\varepsilon_2(t)|^{\gamma_2-1}k_ls^2(t).\end{aligned} \tag{17.50}$$

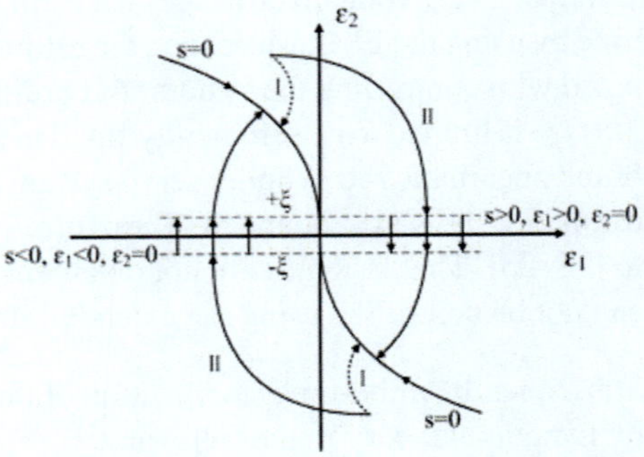

Fig. 17.2 Phase-trajectory plot of pneumatic rod cylinder servo system

When $(L(t) - z_3(t))s(t) \leq k_d s^2(t)$, the following relationship is obtained as

$$\begin{aligned}\dot{V}_1(s(t)) &\leq -k_2\gamma_2|\varepsilon_2(t)|^{\gamma_2-1}k_3|s(t)| - k_2\gamma_2|\varepsilon_2(t)|^{\gamma_2-1}k_l s^2(t)\\ &\leq -\sqrt{2}k_2\gamma_2|\varepsilon_2(t)|^{\gamma_2-1}k_3 V_1(t)^{1/2}\\ &\quad -2k_2\gamma_2|\varepsilon_2(t)|^{\gamma_2-1}k_l V_1(t)\\ &\leq -\psi_1(t)V_1(t)^{1/2} - \psi_2(t)V_1(t),\end{aligned} \tag{17.51}$$

where $\psi_1(t) = \sqrt{2}k_2\gamma_2|\varepsilon_2(t)|^{\gamma_2-1}k_3$, $\psi_2(t) = 2k_2\gamma_2|\varepsilon_2(t)|^{\gamma_2-1}k_l$. In addition, the phase-trajectory plot of the pneumatic rod cylinder servo system (15.18) is shown in Fig. 17.2. Subsequently, two cases are analyzed in the following.

In the first case, the reachability of NFTSM control is not affected under the condition of $\varepsilon_2(t) \neq 0$, please see path I of Fig. 17.2. When $\varepsilon_2(t) \neq 0$, $\psi_1(t) > 0$ and $\psi_2(t) > 0$ are established. According to Lemma 17.2 and inequality (17.51), the tracking error (17.41) moves to the sliding model surface $s(t) = 0$ in finite time and the reaching time T_r is expressed as follows:

$$T_r \leq \frac{1}{\psi_2(t)/2}\ln\frac{\psi_2(t)V_1(s(t_0))^{1/2} + \psi_1(t)}{\psi_1(t)}, \tag{17.52}$$

where $V_1(t_0)$ is the initial value of $V_1(t)$.

In another case, $\varepsilon_2(t) = 0$ is considered in the pneumatic rod cylinder servo system (15.18), please see path II of Fig. 17.2. Substituting Eq. (17.46) into differential formula of the tracking velocity error $\varepsilon_2(t)$ in Eq. (17.42), it follows that

$$\begin{aligned}\dot{\varepsilon}_2(t) &= L(t) - z_3(t) - \frac{1}{k_2\gamma_2}\text{sign}^{2-\gamma_2}\varepsilon_2(t)\\ &\quad\times(1 + k_1\gamma_1|\varepsilon_1(t)|^{\gamma_1-1}) - k_3\text{sign}(s(t)) - k_4 s(t).\end{aligned} \tag{17.53}$$

When $\varepsilon_2(t) = 0$, the Eq. (17.53) is rewritten as follows:

$$\dot{\varepsilon}_2(t) = L(t) - z_3(t) - k_3\text{sign}(s(t)) - k_4 s(t). \tag{17.54}$$

By choosing appropriate parameter k_4, the following relationship is established as

$$|L(t) - z_3(t)| \leq k_4 |s(t)|, \tag{17.55}$$

where $e_3(t) = L(t) - z_3(t)$ is the observer error. According to formulas (17.54) and (17.55), when $s(t) > 0$ or $s(t) < 0$, $\dot{\varepsilon}_2(t) \leq -k_3$ or $\dot{\varepsilon}_2(t) \geq k_3$ is set up, respectively. Therefore, it is known that $\varepsilon_2(t) = 0$ is not an attractor. It also means that there exists a small enough positive constant ξ such that

$$|\varepsilon_2(t)| \leq \xi. \tag{17.56}$$

Then there still holds $\dot{\varepsilon}_2(t) \leq -k_3$ or $\dot{\varepsilon}_2(t) \geq k_3$ under the conditions of $s(t) > 0$ or $s(t) < 0$, respectively [14]. Moreover, the trajectories cross, which include from $\varepsilon_2 \leq \xi$ to $\varepsilon_2(t) \geq -\xi$ for $s(t) > 0$ and from $\varepsilon_2(t) \geq -\xi$ to $\varepsilon_2(t) \leq \xi$ for $s(t) < 0$, are realized in finite time. In addition, the trajectories in the region $|\varepsilon_2(t)| > \xi$ also converge to the boundaries $\pm\xi$ in finite time [26].

To sum up, wherever trajectories of the pneumatic rod cylinder servo system (15.18) are, the sliding mode surface $s(t) = 0$ is reached in finite time. Moreover, according to Lemma 17.3 and the sliding mode surface (17.43), trajectories of the pneumatic servo system (15.18) converge to zero along the sliding mode surface $s(t) = 0$ in finite time and the sliding time T_s is expressed as follows:

$$\begin{aligned} T_s &= \int_0^{|\varepsilon_1(t_0)|} \frac{k_2^{1/\gamma_2}}{(\varepsilon_1(t) + k_1 \varepsilon_1^{\gamma_1}(t))^{1/\gamma_2}} dx \\ &= \frac{\gamma_2 |\varepsilon_1(t_0)|^{1-1/\gamma_2}}{k_2^{-1/\gamma_2}(\gamma_2 - 1)} F\Big(\frac{1}{\gamma_2}, \frac{\gamma_2 - 1}{(\gamma_1 - 1)\gamma_2}; 1 + \frac{\gamma_2 - 1}{(\gamma_1 - 1)\gamma_2}; \\ &\quad -k_1 |\varepsilon_1(t_0)|^{\gamma_1 - 1}\Big), \end{aligned} \tag{17.57}$$

where $\varepsilon_1(t_0)$ is the initial value of $\varepsilon_1(t)$. That is, by choosing appropriate parameters $k_4 = k_l + k_d$ such that $(L(t) - z_3(t))s(t) \leq k_d s^2(t)$ and $|L(t) - z_3(t)| \leq k_4 |s(t)|$, tracking error of the closed-loop system (17.42) converges to zero in finite time. This completes the proof.

17.3 Experiments and Results

In this section, two experiments with variable load, which include a step signal at 100 mm and a sinusoidal signal at 0.4 Hz, 40 mm amplitude, are used to verify the effectiveness of extended state observer-based NFTSM controller for the pneumatic rod cylinder servo system (15.18). The pneumatic rod cylinder servo platform with variable load is shown in Fig. 17.3.

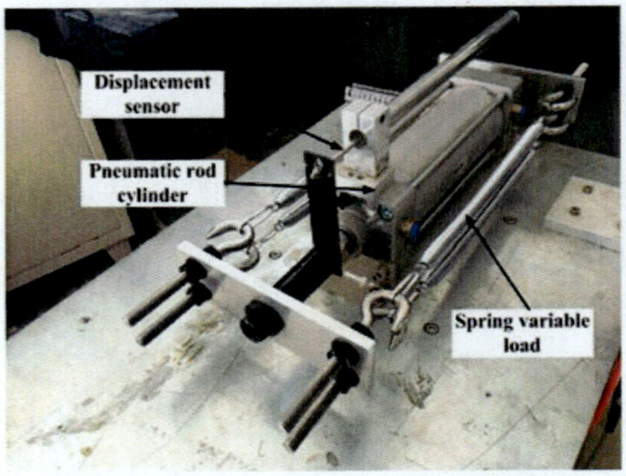

Fig. 17.3 Experimental platform with variable load

Table 17.1 Various parameters in experiment

Parameters			
$\beta_1 = 3800$	$b_0 = 1850$	$\alpha = 1/2$	$\gamma_2 = 9/7$
$\beta_2 = 4500$	$\beta_3 = 2740$	$\gamma_1 = 1.4$	

Note that maximum stroke of the pneumatic cylinder is 200 mm, the spring coefficient is 0.4 N/mm and the supply absolute pressure is 0.6 MPa. Meanwhile, adjustable parameters of the proposed method are listed in Table 17.1. In addition, the experimental results are shown in Figs. 17.4 and 17.5. Moreover, r is the given tracking signal, z_1 is the observation value of x_1, z_3 is the estimate of nonlinearity, x_1, and L are the tracking trajectories of the pneumatic rod cylinder servo system (15.18) based on the proposed method and the linear active disturbance rejection control (LADRC) method, respectively. u_1 and u_2 are the control inputs of the pneumatic rod cylinder servo system (15.18) based on the proposed method and the LADRC method, respectively. e_1 and e_2 are the tracking errors of the pneumatic rod cylinder servo system (15.18) based on the proposed method and the LADRC method, respectively.

In the first experiment, a step signal at 100 mm is tracked as shown in Fig. 17.4. Moreover, parameters of the NFTSM controller (17.46) are set as $k_1 = 4.29$, $k_2 = 7.8$, $k_3 = 9.7$ and $k_4 = 0.86$. Figure 17.4a shows the displacement for tracking the given step signal via the extended state observer (17.9) based NFTSM method. Meanwhile, the observation value z_1 is almost coincident with x_1. Figure 17.4b shows that the piston moves to the given tracking position at about 0.69 s and 0.85 s based on the proposed method and the LADRC method, respectively. In Fig. 17.4c, due to the control inputs u_1 and u_2 eventually tend to be stable, the pneumatic rod cylinder servo system (15.18) is able to stay at the given tracking position based on the proposed method and the LADRC method, respectively. Figure 17.4d shows that the tracking error of the proposed method is less than 0.2 mm and the tracking error of the LADRC method is less than 0.42 mm. In Fig. 17.4e, the estimate of nonlinearity z_3 reaches a steady state finally under the condition of variable load.

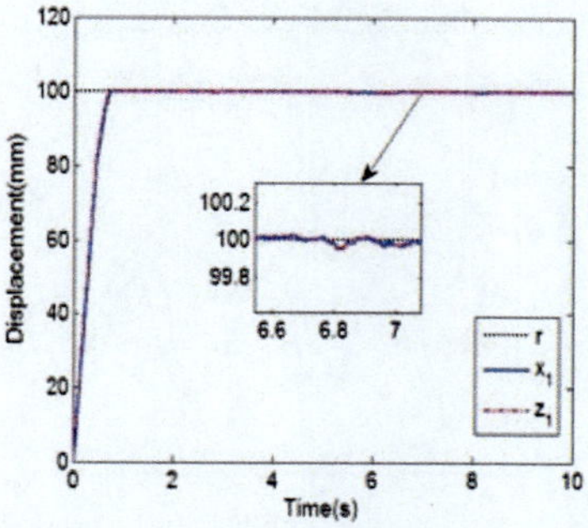

(a) Tracking trajectory of the proposed method

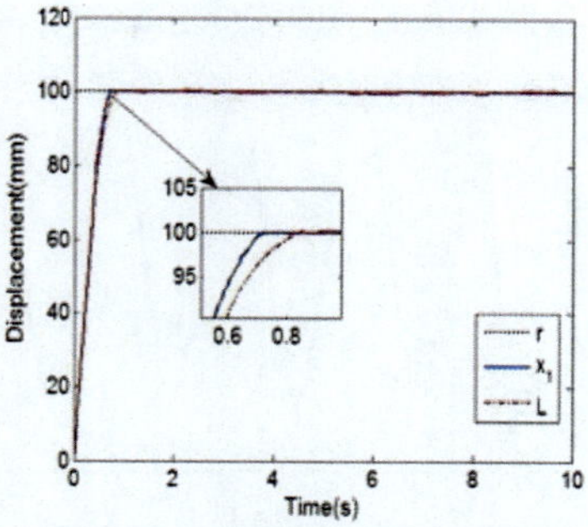

(b) Comparison of tracking trajectories

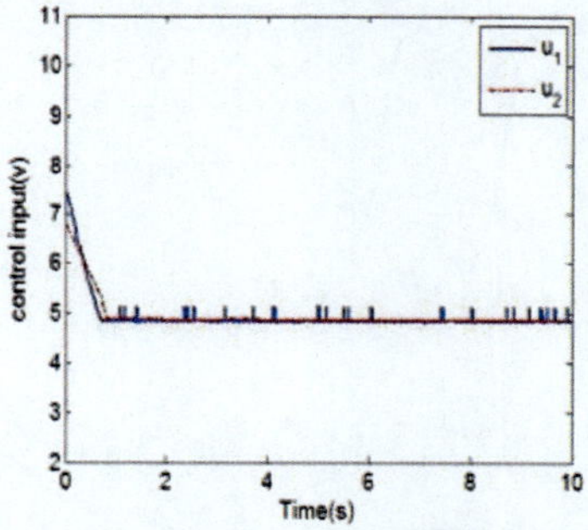

(c) Comparison of control inputs

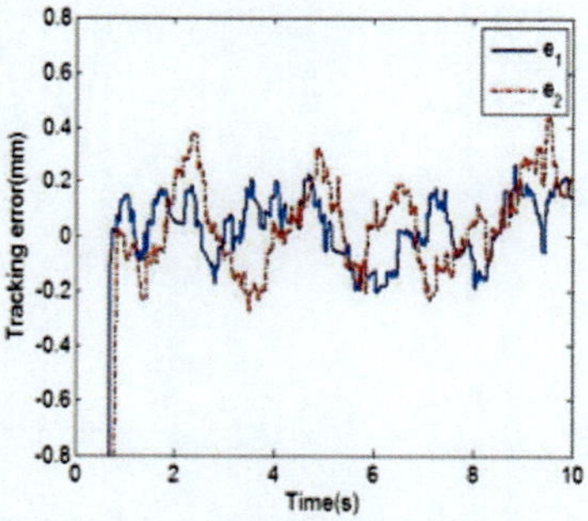

(d) Comparison of tracking errors

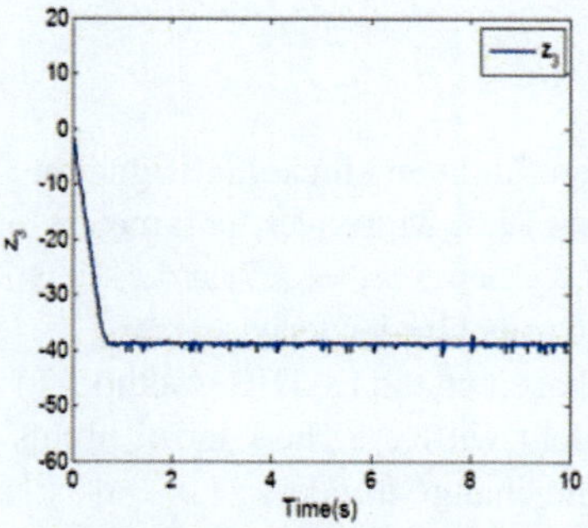

(e) Estimate of nonlinearity

Fig. 17.4 Experimental results of the proposed method and the LADRC method for tracking a step signal

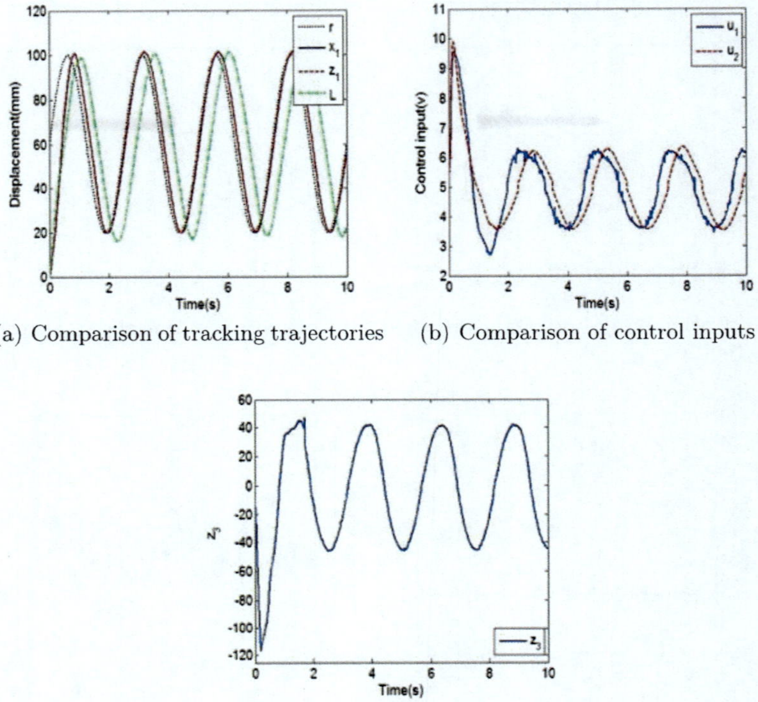

(a) Comparison of tracking trajectories (b) Comparison of control inputs

(c) Estimate of nonlinearity

Fig. 17.5 Experimental results of the proposed method and the LADRC method for tracking a sinusoidal signal

In the second experiment, a given sinusoidal signal at 0.4 Hz, 40 mm amplitude is tracked as shown in Fig. 17.5. Moreover, parameters of the NFTSM controller (17.46) are set as $k_1 = 2.5$, $k_2 = 4.5$, $k_3 = 2.7$, and $k_4 = 3.36$. In Fig. 17.5a, tracking trajectory of the pneumatic rod cylinder servo system (15.18) based on the proposed method is better than that based on the LADRC method. In addition, the observation value z_1 is almost coincident with x_1. The control inputs u_1 and u_2 are shown in Fig. 17.5b. In Fig. 17.5c, the change trend of z_3 is consistent with the fact under the condition of variable load.

According to Figs. 17.4 and 17.5, comparisons of experimental results between the proposed method and the LADRC method are clearly arranged in Table 17.2. From Table 17.2, it is known that the pneumatic rod cylinder servo system (15.18) based on the proposed method in this paper has better response rapidity and control precision compared with the LADRC method. Therefore, the effectiveness of the extended state observer-based NFTSM method is confirmed in the experimental platform with variable loads.

Table 17.2 Comparison of experimental results

Controller	Step signal		Sinusoidal signal	
	Convergence time (s)	Tracking error (mm)	Delay time (s)	Amplitude error (mm)
LADRC	0.85	0.42	0.42	3.1
In this paper	0.69	0.2	0.11	1.2

17.4 Conclusion

A NFTSM finite-time tracking control strategy has been designed in this paper to guarantee the response rapidity and control precision of pneumatic rod cylinder servo system via an ESO. The ESO has been introduced to estimate the total disturbances which include strong nonlinearity, unmodeled dynamics and external disturbances. The NFTSM controller has been designed by combining with the estimation value of ESO. Moreover, the corresponding theoretical analyses on sufficiently small observation error and stabilization of the closed-loop system have also been proved in finite time, respectively. Finally, by comparing the experimental results, the proposed method has obvious improvement for performances of the pneumatic rod cylinder servo system.

References

1. Linares-Flores J, Barahona-Avalos J, Sira-Ramirez H, Contreras-Ordaz MA (2012) Robust passivity-based control of a buckĺcboost-converter/dc-motor system: an active disturbance rejection approach. IEEE Trans Ind Appl 48(6):2362–2371
2. Gao X, Feng Z-J (2005) Design study of an adaptive Fuzzy-PD controller for pneumatic servo system. Control Eng Pract 13(1):55–65
3. Cho SH (2009) Trajectory tracking control of a pneumatic X-Y table using neural network based PID control. Int J Precis Eng Manuf 10(5):37–44
4. Shen X, Zhang J, Barth EJ, Goldfarb M (2006) Nonlinear model-based control of pulse width modulated pneumatic servo systems. J Dyn Syst Meas Control 128(3):663–669
5. Tu DCT, Ahn KK (2006) Nonlinear PID control to improve the control performance of 2 axes pneumatic artificial muscle manipulator using neural network. Mechatronics 16(9):577–587
6. Boiko I, Fridman L, Iriarte R, Pisano A, Usai E (2006) Parameter tuning of second-order sliding mode controllers for linear plants with dynamic actuators. Automatica 42(5):833–839
7. Chen M, Wu Q, Cui R (2013) Terminal sliding mode tracking control for a class of siso uncertain nonlinear systems. ISA Trans 52(2):198–206
8. Wang L, Chai T, Zhai L (2009) Neural-network-based terminal sliding-mode control of robotic manipulators including actuator dynamics. IEEE Trans Ind Electron 56(9):3296–3304
9. Chen S, Lin F (2011) Nonsingular terminal sliding-mode control for nonlinear magnetic bearing system. IEEE Trans Control Syst Technol 19(3):636–643
10. Yang J, Li S, Su J, Yu X (2013) Continuous nonsingular terminal sliding mode control for systems with mismatched disturbances. Automatica 49(7):2287–2291
11. Xu S, Chen C, Wu Z (2015) Study of nonsingular fast terminal sliding-mode fault-tolerant control. IEEE Trans Ind Electron 62(6):3906–3913

12. Li S, Wang X (2013) Finite-time consensus and collision avoidance control algorithms for multiple AUVs. Automatica 49(11):3359–3367
13. Yang Y, Hua C, Guan X (2014) Adaptive fuzzy finite-time coordination control for networked nonlinear bilateral teleoperation system. IEEE Trans Fuzzy Syst 22(3):631–641
14. Yang L, Yang J (2011) Nonsingular fast terminal sliding-mode control for nonlinear dynamical systems. Int J Robust Nonlinear Control 21(16):1865–1879
15. Beukers F (2007) Gauss' hypergeometric function. Prog Math 260:23–42
16. Abramowitz M, Stegun I (1972) Handbook of mathematical functions: with formulas, graphs, and mathematical tables. Dover, New York
17. Wang X, Li G, Li S, Song A (2016) Finite-time output feedback control for a pneumatic servo system. Trans Inst Meas Control 38(12):1520–1534
18. Han J (1998) Auto-disturbances-rejection controller and its applications. Control Dec 13(1):19–23
19. Zhang C, Yang J, Li S, Yang N (2016) A generalized active disturbance rejection control method for nonlinear uncertain systems subject to additive disturbance. Nonlinear Dyn 83(4):2361–2372
20. Shen Y, Shao K, Ren W, Liu Y (2016) Diving control of autonomous underwater vehicle based on improved active disturbance rejection control approach. Neurocomputing 173:1377–1385
21. Su Y, Zheng C, Duan B (2005) Automatic disturbances rejection controller for precise motion control of permanent-magnet synchronous motors. IEEE Trans Ind Electron 52(3):814–823
22. Zhu Z, Xu D, Liu J, Xia Y (2013) Missile guidance law based on extended state observer. IEEE Trans Ind Electron 60(12):5882–5891
23. Nuchkrua T, Leephakpreeda T (2013) Fuzzy self-tuning PID control of hydrogen-driven pneumatic artificial muscle actuator. J Bionic Eng 10(3):329–340
24. Yang H, You X, Xia Y, Li H (2014) Adaptive control for attitude synchronisation of spacecraft formation via extended state observer. IET Control Theory Appl 8(18):2171–2185
25. Guo B-Z, Wu Z-H, Zhou H-C (2016) Active disturbance rejection control approach to output-feedback stabilization of a class of uncertain nonlinear systems subject to stochastic disturbance. IEEE Trans Autom Control 61(6):1613–1618
26. Feng Y, Yu X, Man Z (2002) Non-singular terminal sliding mode control of rigid manipulators. Automatica 38(12):2159–2167

Chapter 18
Integral Sliding Mode Control

18.1 Introduction

Tracking control for the pneumatic rod cylinder is an important part in researches and applications of pneumatic servo systems [1]. However, it is difficult to achieve good performances for pneumatic rod cylinder servo systems mainly due to their inherent strong nonlinearity and low stiffness caused by various friction forces, compressibility of air, and nonlinearity of servo valves [2]. A generalized linear extended state observer has an advantage to estimate these uncertain factors [3]. Performance analysis of generalized linear extended state observer has been presented in handling fast-varying sinusoidal disturbances in [4]. It is also necessary to design an excellent controller in order to obtain good performances of complicated systems with a suitable observer. The sliding mode control method has been widely used in various complicated systems because of strong robustness and fast response, such as semi-Markovian jump systems with mismatched uncertainties [5, 6] and hybrid synchronization of identical chaotic systems [7]. In [8], an integral sliding mode (ISM) control method has been presented to get better performances than the normal sliding mode control method. The ISM control method has also been widely used in various complex systems, such as a chain of uncertain fractional order integrator [9] and second-order multi-agent systems with disturbances [10]. To the best of our knowledge, very few results are available on the generalized nonlinear extended state observer (GNESO)-based ISM finite-time tracking control for pneumatic servo systems with varying loads. This work is important and challenging in both theory and practice, which motivates us to carry on this research work.

In this chapter, an ISM tracking control method is investigated for a pneumatic rod cylinder servo system with varying loads based on the GNESO to improve control precise and response rapidity. The GNESO is designed to estimate the total disturbance and its derivative for the system. The ISM controller is given to obtain good control performances of the system. Moreover, both finite-time convergence of the GNESO and finite-time stabilization of the closed-loop system are realized using the

L. Zhao et al., *Pneumatic Servo Systems Analysis*, Advances in Industrial Control,
https://doi.org/10.1007/978-981-16-9515-5_18

Lyapunovąŕs stability theory. Lastly, experiments for tracking control are carried out to verify the validity of the proposed method.

Three contribution points of the paper are summarized as follows:

i A GNESO is proposed to remove the requirement of highly integrated mathematical model which is hardly obtained for a pneumatic cylinder in servo system.
ii An ISM controller is designed to obtain good control precision and high response rapidity through a combination of an estimation value of the GNESO.
iii Both finite-time convergence of the GNESO and finite-time stabilization of the closed-loop system are realized by employing the Lyapunov's stability theory.

18.2 Main Results

18.2.1 Schematic Diagram of Control Method

Considering inherent strong nonlinearity and low stiffness of the pneumatic rod cylinder servo system with varying loads, an ISM control method is designed to study the finite-time tracking problem via a GNESO. The schematic diagram of the proposed method is shown in Fig. 18.1. Moreover, in order to make this chapter express clearly, a lemma is shown in the following.

Lemma 18.1 ([11]) *Let* $\kappa_1, \ldots, \kappa_n > 0$ *be such that the polynomial* $s^n + \kappa_n s^{n-1} + \cdots + \kappa_2 s + \kappa_1$ *is Hurwitz and consider the system*

$$\begin{cases} \dot{\chi}_1(t) = \chi_2(t) \\ \quad \vdots \\ \dot{\chi}_{n-1}(t) = \chi_n(t) \\ \dot{\chi}_n(t) = o(t) \,. \end{cases} \tag{18.1}$$

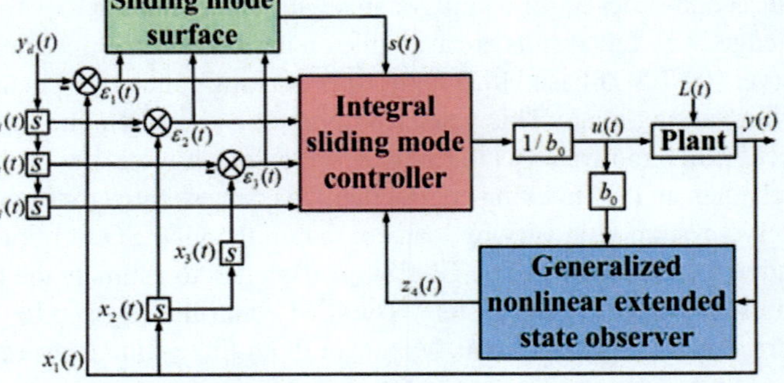

Fig. 18.1 Schematic diagram of the proposed method

There exists $\epsilon \in (0, 1)$ *such that, for every* $\zeta \in (1-\epsilon, 1)$, *the origin is a globally finite-time stable equilibrium for the system (18.1) under the feedback*

$$o(t) = -\kappa_1 sign^{\zeta_1}(\chi_1(t)) - \cdots - \kappa_n sign^{\zeta_n}(\chi_n(t)). \tag{18.2}$$

where $\zeta_1, \ldots, \zeta_n$ *satisfy*

$$\zeta_{i-1} = \frac{\zeta_i \zeta_{i+1}}{2\zeta_{i+1} - \zeta_i}, \; i = 2, \ldots, n, \tag{18.3}$$

with $\zeta_{n+1} = 1$ *and* $\zeta_n = \zeta$.

18.2.2 Generalized Nonlinear Extended State Observer

In the section, the GNESO is used to estimate various uncertainties and their derivatives for the pneumatic rod cylinder servo system with varying loads which is the same as (15.15) in addition to observe the state varyings. The generalized disturbance $L(t)$ is regarded as an extended state, i.e., $L(t) = x_4(t)$. Note that the generalized disturbance $L(t)$ is the nonlinear time-varying function. Moreover, $L(t)$ is continuously differentiable and bounded in practice, i.e., $|\ddot{L}(t)| \leq M$, where M is the positive constant, please refer to [2]. Then the pneumatic rod cylinder servo system (15.15) is rewritten as follows:

$$\begin{cases} \dot{x}_1(t) = x_2(t) \\ \dot{x}_2(t) = x_3(t) \\ \dot{x}_3(t) = x_4(t) + b_0 u(t) \\ \dot{x}_4(t) = x_5(t) \\ \dot{x}_5(t) = \ddot{L}(t), \end{cases} \tag{18.4}$$

where $x_5(t)$ is the derivative of $L(t)$. Considering Eq. (18.4), the GNESO is presented as follows:

$$\begin{cases} \dot{z}_1(t) = z_2(t) - \vartheta_1 \mathrm{sign}^{\frac{\alpha+1}{2}}(z_1(t) - x_1(t)) \\ \dot{z}_2(t) = z_3(t) - \vartheta_2 \mathrm{sign}^{\frac{\alpha+1}{2}}(z_1(t) - x_1(t)) \\ \dot{z}_3(t) = z_4(t) - \vartheta_3 \mathrm{sign}^{\frac{\alpha+1}{2}}(z_1(t) - x_1(t)) + b_0 u(t) \\ \dot{z}_4(t) = z_5(t) - \vartheta_4 \mathrm{sign}^{\frac{\alpha+1}{2}}(z_1(t) - x_1(t)) \\ \dot{z}_5(t) = -\vartheta_5 \mathrm{sign}^{\alpha}(z_1(t) - x_1(t)), \end{cases} \tag{18.5}$$

where α is the given parameter with $\alpha \in (0, 1)$, $z_i(t)$ is the observation value of $x_i(t)$, ϑ_i is the observation gain with $i = 1, \ldots, 5$. Let $e_i(t) = z_i(t) - x_i(t)$, where $e_i(t)$ is the observation error. According to systems (18.4) and (18.5), the observation error system is obtained as follows:

$$\begin{cases} \dot{e}_1(t) = e_2(t) - \vartheta_1 \text{sign}^{\frac{\alpha+1}{2}}(e_1(t)) \\ \dot{e}_2(t) = e_3(t) - \vartheta_2 \text{sign}^{\frac{\alpha+1}{2}}(e_1(t)) \\ \dot{e}_3(t) = e_4(t) - \vartheta_3 \text{sign}^{\frac{\alpha+1}{2}}(e_1(t)) \\ \dot{e}_4(t) = e_5(t) - \vartheta_4 \text{sign}^{\frac{\alpha+1}{2}}(e_1(t)) \\ \dot{e}_5(t) = -\ddot{L}(t) - \vartheta_5 \text{sign}^{\alpha}(e_1(t)). \end{cases} \tag{18.6}$$

Subsequently, both Lyapunov theory and finite-time lemmas are used to achieve the sufficiently small observation error in the following.

Theorem 18.2 *For the observation error system (18.6), there exist $\vartheta_1, \ldots, \vartheta_5 > 0$ and $0 < \alpha < 1$ such that the observation error is bounded in finite-time $t_r > 0$, i.e.,*

$$\|\xi(t)\|_2 \le \frac{JM}{\sigma_{\min}\{A_2\}\sigma_{\min}\{P\}},$$

where

$$\xi(t) = [sign^{(\alpha+1)/2}(e_1(t)), e_2(t), e_3(t), e_4(t), e_5(t)]^T, \; J = \sqrt{\vartheta_5^2 + 4},$$

$$A_2 = \begin{bmatrix} \vartheta_1 & -1 & 0 & 0 & 0 \\ \vartheta_2 & 0 & -1 & 0 & 0 \\ \vartheta_3 & 0 & 0 & -1 & 0 \\ \vartheta_4 & 0 & 0 & 0 & -1 \\ \vartheta_5 & 0 & 0 & 0 & 0 \end{bmatrix},$$

$$P = \begin{bmatrix} 2\vartheta_1/(\alpha+1) + \vartheta_2^2 + \vartheta_3^2 + \vartheta_4^2 + \vartheta_5^2 & -\vartheta_2 & -\vartheta_3 & -\vartheta_4 & -\vartheta_5 \\ -\vartheta_2 & 2 & 0 & 0 & 0 \\ -\vartheta_3 & 0 & 2 & 0 & 0 \\ -\vartheta_4 & 0 & 0 & 2 & 0 \\ -\vartheta_5 & 0 & 0 & 0 & 2 \end{bmatrix}.$$

Moreover, select the suitable parameters $\vartheta_1, \ldots, \vartheta_5$ and α to meet $JM \ll \sigma_{\min}\{A_2\}\sigma_{\min}\{P\}$, then $\|\xi(t)\|_2$ is limited to be sufficiently small in finite-time t_r.

Proof A Lyapunov function is designed as follows:

$$V(\xi(t)) = \xi^T(t)P\xi(t), \tag{18.7}$$

Considering Eq. (18.7), the following formula is get as

$$\begin{aligned}V(\xi(t)) = &\frac{2\vartheta_1}{\alpha+1}|e_1(t)|^{\alpha+1} + e_2^2(t) + e_3^2(t) + e_4^2(t) + e_5^2(t)\\ &+ \left[e_2(t) - \vartheta_2 \text{sign}^{\frac{\alpha+1}{2}}(e_1(t))\right]^2 + \left[e_3(t) - \vartheta_3 \text{sign}^{\frac{\alpha+1}{2}}(e_1(t))\right]^2\\ &+ \left[e_4(t) - \vartheta_4 \text{sign}^{\frac{\alpha+1}{2}}(e_1(t))\right]^2 + \left[e_5(t) - \vartheta_5 \text{sign}^{\frac{\alpha+1}{2}}(e_1(t))\right]^2\\ \geq\, &0 \end{aligned} \tag{18.8}$$

According to formula (18.8), when $e_1(t) \neq 0$, $V(\xi(t))$ is continuous and differentiable everywhere. Moreover, before reaching the origin, the observation error system (18.6) is not possible to stay on $e_1(t) = 0$. Therefore, $\dot{V}(\xi(t))$ is calculated in accordance with the conventional way when $e_1(t) \neq 0$. Then the following relationship is obtained as

$$\frac{d\left[\text{sign}^{(\alpha+1)/2}(e_1(t))\right]}{dt} = \frac{(\alpha+1)|e_1(t)|^{(\alpha-1)/2}\dot{e}_1(t)}{2}, \; e_1(t) \neq 0. \tag{18.9}$$

The derivative of $\xi(t)$ is shown as follows:

$$\begin{aligned}\dot{\xi}(t) &= \begin{bmatrix} \frac{\alpha+1}{2}|e_1(t)|^{\frac{\alpha-1}{2}}\left(e_2(t) - \vartheta_1 sig^{\frac{\alpha+1}{2}}(e_1(t))\right) \\ e_3(t) - \vartheta_2 \text{sign}^{\frac{\alpha+1}{2}}(e_1(t)) \\ e_4(t) - \vartheta_3 \text{sign}^{\frac{\alpha+1}{2}}(e_1(t)) \\ e_5(t) - \vartheta_4 \text{sign}^{\frac{\alpha+1}{2}}(e_1(t)) \\ -\vartheta_5 \text{sign}^{\alpha}(e_1(t)) - \ddot{L}(t) \end{bmatrix}\\ &= \begin{bmatrix} -r\mu\vartheta_1 & r\mu & 0 & 0 & 0 \\ -\vartheta_2 & 0 & 1 & 0 & 0 \\ -\vartheta_3 & 0 & 0 & 1 & 0 \\ -\vartheta_4 & 0 & 0 & 0 & 1 \\ -\mu\vartheta_5 & 0 & 0 & 0 & 0 \end{bmatrix} \xi(t) - \begin{bmatrix} 0 \\ 0 \\ 0 \\ 0 \\ 1 \end{bmatrix} \ddot{L}(t)\\ &= A\xi(t) - B\ddot{L}(t), \end{aligned} \tag{18.10}$$

where $1/2 < r = (\alpha+1)/2 < 1$, $\mu = |e_1(t)|^{(\alpha-1)/2} > 0$. The characteristic equation of matrix A is given as follows:

$$G(\lambda) = \lambda^5 + r\mu\vartheta_1\lambda^4 + r\mu\vartheta_2\lambda^3 + r\mu\vartheta_3\lambda^2 + r\mu\vartheta_4\lambda + r\mu^2\vartheta_5. \tag{18.11}$$

If there exist $\vartheta_i > 0$ and $e_1 \neq 0$, then all coefficients of $G(\lambda)$ is positive. This means that the matrix A is Hurwitz, i.e., A is stable. Therefore, there exists a corresponding symmetrical positive definite matrix $Q > 0$ such that P is the solution of Lyapunov equation. It follows that

$$A^T P + PA = -Q, \; Q > 0. \tag{18.12}$$

Considering Eq. (18.7), the following relationship is given as

$$\lambda_{\min}\{P\}\|\xi(t)\|_2^2 \leq V(\xi(t)) \leq \lambda_{\max}\{P\}\|\xi(t)\|_2^2, \tag{18.13}$$

where $\|\xi(t)\|_2^2 = |e_1(t)|^{\alpha+1} + e_2^2(t) + e_3^2(t) + e_4^2(t) + e_5^2(t)$. Then the inequality is written as follows:

$$\|\xi(t)\|_2 \geq |e_1(t)|^{\frac{\alpha+1}{2}}. \tag{18.14}$$

The derivative of Eq. (18.7) is obtained as follows:

$$\begin{aligned}\dot{V}(\xi(t)) &= \dot{\xi}^T(t)P\xi(t) + \xi(t)^T P\dot{\xi}(t)\\ &= (A\xi(t) - B\ddot{L}(t))^T P\xi(t) + \xi^T(t)P(A\xi(t) - B\ddot{L}(t))\\ &= \xi^T(t)(A^T P + PA)\xi(t) + 2\ddot{L}(t)\hat{B}^T\xi(t),\end{aligned} \tag{18.15}$$

where $\hat{B}^T = -B^T P = [\vartheta_5\ 0\ 0\ 0\ -2]$, then $J = \|\hat{B}\|_2 = \sqrt{\vartheta_5^2 + 4}$. According to Eqs. (18.12) and (18.15), it follows that

$$\begin{aligned}\dot{V}(\xi(t)) &= -\xi^T(t)Q\xi(t) + 2\ddot{L}(t)\hat{B}\xi(t)\\ &\leq -\lambda_{\min}\{Q\}\|\xi(t)\|_2^2 + 2JM\|\xi(t)\|_2\\ &= -(\lambda_{\min}\{Q\}\|\xi(t)\|_2 - 2JM)\|\xi(t)\|_2.\end{aligned} \tag{18.16}$$

Note that there exists $\sigma_{\min}\{Q\} = \lambda_{\min}\{Q\}$ for the symmetrical positive definite matrix Q. Considering Eq. (18.12), the following relationship is given as

$$\sigma_{\min}\{Q\} = 2\sigma_{\min}\{-AP\} \geq 2\sigma_{\min}\{-A\}\sigma_{\min}\{P\}. \tag{18.17}$$

Then A is rewritten as follows:

$$-A = \begin{bmatrix} r\mu & 0 & 0 & 0 & 0\\ 0 & 1 & 0 & 0 & 0\\ 0 & 0 & 1 & 0 & 0\\ 0 & 0 & 0 & 1 & 0\\ 0 & 0 & 0 & 0 & \mu \end{bmatrix} \begin{bmatrix} \vartheta_1 & -1 & 0 & 0 & 0\\ \vartheta_2 & 0 & -1 & 0 & 0\\ \vartheta_3 & 0 & 0 & -1 & 0\\ \vartheta_4 & 0 & 0 & 0 & -1\\ \vartheta_5 & 0 & 0 & 0 & 0 \end{bmatrix} = A_1 A_2 \tag{18.18}$$

One has that

$$\sigma_{\min}\{-A\} = \sigma_{\min}\{A_1 A_2\} \geq \sigma_{\min}\{A_1\}\sigma_{\min}\{A_2\}. \tag{18.19}$$

Since A_1 is a diagonal matrix and $r\mu < \mu$, there exists

$$\sigma_{\min}\{A_1\} = \begin{cases} 1, & |e_1(t)| < \left(\dfrac{2}{\alpha+1}\right)^{\frac{2}{\alpha-1}} \\ r\mu, & |e_1(t)| \geq \left(\dfrac{2}{\alpha+1}\right)^{\frac{2}{\alpha-1}} \end{cases}. \tag{18.20}$$

According to inequality (18.14), when $|e_1(t)| \geq \left(\frac{2}{\alpha+1}\right)^{\frac{2}{\alpha-1}}$, then $\|\xi(t)\|_2 \geq \left(\frac{2}{\alpha+1}\right)^{\frac{\alpha+1}{\alpha-1}}$. The relationship (18.17) is rewritten as follows:

$$\lambda_{\min}\{Q\} = \sigma_{\min}\{Q\} \geq 2r\mu\sigma_{\min}\{A_2\}\sigma_{\min}\{P\} \tag{18.21}$$

Considering expressions (18.14), (18.16) and (18.21), it follows that

$$\begin{aligned} \eta_1 &= \lambda_{\min}\{Q\}\|\xi(t)\|_2 - 2JM \\ &\geq 2r\mu\sigma_{\min}\{A_2\}\sigma_{\min}\{P\}|e_1(t)|^{\frac{\alpha+1}{2}} - 2JM \\ &= (\alpha+1)\sigma_{\min}\{A_2\}\sigma_{\min}\{P\}|e_1(t)|^{\alpha} - 2JM \\ &\geq (\alpha+1)\left(\frac{2}{\alpha+1}\right)^{\frac{2\alpha}{\alpha-1}}\sigma_{\min}\{A_2\}\sigma_{\min}\{P\} - 2JM. \end{aligned} \tag{18.22}$$

Since α is the given parameter with $\alpha \in (0, 1)$, there exists a minimum value Υ such that

$$(\alpha+1)\left(\frac{2}{\alpha+1}\right)^{\frac{2\alpha}{\alpha-1}} \geq \Upsilon. \tag{18.23}$$

Therefore, inequality (18.22) is rewritten as follows:

$$\eta_1 \geq \Upsilon\sigma_{\min}\{A_2\}\sigma_{\min}\{P\} - 2JM = \eta_{1,\min}. \tag{18.24}$$

The relationship $\Upsilon\sigma_{\min}\{A_2\}\sigma_{\min}\{P\} > 2JM$ is established by selecting the suitable parameters $\vartheta_1, \ldots, \vartheta_5$, i.e., $\dot{V}(\xi(t)) < 0$. For expressions (18.16) and (18.22), the following inequality is given as

$$\dot{V}(\xi(t)) \leq -\eta_1\|\xi(t)\|_2 \leq -\frac{\eta_{1,\min}}{\sqrt{\lambda_{\max}\{P\}}}V^{\frac{1}{2}}(\xi(t)) = -C_1V^{\frac{1}{2}}(\xi(t)). \tag{18.25}$$

According to Lemma 17.1 in this chapter, $\|\xi(t)\|_2 < \left(\frac{2}{\alpha+1}\right)^{\frac{\alpha+1}{\alpha-1}}$ is obtained in finite-time t_{r1}. Meanwhile, t_{r1} is expressed as follows:

$$t_{r1} \leq \frac{1}{C_1/2}V^{\frac{1}{2}}(\xi(t)) \leq \frac{1}{C_1/2}V^{\frac{1}{2}}(\xi(t_0)). \tag{18.26}$$

Considering inequality (18.14), when $\|\xi(t)\|_2 < \left(\frac{2}{\alpha+1}\right)^{\frac{\alpha+1}{\alpha-1}}$, then $|e_1(t)| < \left(\frac{2}{\alpha+1}\right)^{\frac{2}{\alpha-1}}$. One has that

$$\lambda_{\min}\{Q\} = \sigma_{\min}\{Q\} \geq 2\sigma_{\min}\{A_2\}\sigma_{\min}\{P\}. \tag{18.27}$$

For expressions (18.14), (18.16) and (18.27), it follows that

$$\eta_2 = \lambda_{\min}\{Q\}\|\xi(t)\|_2 - 2JM \geq 2\sigma_{\min}\{A_2\}\sigma_{\min}\{P\}\|\xi(t)\|_2 - 2JM. \tag{18.28}$$

If the relationship

$$\left(\frac{2}{\alpha+1}\right)^{\frac{\alpha+1}{\alpha-1}} > \|\xi(t)\|_2 > \frac{JM}{\sigma_{\min}\{A_2\}\sigma_{\min}\{P\}} = W \tag{18.29}$$

is satisfied, then $\eta_2 > 2\sigma_{\min}\{A_2\}\sigma_{\min}\{P\}W - 2JM = 0$ and $\dot{V}(\xi(t)) < 0$. Similar to expressions (18.25) and (18.26), there exist

$$\dot{V}(\xi(t)) \leq -\eta_2\|\xi(t)\|_2 \leq -\frac{\eta_2}{\sqrt{\lambda_{\max}\{P\}}}V^{\frac{1}{2}}(\xi(t)) = -C_2V^{\frac{1}{2}}(\xi(t)) \tag{18.30}$$

and

$$t_{r2} \leq \frac{1}{C_2/2}V^{\frac{1}{2}}(\xi(t)) \leq \frac{1}{C_2/2}V^{\frac{1}{2}}(\xi(t_1)). \tag{18.31}$$

Therefore, the following inequality is given in finite time $t_r = t_{r1} + t_{r2}$ as

$$\|\xi(t)\|_2 \leq \frac{JM}{\sigma_{\min}\{A_2\}\sigma_{\min}\{P\}}. \tag{18.32}$$

Obviously, sufficiently small observation error is obtained in finite time by selecting suitable parameters $\vartheta_1, \ldots, \vartheta_5$, and α to make $\sigma_{\min}\{A_2\}\sigma_{\min}\{P\}$ large enough. This completes the proof.

18.2.3 Integral Sliding Mode Controller

In order to obtain good performances of the pneumatic rod cylinder servo system with varying loads, an ISM controller is designed in the following. The tracking error is presented as follows:

$$\begin{cases} \varepsilon_1(t) = x_1(t) - y_d(t) \\ \varepsilon_2(t) = x_2(t) - \dot{y}_d(t) \\ \varepsilon_3(t) = x_3(t) - \ddot{y}_d(t), \end{cases} \tag{18.33}$$

where $y_d(t)$ is the reference signal. Considering systems (18.4) and (18.33), the following tracking error system is given as

$$\begin{cases} \dot{\varepsilon}_1(t) = \varepsilon_2(t) \\ \dot{\varepsilon}_2(t) = \varepsilon_3(t) \\ \dot{\varepsilon}_3(t) = x_4(t) + b_0 u(t) - \dddot{y}_d(t). \end{cases} \tag{18.34}$$

According to expression (18.33), the sliding mode surface $s(t)$ of controller is designed as follows:

$$\begin{aligned} s(t) = \varepsilon_3(t) + \int_0^t & (\kappa_1 \text{sign}^{\delta_1}(\varepsilon_1(\tau)) s(t) + \kappa_2 \text{sign}^{\delta_2}(\varepsilon_2(\tau)) \\ & + \kappa_3 \text{sign}^{\delta_3}(\varepsilon_3(\tau))) d\tau, \end{aligned} \tag{18.35}$$

where $\kappa_i > 0$ and $\delta_i > 0$ are the adjustable parameters with $i = 1, 2, 3$. Moreover, there exists κ_i such that the polynomial $s^3 + \kappa_3 s^2 + \kappa_2 s + \kappa_1$ is Hurwitz, where s is the differential operator. δ_i is selected according to Eq. (18.3). Considering expressions (18.34) and (18.35), the derivative of $s(t)$ is gotten as follows:

$$\begin{aligned} \dot{s}(t) &= \dot{\varepsilon}_3(t) + \kappa_1 \text{sign}^{\delta_1}(\varepsilon_1(t)) + \kappa_2 \text{sign}^{\delta_2}(\varepsilon_2(t)) + \kappa_3 \text{sign}^{\delta_3}(\varepsilon_3(t)) \\ &= x_4(t) + b_0 u(t) - \dddot{y}_d(t) + \kappa_1 \text{sign}^{\delta_1}(\varepsilon_1(t)) \\ &\quad + \kappa_2 \text{sign}^{\delta_2}(\varepsilon_2(t)) + \kappa_3 \text{sign}^{\delta_3}(\varepsilon_3(t)). \end{aligned} \tag{18.36}$$

Then the following ISM controller is designed as

$$\begin{aligned} u(t) = -\frac{1}{b_0} & (z_4(t) - \dddot{y}_d(t) + \kappa_1 \text{sign}^{\delta_1}(\varepsilon_1(t)) + \kappa_2 \text{sign}^{\delta_2}(\varepsilon_2(t)) \\ & + \kappa_3 \text{sign}^{\delta_3}(\varepsilon_3(t)) + \kappa_4 s(t) + \kappa_5 \text{sign}^{\varphi}(s(t))), \end{aligned} \tag{18.37}$$

where $\kappa_4 = \kappa_l + \kappa_d > 0$, $\kappa_5 > 0$, and $0 < \varphi < 1$ are adjustable parameters of the controller. Subsequently, finite-time stabilization of the closed-loop system (18.34) is proven by Lyapunov theory in the following.

Theorem 18.3 *Consider the closed-loop system (18.34) with the ISM controller (18.37). When the sliding mode surface is selected as (18.35), and by choosing suitable parameter $\kappa_4 = \kappa_l + \kappa_d$ such that $(x_4(t) - z_4(t))s(t) \leq \kappa_d s^2(t)$, then tracking error of the closed-loop system (18.34) converges to zero in finite time.*

Proof According to the closed-loop system (18.34), the following Lyapunov function is given as

$$V_1(s(t)) = \frac{1}{2}s^2(t). \tag{18.38}$$

Differentiating Eq. (18.38), it follows that

$$\begin{aligned}\dot{V}_1(s(t)) &= s(t)\dot{s}(t)\\ &= s(t)(x_4(t) + b_0u(t) - \dddot{y}_d(t) + \kappa_1\text{sign}^{\delta_1}(\varepsilon_1(t))\\ &\quad+\kappa_2\text{sign}^{\delta_2}(\varepsilon_2(t)) + \kappa_3\text{sign}^{\delta_3}(\varepsilon_3(t)).\end{aligned} \tag{18.39}$$

Considering Eqs. (18.37)–(18.39), when $(x_4(t) - z_4(t))s(t) \leq \kappa_d s^2(t)$, the following relationship is obtained as

$$\begin{aligned}\dot{V}_1(s(t)) &= s(t)(x_4(t) - z_4(t) - \kappa_4 s(t) - \kappa_5\text{sign}^{\varphi}(s(t)))\\ &= (x_4(t) - z_4(t))s(t) - \kappa_4 s^2(t) - \kappa_5\text{sign}^{\varphi}(s(t))s(t)\\ &\leq -\kappa_l s^2(t) - \kappa_5\text{sign}(s(t))|s(t)|^{\varphi}s(t)\\ &= -\kappa_l s^2(t) - \kappa_5|s(t)|^{\varphi+1}\\ &= -2\kappa_l V_1(s(t)) - 2^{\frac{\varphi+1}{2}}\kappa_5 V_1^{\frac{\varphi+1}{2}}(s(t)).\end{aligned} \tag{18.40}$$

According to Lemma 17.2 and relationship (18.40), the sliding mode surface $s(t)$ and its derivative $\dot{s}(t)$ converge to zero in finite-time t_s. Then Eq. (18.36) is expressed as follows:

$$\dot{\varepsilon}_3(t) = -\kappa_1\text{sign}^{\delta_1}(\varepsilon_1(t)) - \kappa_2\text{sign}^{\delta_2}(\varepsilon_2(t)) - \kappa_3\text{sign}^{\delta_3}(\varepsilon_3(t)). \tag{18.41}$$

Therefore, the tracking error system (18.34) is rewritten as follows:

$$\begin{cases}\dot{\varepsilon}_1(t) = \varepsilon_2(t)\\ \dot{\varepsilon}_2(t) = \varepsilon_3(t)\\ \dot{\varepsilon}_3(t) = -\kappa_1\text{sign}^{\delta_1}(\varepsilon_1(t)) - \kappa_2\text{sign}^{\delta_2}(\varepsilon_2(t)) - \kappa_3\text{sign}^{\delta_3}(\varepsilon_3(t)).\end{cases} \tag{18.42}$$

According to Lemma 18.1, $\varepsilon_1(t)$, $\varepsilon_2(t)$, and $\varepsilon_3(t)$ converge to equilibrium point in finite-time t_s, which means that tracking error of the closed-loop system (18.34) converges to zero in finite time.

18.3 Experiments and Results

In this section, experimental results are presented to demonstrate the effectiveness of the proposed method. Adjustable parameters of the proposed method are listed in Table 18.1. Experimental results of the proposed method and the ISM control

Table 18.1 Adjustable parameters of the proposed method

Parameters

Symbol	Value	Symbol	Value	Symbol	Value	Symbol	Value
ϑ_1	420	κ_1	9.4	δ_1	2/11	ϑ_5	1000
ϑ_2	8500	κ_2	6	δ_2	1/4	κ_5	0.01
ϑ_3	16000	κ_3	1.1	δ_3	0.4	b_0	60
ϑ_4	1500	κ_4	0.01	φ	0.3	α	1/2

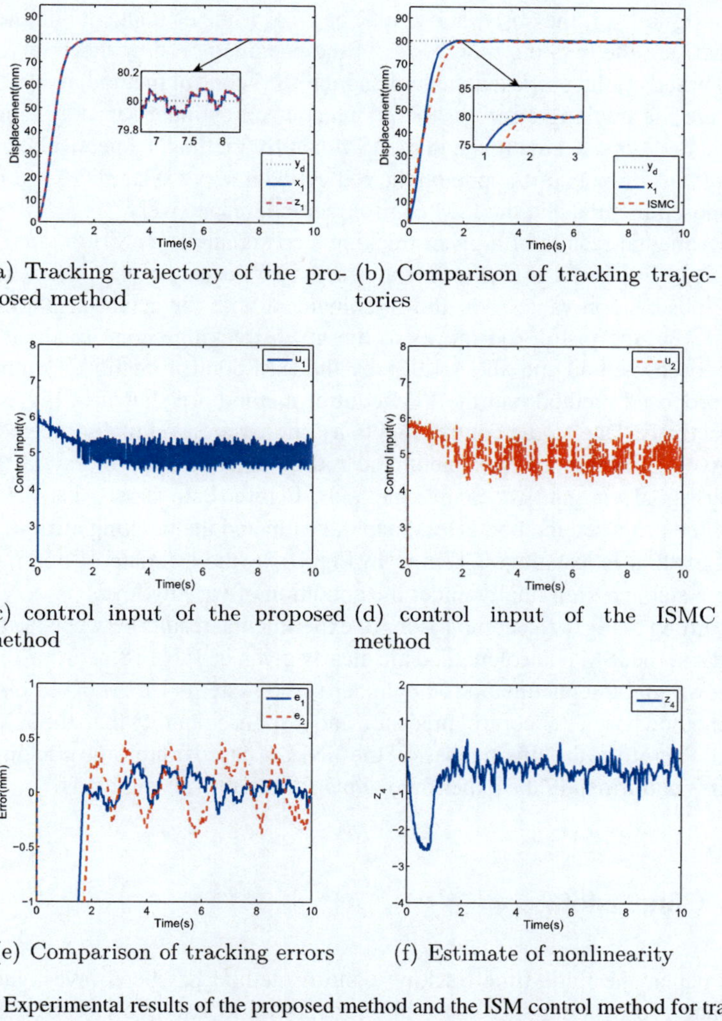

(a) Tracking trajectory of the proposed method
(b) Comparison of tracking trajectories
(c) control input of the proposed method
(d) control input of the ISMC method
(e) Comparison of tracking errors
(f) Estimate of nonlinearity

Fig. 18.2 Experimental results of the proposed method and the ISM control method for tracking a step signal

Table 18.2 Comparison of experimental results

Controller	Step signal	
	Convergence time (s)	Tracking error (mm)
ISM control method	1.95	0.47
The proposed method	1.7	0.25

method for tracking a step signal are shown in Fig. 18.2. Moreover, y_d is the given tracking signal, z_1 is the observation value of x_1, z_4 is the estimate of nonlinearity, x_1 and ISMC are the tracking trajectories of the pneumatic rod cylinder servo system (15.15) based on the proposed method and the ISM control method, respectively. e_1 and e_2 are the tracking errors of the pneumatic rod cylinder servo system (15.15) based on the proposed method and the ISM control method, respectively. u_1 and u_2 are the control inputs of the pneumatic rod cylinder servo system (15.15) based on the proposed method and the ISM control method, respectively.

Experimental results of a given tracking step signal set as 80 mm are shown in Fig. 18.2. In Fig. 18.2a, it is shown from tracking trajectory of the proposed method that the observation value z_1 is almost coincident with the actual displacement x_1. In Fig. 18.2b, the piston rod moves to the given tracking signal at about 1.7 s by the proposed method and about 1.95 s by the ISM control method. Control inputs of the proposed method and the ISM control method are shown in Fig. 18.2c and d, respectively. Due to the control inputs u_1 and u_2 are eventually located in their respective stable regions, the pneumatic rod cylinder is able to stay at the given tracking signal via the two control methods. Figure 18.2e shows that the tracking error of the proposed method is less than 0.25 mm and the tracking error of the ISM control method is less than 0.47 mm. In Fig. 18.2f, the estimate of nonlinearity z_4 reaches a stable region finally under the condition of varying loads.

According to Fig. 18.2, comparisons of experimental results between the proposed method and the ISM control method are clearly given in Table 18.2. From Table 18.2, it is known that the pneumatic rod cylinder servo system (15.15) based on the proposed method has better control precision and response rapidity than the ISM control method. Therefore, the effectiveness of the GNESO based finite-time tracking control method is confirmed in the experimental platform with varying loads.

18.4 Conclusion

In this paper, the finite-time tracking control method has been investigated for a pneumatic rod cylinder servo system with varying loads via a GNESO. Both various uncertainties and their derivatives of the system have been estimated by the observer. An ISM controller has been designed to ensure good performances for the closed-loop system by combining with estimation value of the observer. Moreover, the corresponding theoretical analyses on sufficiently small observation error and stabilization of the closed-loop system have also been proved in finite time, respectively. Finally, it is shown from experimental results that the proposed method has better performances for the pneumatic rod cylinder servo system with varying loads compared with ISM control method.

References

1. Taheri B, Case D, Richer E (2014) Force and stiffness backstepping-sliding mode controller for pneumatic cylinders. IEEE-ASME Trans Mechatron 19(6):1799–1809
2. Tang Z, Ge SS, Lee KP, He W (2016) Adaptive neural control for an uncertain robotic manipulator with joint space constraints. Int J Control 89(7):1428–1446
3. Nuchkrua T, Leephakpreeda T (2013) Fuzzy self-tuning PID control of hydrogen-driven pneumatic artificial muscle actuator. J Bionic Eng 10(3):329–340
4. Godbole AA, Kolhe JP, Talole SE (2013) Performance analysis of generalized extended state observer in tackling sinusoidal disturbances. IEEE Trans Control Syst Technol 21(6):2212–2223
5. Haroun R, Aroudi A, Cid-Pastor A, Garcia G (2015) Impedance matching in photovoltaic systems using cascaded boost converters and sliding-mode control. IEEE Trans Power Electron 30(6):3185–3199
6. Li F, Shi P, Wu L (2015) State estimation and sliding mode control for semi-markovian jump systems. Automatica 51:385–393
7. Vaidyanathan S, Azar A (2015) Hybrid synchronization of identical chaotic systems using sliding mode control and an application to vaidyanathan chaotic systems. IEEE Trans Med Imag 17(1):549–569
8. Xue W, Bai W, Yang S, Song K, Huang Y, Xie H (2015) ADRC with adaptive extended state observer and its application to air-fuel ratio control in gasoline engines. IEEE Trans Ind Electron 62(9):5847–5857
9. Kamal S, Raman A, Bandyopadhyay B (2012) Finite-time stabilization of fractional order uncertain chain of integrator: an integral sliding mode approach. IEEE Trans Autom Control 58(6):1597–1602
10. Yu S, Long X (2015) Finite-time consensus for second-order multi-agent systems with disturbances by integral sliding mode. Automatica 54:158–165
11. Qiu Y, Hu J, Lu H, Liu L (2016) Finite time motion control for a launching platform based on sliding mode disturbance observer. Adv Mech Eng 8(10):1–10

Part V
Pneumatic Right Angle Composite Motion System

Chapter 19
Platform Introduction

19.1 Application Background

With the advantages of clean and pollution-free, rodless cylinders have been widely used in microelectronics industry, automobile industry, robot automation, and other industries. In the microelectronics industry with high requirements of cleanliness, the rodless cylinder can be used to meet needs of dust-free applications, realize transmission of car accessories into different positions, and achieve accurate position. In the automotive industry, when an engine is moved from a load bench to a test bench, the rodless cylinder is used to solve the problem of space limited, which can greatly save labor and improve production efficiency. When the rodless cylinder is used in the ultrasonic welding machine, the machine pressure becomes stable, and control accuracy and production efficiency are both improved. In silk-screen printing, silica gel is used instead of ink. The problem is that silica gel is viscous and must be used in a very clean environment, and driving force of the printing head must be large, which can be solved well by an application of the rodless cylinder. In these applications, different requirements are put forward on the control of position, output force, and moving speed for the rodless cylinder. In order to make the rodless cylinder get better application in these industries, researches on the servo control technology of the rodless cylinder are particularly important. Therefore, the research and control of the pneumatic right angle composite motion system is an important topic.

19.2 Platform Structure

An experiment platform for pneumatic right angle composite motion system is depicted in Fig. 19.1, and it is an pneumatic right angle composite motion platform. In the experiment platform, important mechanism parts are a pneumatic manipulator and 12 small cylinders. The pneumatic manipulator consists of a cylinder with vacuum chuck (SMC, ZCDUKD20-40D) in Z-axis and two rodless pneumatic

L. Zhao et al., *Pneumatic Servo Systems Analysis*, Advances in Industrial Control,
https://doi.org/10.1007/978-981-16-9515-5_19

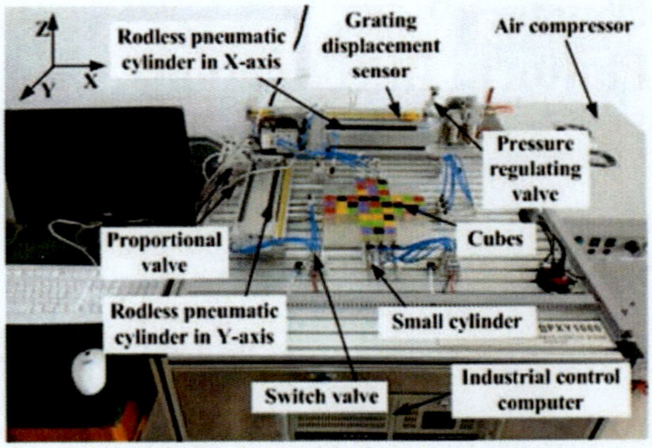

Fig. 19.1 Pneumatic right angle composite motion platform

cylinders (SMC, CY1H20-300-Y7BWS) with the stroke of 300 mm in X-axis and Y-axis, respectively. Moreover, each rodless pneumatic cylinder equipped with displacement sensor is driven by the proportional valve (Festo, MPYE-5-M5-010-B). Besides the above components, there are an industrial control computer (Advantech, 610H), a counting card (Advantech, PCL-833), and a D/A card (Advantech, PCL-726). The operated objects are 45 cubes with 5 different colors. Each cube is pushed by a small cylinder (SMC, CJ2B10-45). In order to grasp the cubes accurately, high-precision position control of the rodless cylinder is a key for the pneumatic right angle composite motion system. Therefore, the servo position control of the single rodless pneumatic cylinder is studied in this chapter, and an experimental platform with rodless pneumatic cylinder is shown in Fig. 19.1.

In the study of the pneumatic right angle composite motion system, main contents include three parts: image identification, path planning, and position control of rodless pneumatic cylinder. The hardware composition and working principle of the experiment platform are introduced in this chapter from the following three aspects: platform components, an electrical structure, and a pneumatic structure.

19.2.1 Platform Components

As the main body of the experimental platform, the pneumatic manipulator is shown in Fig. 19.2. The pneumatic manipulator is mainly composed of two magnetic rodless cylinders (SMC, CY1H20-300-Y7BWS, stroke 300 mm, bore 20 mm) in X-axis and Y-axis, and a vacuum cylinder (SMC, ZCDUKD20-40D) in Z-axis. The two magnetic rodless cylinders are controlled by two proportional directional control valves (FESTO, MPYE-5-M5-010-B, input range 0–10 V), which are used to move the vacuum cylinder to the position above cubes in the platform. Furthermore, the vacuum cylinder on the Z-axis is used to flexibly extend or retract and absorb the cube to complete the grasping action. So the pneumatic manipulator has three degrees of freedom. The structural design drawing is shown in Fig. 19.2.

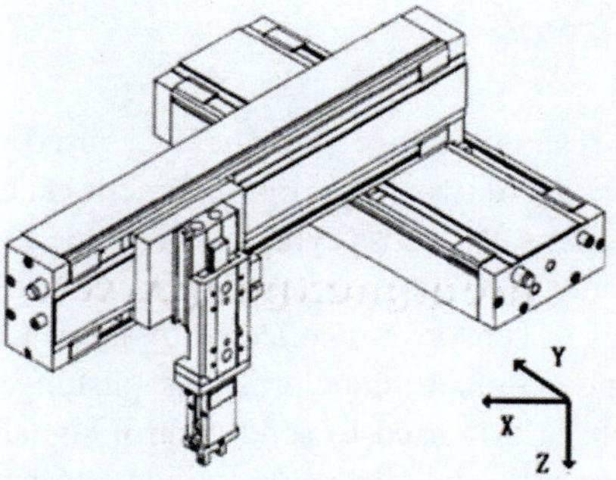

Fig. 19.2 The structural design drawing of the pneumatic manipulator

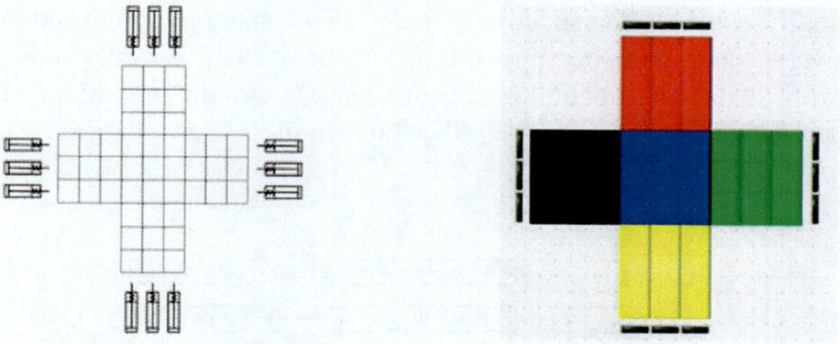

Fig. 19.3 The sketch map of the cubes

There are 45 cubes on the platform with 5 colors, and there are 9 cubes in each color, as shown in Fig. 19.3. There are 12 small telescopic cylinders on the four sides of the cubes, which is used to push the corresponding cubes. The telescopic cylinders on each side are controlled by integrated valves(SMC, 3-SY3000). The pneumatic manipulator can capture cubes at any position on the platform. When a cube is captured, a vacancy exists in the position of the cube. The small telescopic cylinders will push corresponding cubes, and a new vacancy will appear at the same time. The captured cube will be placed in the new vacancy.

The motion steps of the pneumatic manipulator is shown as follows:

i X-axis and Y-axis rodless cylinder move to the position above the selected cube.
ii The vacuum cylinder drives the vacuum chuck to move down and the vacuum generator generates vacuum. Then the vacuum chuck sucks up a cube and leave a vacancy.
iii The vacuum cylinder drives vacuum chuck upward, and a telescopic cylinder pushes the whole row of cubes to create a new vacancy.
iv X-axis and Y-axis rodless cylinder move to the position above new vacancy. Put down the selected cube.
v X-axis and Y-axis rodless cylinder move to the position above another selected cube. Return to step ii.

19.2.2 Control Circuit

The electrical structure is shown in Fig. 19.4. The IPC is used as a hardware controller. It sends two voltage signals to two proportional directional control valves. Pressure difference for two chambers of rodless cylinder drives the piston move as expected. The displacement sensors (ZK-200) are installed to measure displacements of the rodless cylinders. Dates of sensors are collected by the IPC equipped with an data acquisition card. During actions of cube grasping, pushing, and putting down, the PCL-726 digital output card is used to send control signal for the motion of the cylinders, vacuum generators, and telescopic cylinders. A camera is used to capture the cube image to the IPC, which is used to supply position information in the system.

The pneumatic right angle composite motion system is mainly composed by air supply, control valves, and executive cylinders. The air supply produced by the air compressor is the power source of the pneumatic system. In order to ensure the normal work of the pneumatic system, the compressed air must be processed as cooled, purified, and stabilized. The pneumatic circuit is shown in Fig. 19.5.

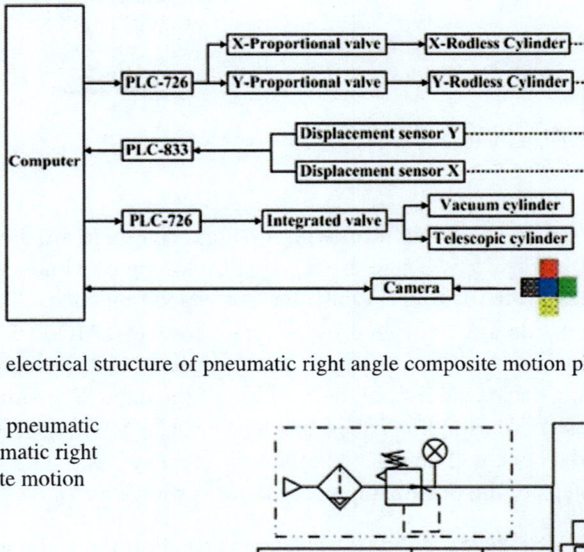

Fig. 19.4 The electrical structure of pneumatic right angle composite motion platform

Fig. 19.5 The pneumatic circuit of pneumatic right angle composite motion platform

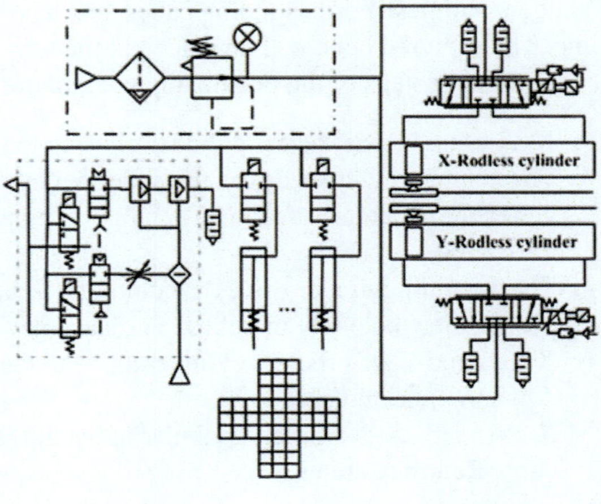

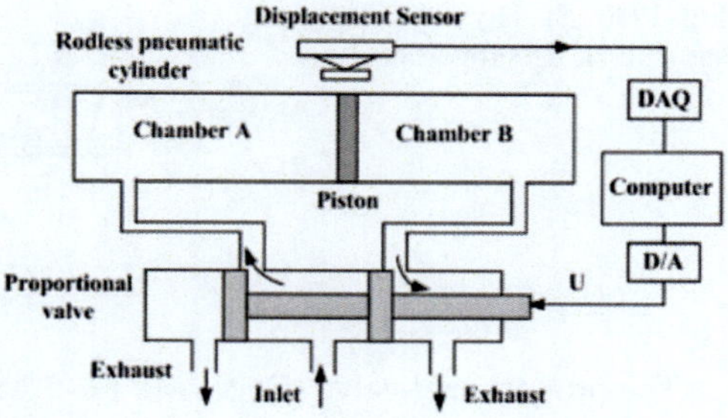

Fig. 19.6 The hardware block diagram of pneumatic rodless cylinder system

Fig. 19.7 The physical structures of CY1H series cylinder

A single rodless cylinder pneumatic control loop is shown in Fig. 19.6. The rodless cylinder is SMC CY1H-type rodless cylinder with the stroke of 300 mm and the bore of 20 mm. The proportional valve is FESTO MPYE-type electric proportional flow valve. The control input signal is the voltage signal with the range of 0–10 V, and +5 V is the valve mid-position. The displacement sensor is ZK type grating displacement sensor with the precision of ±0.005 mm. The PCL-833 receives two quadrature coded square wave signals from the displacement sensor, and the feedback of the position signal is realized to form a closed-loop control system.

CY1H series cylinder is a type of rodless cylinder produced by SMC. Characteristics of the cylinder include high precision guide bearing, which ensures high linearity and repeatability of the cylinder, no air leakage, centralized air supply, and hydraulic bumper. Physical structures of CY1H series cylinder are shown in Fig. 19.7. The rodless cylinders (SMC, CY1H20-300-Y7BWS) with the stroke of 300 mm are used in the experiment platform. The parameters of the CY1H series cylinder are shown in Table 19.1.

Table 19.1 Parameters of the CY1H series cylinder

Bore size (mm)	20
Fluid	Air
Action	Double acting
Maximum operating pressure (Mpa)	0.7
Minimum operating pressure (Mpa)	0.2
Piping	Gentralized piping type
Piping port size	$M5 \times 0.8$

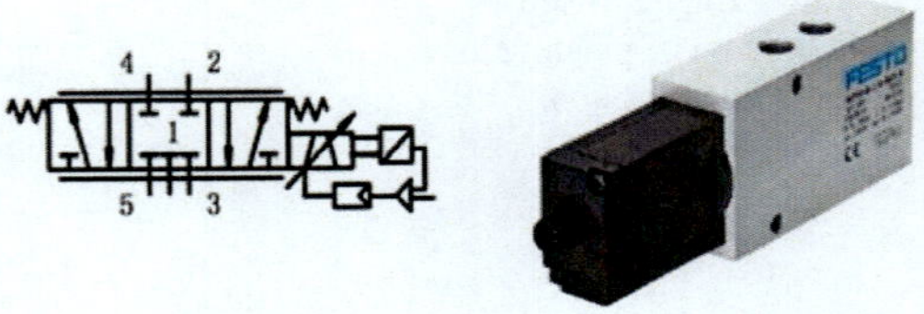

Fig. 19.8 The physical structure of the proportional valve

The proportional valve (Festo, MPYE-5-M5-010-B) used in this pneumatic system is shown in Fig. 19.8. The speed of the cylinder is controlled by the flow of compressed air into the cylinder chamber. The main technical parameters of the proportional valve are shown in Table 19.2. The characteristic curve of the setpoint voltage U with output flow q is shown in Fig. 19.9.

The displacement sensor (ZK-200) with the precision of ±0.005 mm is installed to measure the displacement of the rodless cylinder. The physical structure of the displacement sensor is shown in Fig. 19.10. The main technical parameters of the displacement sensor are shown in Table 19.3. The output signal of the sensor is two quadrature coded square wave signals with amplitude 5 V, phase angle 90°.

Table 19.2 Parameters of the MPYE proportional valve

Valve function	5/3-way
Flow rate	100–2000 L/min
Pressure	0–10 bar
Setpoint value (Voltage type)	0–10 V
Valve mid-position (Voltage type)	5 V
Minimum operating pressure	0.2 Mpa
Power consumption	24 V
Critical frequency	125 Hz
Max. hysteresis	0.4%

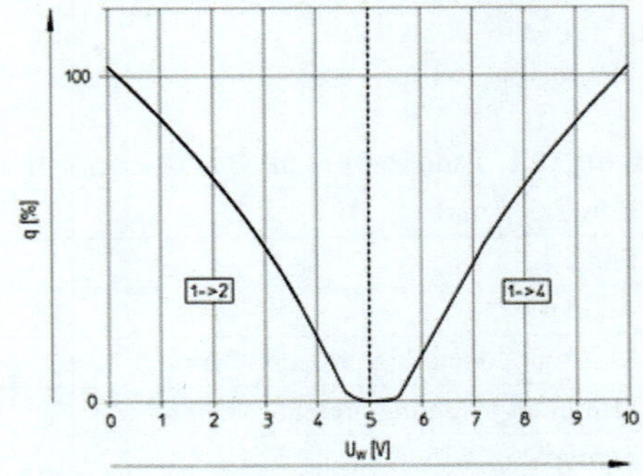

Fig. 19.9 The characteristic curve of the setpoint voltage-output flow of the proportional valve

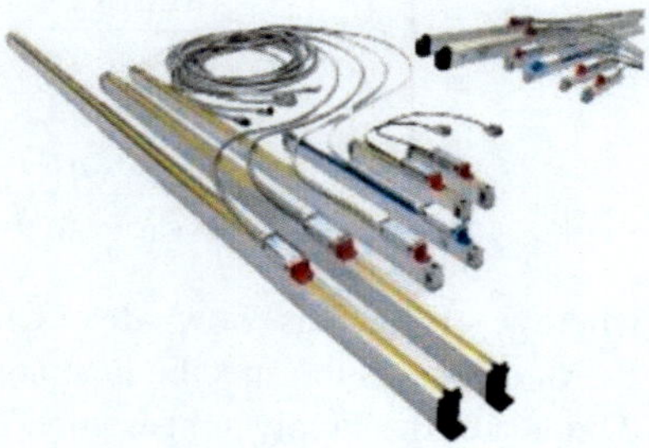

Fig. 19.10 The physical structure of the displacement sensor

Table 19.3 Parameters of the displacement sensor

Grid pitch (mm)	Range (m)	Accuracy (μm)	Working temperature (°C)
0.02	30	±5	0–45

19.3 System Model

19.3.1 Rodless Cylinder Model

An accurate mathematical model of the pneumatic rodless cylinder system is needed to achieve well control performances for the pneumatic right angle composite motion platform. The piston in rodless cylinder is driven by differential pressures between Chamber A and Chamber B. According to [1], the following expression is obtained as

$$Ma(t) = F_w(t) + A(p_1(t) - p_2(t)), \tag{19.1}$$

where M describes the mass of piston and load, $a(t)$ represents the acceleration of piston, $F(t)$ stands for the nonlinearity part mainly including friction of rodless cylinder, A denotes piston area, $p_1(t)$ and $p_2(t)$ are pressures of Chamber A and Chamber B, respectively. Equation (19.1) is rewritten as follows:

$$\begin{cases} \dot{x}_1(t) = x_2(t) \\ \dot{x}_2(t) = F_w(t)/M + A(p_1(t) - p_2(t))/M, \end{cases} \tag{19.2}$$

where $x_1(t)$ and $x_2(t)$ represent the position and velocity of piston, respectively.

Based on the first law of thermodynamics and results in [1], the thermodynamic model is obtained as

$$\begin{cases} \dot{p}_1(t) = \dfrac{K_0 RT}{A(x_0 + x_1(t))} q_{m1} - \dfrac{K_0 p_1(t) x_2(t)}{x_0 + x_1(t)} \\ \dot{p}_2(t) = \dfrac{K_0 RT}{A(L_0 - x_0 - x_1(t))} q_{m2} + \dfrac{K_0 p_2 x_2(t)}{L_0 - x_0 - x_1(t)}, \end{cases} \tag{19.3}$$

where q_{m1} is the mass flow rate of Chamber A, q_{m2} is the mass flow rate of Chamber B, K_0 describes the specific heat constant, R stands for the universal gas constant, T indicates the supply temperature, L_0 represents the piston stroke with the value of 0.3 m, and x_0 is the starting position of piston. Parameters in this subsection are set $M = 2\,\text{kg}$, $A = 3.14159 \times 10^{-4}\,\text{m}^2$, $K_0 = 1.4$, $R = 287\,\text{N}\cdot\text{m/(kg}\cdot\text{K)}$, $T = 300\,\text{K}$, and $x_0 = 0.001\,\text{m}$.

19.3.2 Five-Way Proportional Valve Model

It is hard to get a model of the proportional valve because of complex mechanical structures and pressure changes. In [2], the five-way proportional valve model has been given as follows:

$$q_{mi}(t) = A(u(t))C_1(p_1(t), p_2(t))C_2(p_1(t), p_2(t))C_3 p_2/\sqrt{T_i}, \tag{19.4}$$

where $i = 1, 2$, $A(u(t))$ represents the effective valve orifice area, $C_1(p_1(t), p_2(t))$ describes a subsection function, $p_1(t)$ is the upstream pressure, $p_2(t)$ is the downstream pressure, $C_2(p_1(t), p_2(t))$ describes the discharge coefficient, C_3 indicates a constant with the value of 0.0404, and T_i is the temperature of upstream air. Simply for calculating, set $T_i = 300\,\text{K}$.

To describe the relationship between $u(t)$ and $A(u(t))$ precisely, $A(u(t))$ is set as the following quadratic function:

$$A(u(t)) = a_1 u(t)^2 + a_2 u(t) + a_3, \tag{19.5}$$

where a_1, a_2, and a_3 are three coefficients of function (19.5). Note that $C_1(p_1(t), p_2(t))$ in (19.4) is written in the following form:

$$C_1(p_1(t), p_2(t)) = \begin{cases} 1, & \dfrac{p_1(t)}{p_2(t)} \le p^* \\ \sqrt{1 - \left(\dfrac{\frac{p_1(t)}{p_2(t)} - p^*}{1 - p^*} \right)^2}, & \dfrac{p_1(t)}{p_2(t)} > p^*, \end{cases}$$

where p^* is the critical pressure ratio. Moreover, the expression of $C_2(p_1(t), p_2(t))$ is given as

$$C_2(p_1(t), p_2(t)) = b_1\left(\frac{p_1(t)}{p_2(t)}\right)^2 + b_2\left(\frac{p_1(t)}{p_2(t)}\right) + b_3, \tag{19.6}$$

where b_1, b_2, and b_3 are three coefficients of function $C_2(p_1(t), p_2(t))$. Let

$$p_{12}(t) = \frac{p_1(t)}{p_2(t)},\ C^* = C_1(p_1(t), p_2(t))C_3 p_1(t)/\sqrt{T_i}.$$

Moreover, the model of proportional valve (19.4) is rewritten as follows:

$$q_{mi}(t) = (a_1 u(t)^2 + a_2 u(t) + a_3)(b_1 p_{12}(t)^2 + b_2 p_{12}(t) + b_3)C^*. \tag{19.7}$$

Remark 19.1 In the relational expression C_1, $p_1(t)$ and $p_2(t)$ represent the upstream and downstream pressures, respectively. Moreover, the upstream pressure $p_1(t)$ is always greater than the downstream pressure $p_2(t)$. Therefore, $p^* < 1$ holds when $\frac{p_1(t)}{p_2(t)} > p^*$, namely, p^* is always different than 1.

19.3.3 Parameter Identification

Restricted by realistic conditions, some parameters in systems (19.7) can't be obtained directly. Note that a_i and b_i $(i = 1, 2, 3)$ are unknown parameters in Eq. (19.7). Therefore, a parameter identification approach is used to deal with this problem in this chapter as in [3, 4]. It is known from [5] that least-square algorithms are simple and effective in identification parameters. In the following, a recursive least-square algorithm is given to obtain an accurate mathematical model of the five-way proportional valve.

Consider the following linear identification model as

$$y(t) = \varphi^T\theta + v(t), \tag{19.8}$$

where $y(t)$ is the output sequence, v is the stochastic error, θ is the parameter vector, and φ^T $(\varphi \in R^m)$ is the information vector which is comprised of input sequences. Supposing that there are n arrays and letting

$$Y = \begin{bmatrix} y(1) \\ y(2) \\ \vdots \\ y(n) \end{bmatrix} \in R^n,\ H = \begin{bmatrix} \varphi^T(1) \\ \varphi^T(2) \\ \vdots \\ \varphi^T(n) \end{bmatrix} \in R^{n\times m}$$

the least squares estimate of θ is obtained as

$$\widehat{\theta} = (H^T H)^{-1} H^T Y, \tag{19.9}$$

where matrix $H^T H$ is nonsingular, $\widehat{\theta}$ is the estimated vector of θ. Let

$$\widehat{\theta} = \left[b_1a_1\ b_1a_2\ b_1a_3\ b_2a_1\ b_2a_2b_2a_3\ b_3a_1\ b_3a_2\ b_3a_3 \right]^T$$

It is well known that the recursive least-square method is suitable for linear systems [6]. However, the proportional valve model (19.7) is a nonlinear system. In the next, system (19.7) is changed to the following linear form as

$$\begin{aligned} q_{mi}(t) = {} & b_1a_1u(t)^2p_{12}(t)^2C^* + b_1a_2u(t)p_{12}(t)^2C^* + b_1a_3p_{12}(t)^2C^* \\ & + b_2a_1u(t)^2p_{12}(t)C^* + b_2a_2u(t)p_{12}(t)C^* + b_2a_3p_{12}(t)C^* \\ & + b_3a_1u(t)^2C^* + b_3a_2u(t)C^* + b_3a_3C^*, \end{aligned} \tag{19.10}$$

where $b_{i_1}a_{j_1}(i_1, j_1 = 1, 2, 3)$ are parameters which is needed to estimate, and $u(t)^{i_2}p_{12}(t)^{j_2}C^*(i_2, j_2 = 0, 1, 2)$ are input variables. Therefore, Eq. (19.10) is similar to model (19.8), and $b_{i_1}a_{j_1}$ is obtained by Eq. (19.9). Moreover, b_{i_1} and a_{j_1} are calculated according to some real parameters of a five-way valve [2]. There are nine equations on $b_{i_1}a_{j_1}$ which is obtained by Eq. (19.9). For example, parameter a_1 is gotten from the real parameters of a five-way valve in [7]. Moreover, b_1, b_2, and b_3 are calculated based on three equations in the nine equations on $b_{i_1}a_{j_1}$. Then three parameters on a_2 and three parameters on a_3 are calculated. All the three parameters on a_2 are very closely, so do as for a_3. Finally, both a_2 and a_3 are obtained by calculating an average of all parameters on a_2 and a_3. The six parameters b_{i_1} and a_{j_1} with $i_1, j_1 = 1, 2, 3$ are gotten from the calculation process. Results of parameter identification are shown in Table 19.4.

In Table 19.4, Case 1 and Case 2 represent the condition that the range of upstream pressure is 0.15–0.2 MPa, Case 3 and Case 4 represent that upstream pressure changes from 0.2 to 0.6 MPa.

Moreover, curves in Fig. 19.11 shows the relationship between voltage and mass flow rate under different upstream pressures. In Fig. 19.11, the discrete points are

Table 19.4 Parameters of the five-way valve

	Case 1	Case 2	Case 3	Case 4
Voltage (V)	0–5	0–5	5–10	5–10
b_1	0.6678	−5.6776	−0.4602	6.8879
b_2	−0.2829	5.2761	−0.0531	−6.4772
b_3	−0.2596	−1.4798	0.4326	1.8361
$a_1 \times 10^4$	−0.0007	−0.0007	−0.0001	−0.0001
$a_2 \times 10^4$	−0.0256	−0.0256	0.0324	0.0324
$a_3 \times 10^4$	0.135	0.135	−0.1687	−0.1687

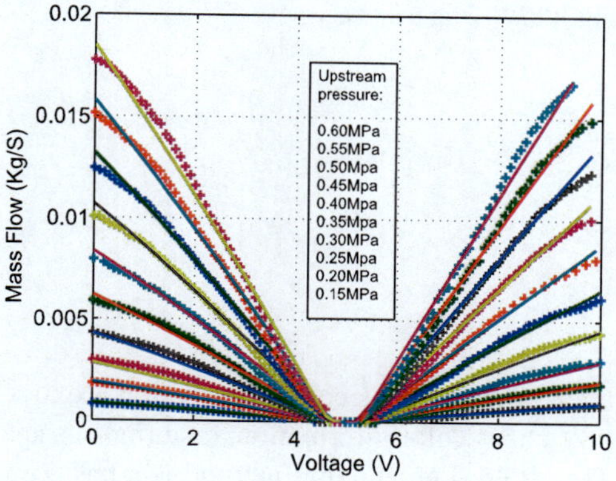

Fig. 19.11 Curve fitting results of the five-way valve

experimental data and the fitting lines are obtained by Eq. (19.10). The results show a good performance in parameter identification. For simplicity, the five-way proportional valve model is rewritten as follows:

$$q_{mi}(t) = A(u(t))g(p_i(t)). \tag{19.11}$$

Note that the five-way valve has null voltage u^*. If voltage $u(t)$ is greater than u^*, there exist $q_{m1}(t) > 0$ and $q_{m2}(t) < 0$ for the experimental platform. On the contrary, there are $q_{m1}(t) < 0$ and $q_{m2}(t) > 0$. Then, the proportional valve model (19.4) is rewritten in the following form:

$$\begin{cases} q_{m1}(t) = \text{sign}(u(t) - u^*)A(u(t))g(p_1(t)) \\ q_{m2}(t) = -\text{sign}(u(t) - u^*)A(u(t))g(p_2(t)). \end{cases} \tag{19.12}$$

19.4 Multi-point Positioning Experimental

19.4.1 Color Recognition of Cubes

The aim of image processing is to identify color information of cubes. Color recognition in this part is prepared for the following path planning design. It is consist of image segmentation and image correction. The corresponding path planning of the cubes will be introduced in the next part.

19.4.1.1 Image Segmentation

Image segmentation is an important preprocessing for image recognition. The image collected by a camera is shown in Fig. 19.12, which is segmented into several sub-

Fig. 19.12 Original image

areas, and the block color information is extracted from the image. In order to recognize cubes color information, disturbances apart from the cubes should be removed. Threshold segmentation method is a basic image processing segmentation method which is studied in this chapter. The threshold segmentation method plays an important role in image analysis and recognition due to its less calculation and simple to achieve. The criteria of threshold segmentation is given as follows:

$$\text{color} = \begin{cases} 1, & \text{if } r \in \tilde{R},\ g \in \tilde{G},\ b \in \tilde{B} \\ 0, & \text{else,} \end{cases} \tag{19.13}$$

where color is the value of pixel in the binary image, r, g, and b are three RGB values of pixel in the collected image, $\tilde{R}$, $\tilde{G}$, and $\tilde{B}$ are three ranges of RGB values of image which is to be retained. Four colors RGB threshold ranges are set for red, blue, yellow, and green, respectively. Each pixel RGB value is compared with the preset threshold. If the pixel RGB value within the setting range, the binary image pixel value is set to 1, else 0. Then a median filter is used to remove the noises. The binary image correspond to the color image is obtained as shown in Fig. 19.13a. The black cubes binary image is also obtained by (19.13), but it needs further processing to remove noises. Figure 19.13a is dilated firstly, then combine the expanded motion part and black cubes binary image by performing a logical 'and' operation. A median filter is deployed to remove the noises, and the result is shown in Fig. 19.13b. Color binary image Fig. 19.13a and black binary image Fig. 19.13b are combined to form a

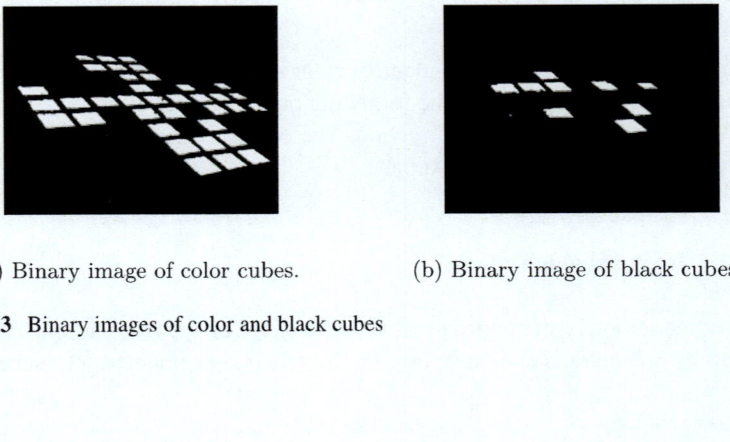

(a) Binary image of color cubes. (b) Binary image of black cubes.

Fig. 19.13 Binary images of color and black cubes

(a) A result of image segmentation. (b) A result of image correction.

Fig. 19.14 Results of image segmentation and correction

complete binary image. Create a three-dimensional image mask for the binary image, and perform a logical 'and' operation with the original image to get the final image, as shown in Fig. 19.14a.

19.4.1.2 Image Correction

It is not convenient for color recognition due to the final image in Fig. 19.14a. Therefore, it is necessary to design an image correction algorithm. A control point transformation method is introduced in this chapter. In Fig. 19.15, the process of the image correction is shown. According to known coordinates, mapping relations of the two images are established as

$$u^* = a^*x + b^*y + c^*xy + d^* \tag{19.14}$$
$$v^* = e^*x + f^*y + g^*xy + h^*, \tag{19.15}$$

where x and y are the coordinates of the image before correction, u^* and v^* are the coordinates of the image after correction, a^*, b^*, c^*, d^*, e^*, f^*, g^*, and h^* are unknown parameters.

Fig. 19.15 Diagram of image correction

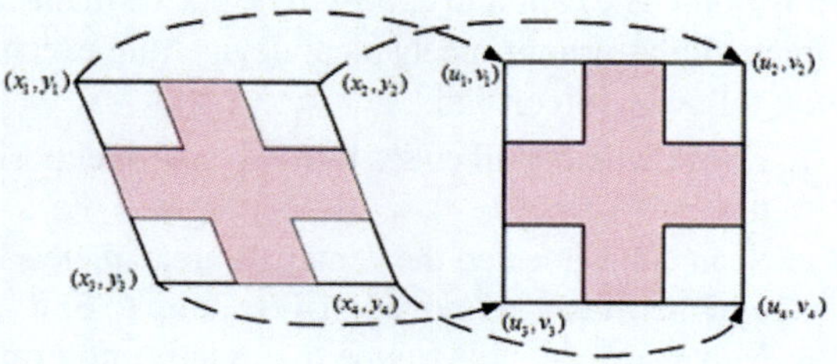

According to Fig. 19.15, equalities (19.14) and (19.15), the mapping relations are expressed as follows:

$$\begin{bmatrix} A^* & 0 \\ 0 & A^* \end{bmatrix} C^* = U^*,$$

where

$$A^* = \begin{bmatrix} x_1 & y_1 & x_1 y_1 & 1 \\ x_2 & y_2 & x_2 y_2 & 1 \\ x_3 & y_3 & x_3 y_3 & 1 \\ x_4 & y_4 & x_4 y_4 & 1 \end{bmatrix}$$

$$C^* = [a^* \; b^* \; c^* \; d^* \; e^* \; f^* \; g^* \; h^*]^T$$

$$U^* = [u_1 \; u_2 \; u_3 \; u_4 \; v_1 \; v_2 \; v_3 \; v_4]^T.$$

Finally, a bilinear interpolation method is employed to confirm gray value of each pixel. The result of image correction is shown in Fig. 19.14b, which meets the practical requirements of the experimental platform.

19.4.2 Design of Path Planning

The path planning is designed on a basis of the image processing technology. The cubes initial color information will be recognized by the image processing technology. Furthermore, the image processing is only done once, which do not require a lot of computation time. There are 45 cubes on the platform with five colors, and each color has nine cubes. These colorful cubes are divided into five regions and each region is numbered consecutively according to the order of top to bottom and left to right, as shown in Fig. 19.16. The pneumatic manipulator can capture an object located arbitrarily in space. At the same time, a vacancy is formed in its former position after the object is captured. The platform is equipped with 12 small telescopic cylinders as auxiliary tools which push corresponding 12 cubes labeled as 0, 1, 2, 9, 12, 15, 29, 32, 35, 42, 43, and 44. Then a new vacancy will appear at above 12 positions. The captured cube will be placed in the new space. According to the experimental platform of mechanical device rule execution, path planning is designed in the following algorithm:

1: Check whether all cubes move to a correct position, if no, jump to 2, if yes, jump to 6.

2: Scan cubes located the center III area labeled 18–26 sequentially, and determine whether it can be moved, if yes, jump to 3, if no, jump to 4.

3: Move corresponding cube to a relative color position such as I, II, IV, or V, and then jump to 1.

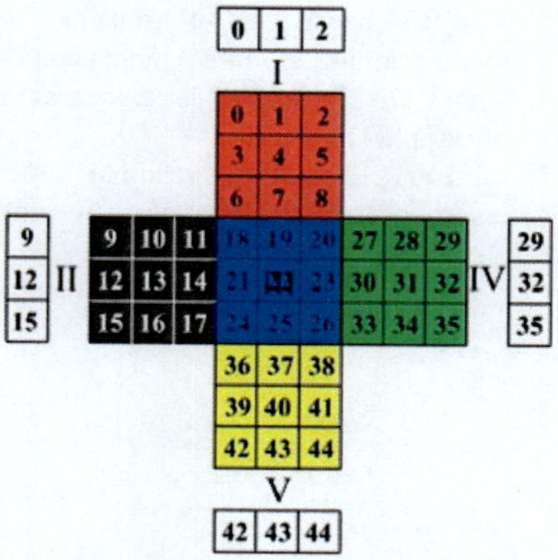

Fig. 19.16 Cubes number

4: Check whether all cubes homing to the correct position again, if yes, jump to 6, if no, jump to 5.
5: Move the no returned cube in I, II, IV, and V area to the specified position in center III area, and then jump to 3.
6: End the path planning.

19.5 Conclusion

In this chapter, a pneumatic right angle composite motion platform has been introduced. Platform components of the pneumatic right angle composite motion platform have been shown and listed in this chapter. Pneumatic circuit and control circuit for the pneumatic right angle composite motion platform have been given to show structure of the pneumatic right angle composite motion platform. Moreover, a dynamic nonlinear system for rodless cylinder model and a dynamic nonlinear system for five-way proportional valve model have been derived to express the pneumatic right angle composite motion platform. Finally, multi-point positioning experimental has been introduced for the pneumatic right angle composite motion platform.

References

1. Gulati N, Barth EJ (2009) A globally stable, load-independent pressure observer for the servo control of pneumatic actuators. IEEE-ASME Trans Mechatron 14(3):295–306
2. Nuchkrua T, Leephakpreeda T (2013) Fuzzy self-tuning PID control of hydrogen-driven pneumatic artificial muscle actuator. J Bionic Eng 10(3):329–340
3. Ding J, Ding F, Liu XP, Liu G (2011) Hierarchical least squares identification for linear SISO systems with dual-rate sampled-data. IEEE Trans Autom Control 56(11):2677–2683
4. Wang D, Ding F (2011) Least squares based and gradient based iterative identification for wiener nonlinear systems. Signal Proc 91(5):1182–1189

5. Wang C, Zhao Y, Lin C-Y, Tomizuka M (2014) Fast planning of well conditioned trajectories for model learning. In: International conference on intelligent robots and systems, pp 1460–1465
6. Ding F (2013) Coupled-least-squares identification for multivariable systems. IET Control Theory Appl 7(1):68–79
7. Moreau R, Pham MT, Tavakoli M, Le MQ, Redarce T (2012) Sliding-mode bilateral teleoperation control design for master-slave pneumatic servo systems. Control Eng Pract 20(6):584–597

Chapter 20
Nonlinear Feedback Control

20.1 Introduction

Pneumatic actuators in servo systems have attracted increasing attention in recent years [1, 2]. Especially, pneumatic cylinder, which is an important class of pneumatic actuators, has wide applications in industrial automation, health care, and other fields [3–5]. It is known that servo position control for a pneumatic cylinder is a key technique in the applications of pneumatic cylinder systems [6]. However, both the compressibility of air and the negative effects of friction force result in the strong nonlinearity of the rodless cylinder, and the nonlinearity makes it difficulty to achieve precise position control [7]. Researchers have adopted many methods to improve the position control performance of pneumatic rodless cylinder systems. Various of improved proportional integral derivative controllers with the simple structure and easy implementation are designed in pneumatic cylinder systems [8, 9]. In addition, an adaptive fuzzy control method has been used to study the position control for a pneumatic rodless cylinder in [10]. However, pressure states or friction model have to be known in most of the control methods for position control of pneumatic rodless cylinder. Design of pressure observer is necessary and difficult in control and friction model is not easy to gain for systems with different pneumatic rodless cylinders. Therefore, active disturbance rejection rodless position control is studied for the pneumatic magnetic rodless cylinder in this chapter.

In this chapter, an active disturbance rejection position control scheme for a pneumatic magnetic rodless cylinder is studied. Then, both the convergence of ESO and the stability of the close-loop pneumatic system are analyzed. Finally, experiments for servo position control are provided to illustrate the effectiveness of the proposed techniques. The main contributions of this chapter are summarized as below:

i An active disturbance rejection position control scheme for a pneumatic magnetic rodless cylinder is used to ensure the control performances.

L. Zhao et al., *Pneumatic Servo Systems Analysis*, Advances in Industrial Control,
https://doi.org/10.1007/978-981-16-9515-5_20

ii The convergence of the ESO is proven by the self-stable region theory, and the Lyapunov stability theory is used to discuss the stability of the pneumatic closed-loop system.
iii The steady-state error is less than 0.05 mm, and the response time is 1 s in step signal experimental results.

20.2 Main Results

20.2.1 Schematic Diagram of Control Method

Considering nonlinear characteristics of the pneumatic system and inaccuracy problem of pneumatic rodless cylinder model, ADRC technique is developed to study servo position control of pneumatic rodless cylinder. The schematic diagram of the active disturbance rejection controller is shown in Fig. 20.1.

In this chapter, tracking differentiator is designed to arrange the transient process which is used to get a continuous smooth signal and its differential signal for the abrupt input. Considering system model (19.2), define $\eta = v_1(t) - v_0(t)$ and the second-order tracking differentiator is designed as follows:

$$\begin{cases} \dot{v}_1(t) = v_2(t) \\ \dot{v}_2(t) = f_{han}(\eta(t), v_2(t), r_0, h_0), \end{cases} \tag{20.1}$$

where $\eta(t)$ is an error, $v_0(t)$ is the given input, $v_1(t)$ is the output tracking $v_0(t)$, and $v_2(t)$ is the differential signal of $v_1(t)$. Both $v_1(t)$ and $v_2(t)$ are regarded as parts of control inputs based on proportional derivative control theory.

Remark 20.1 The parameters r_0 and h_0 are positive numbers. We can get different transient processes by regulating r_0 and h_0. However, if they are out of adjustable ranges, the right transient process can't be obtained. There is not a theoretical approach to calculate the adjustable ranges of r_0 and h_0. Therefore, the regulation of r_0 and h_0 is almost on the basis of experience.

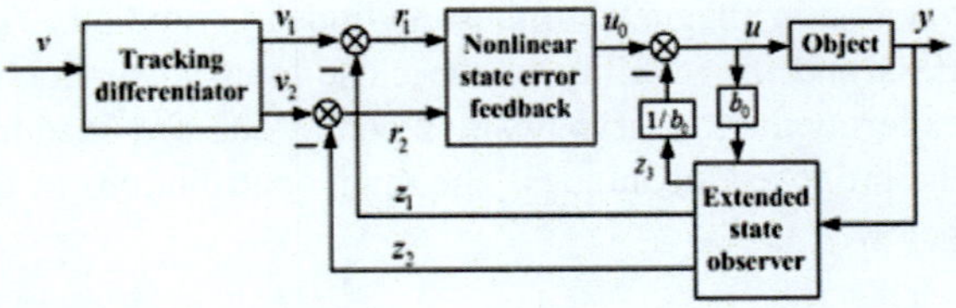

Fig. 20.1 The schematic diagram of the active disturbance rejection controller

20.2.2 Extended State Observer

In this section, ESO is introduced to deal with strong nonlinearity of the pneumatic rodless cylinder system (19.2). The unknown nonlinear term $f(x_1(t), x_2(t))$ is continuously differentiable and bounded, and it is treated as an extended state $x_3(t)$, i.e., $f(x_1(t), x_2(t)) = x_3(t)$. The pneumatic rodless cylinder system (19.2) is rewritten as follows:

$$\begin{cases} \dot{x}_1(t) = x_2(t) \\ \dot{x}_2(t) = x_3(t) + b_0 u(t) \\ \dot{x}_3(t) = w(t), \end{cases} \tag{20.2}$$

where $w(t)$ is the derivative of $x_3(t)$. Note that $w(t)$ is bounded in practice. If $(x_1(t), x_2(t)) = (0, 0)$, then $w(t) = 0$. Defining $e_1(t) = z_1(t) - y(t)$, the ESO of the system is represented as follows:

$$\begin{cases} \dot{z}_1(t) = z_2(t) - \beta_1 e_1(t) \\ \dot{z}_2(t) = z_3(t) - \beta_2 fal(e_1(t), \sigma_1, \delta) + b_0 u(t) \\ \dot{z}_3(t) = -\beta_3 fal(e_1(t), \sigma_2, \delta), \end{cases} \tag{20.3}$$

where $z_1(t)$, $z_2(t)$, and $z_3(t)$ are observations of $x_1(t)$, $x_2(t)$, and $x_3(t)$, and β_1, β_2, and β_3 are observer gains. For simplicity, $fal(e_1(t), \sigma_1, \delta)$ is indicated by $f_1(e_1(t))$ and $fal(e_1(t), \sigma_2, \delta)$ by $f_2(e_1(t))$.

Considering systems (20.2) and (20.3), we get the following error system:

$$\begin{cases} \dot{e}_1(t) = e_2(t) - \beta_1 e_1(t) \\ \dot{e}_2(t) = e_3(t) - \beta_2 f_1(e_1(t)) \\ \dot{e}_3(t) = -w(t) - \beta_3 f_2(e_1(t)), \end{cases} \tag{20.4}$$

where $e_i(t) = z_i(t) - x_i(t)$ and $i = 1, 2, 3$.

Remark 20.2 In this chapter, there exist two points which aren't differentiable for function $fal(\cdot)$. Since the above factor doesn't have an effect on the following theoretical analysis, we define that $f_1'(e_1)$ is the derivative of $f_1(e_1(t))$ and $f_2'(e_1(t))$ is the derivative of $f_2(e_1(t))$. Moreover, $f_1'(e_1(t)) > 0$ and $f_2'(e_1(t)) > 0$ for the reason that $fal(\cdot)$ is also a monotonous increasing function.

In this section, the self-stable region approach is used to analyze the convergence of system (20.4), such as in [11, 12].

Definition 20.3 The region Ω with the origin as the vertex is called a self-stable region of a system, if all state trajectories of the system remaining in Ω after a certain time will eventually converge to the origin.

According to the error system (20.4), we set the following two error functions:

$$h_2(e_1(t), e_2(t)) = e_2(t) - \beta_1 e_1(t) + k_1 g_1(e_1(t))\text{sign}(e_1(t))$$

and

$$\begin{aligned}h_3(e_1(t), e_2(t), e_3(t)) &= e_3(t) - \beta_2 f_1(e_1(t)) - \beta_1(e_2(t) - \beta_1 e_1(t)) \\ &+ k_2 g_2(e_1(t), e_2(t))\text{sat}(h_2(e_1(t), e_2(t))/g_1(e_1(t))),\end{aligned}$$

where $g_1(e_1(t))$ and $g_2(e_1(t), e_2(t))$ are continuous positive definite functions, and $g_1(0) = 0$, $g_2(0, 0) = 0$. Note that k_1 and k_2 are two constants with $k_1 > 1$ and $k_2 > 1$. Let region $\Omega_2 = \{(e_1(t), e_2(t), e_3(t)) : |h_2(e_1(t), e_2(t))| \leq g_1(e_1(t))\}$ and region $\Omega_3 = \{(e_1(t), e_2(t), e_3(t)) : |h_3(e_1(t), e_2(t), e_3(t))| \leq g_2(e_1(t), e_2(t))\}$. For notational simplicity, we denote $g_1(e_1(t))$ by g_1, $g_1\text{sign}(e_1(t))$ by $g_1 s$, $g_2(e_1(t), e_2(t))$ by g_2, $h_2(e_1(t), e_2(t))$ by h_2 and $h_3(e_1(t), e_2(t), e_3(t))$ by h_3.

Theorem 20.4 *Consider the error system (20.4). If the following inequation is satisfied*

$$g_2 > \frac{(k_1 + 1)^2}{k_2 - 1}\left|\frac{dg_1}{de_1(t)}\right||h_2|$$

and there exists appropriate gains β_1, β_2 and β_3 which are satisfied with $\dot{V}_{h_3 g_2}(t) < 0$, then the error system (20.4) is convergent. The observed states $z_1(t)$, $z_2(t)$, and $z_3(t)$ converge to the actual states $x_1(t)$, $x_2(t)$, and $x_3(t)$.

Proof Firstly, consider the condition that there has been a trajectory $(e_1(t), e_2(t), e_3(t)) \in \Omega_3$ after a certain time. From the structure of Ω_3, we get that

$$\begin{aligned}\beta_2 f_1(e_1(t)) + \beta_1(e_2(t) - \beta_1 e_1(t)) - g_2(1 + k_2\text{sat}(h_2/g_1)) &\leq e_3(t) \\ \leq \beta_2 f_1(e_1(t)) + \beta_1(e_2(t) - \beta_1 e_1(t)) + g_2(1 - k_2\text{sat}(h_2/g_1)) &\quad (20.5)\end{aligned}$$

Let $V_{h_2 g_1}(t) = \frac{1}{2}(h_2^2 - g_1^2)$. Consider the situation that $V_{h_2 g_1}(t) \geq 0$. Differentiating $V_{h_2 g_1}(t)$ results in

$$\begin{aligned}\dot{V}_{h_2 g_1}(t) = &+ \left(k_1 h_2 \frac{dg_1 s}{de_1(t)} - g_1 \frac{dg_1}{de_1(t)}\right)(h_2 - k_1 g_1 s) \\ &h_2(h_3 - k_2 g_2 \text{sat}(h_2/g_1)).\end{aligned}$$

In region Ω_3, there exists $|h_3| \leq g_2$. The following relation expression is gotten

$$h_2(h_3 - k_2 g_2\text{sat}(h_2/g_1)) \leq -(k_2 - 1)|h_2| g_2. \quad (20.6)$$

Then it is easy to get the following inequality:

$$\left(h_2 k_1 \frac{dg_1 s}{de_1(t)} - g_1 \frac{dg_1}{de_1(t)}\right)(h_2 - k_1 g_1 s) \le (k_1+1)^2 |h_2|^2 \left|\frac{dg_1}{de_1(t)}\right|. \tag{20.7}$$

Considering (20.6) and (20.7), we get

$$\dot{V}_{h_2 g_1}(t) \le -(k_2-1)|h_2| g_2 + (k_1+1)^2 |h_2|^2 \left|\frac{dg_1}{de_1(t)}\right|.$$

Based on $V_{h_2 g_1}(t) \ge 0$ with $(e_1(t), e_2(t)) \ne (0, 0)$ and

$$g_2 > \frac{(k_1+1)^2}{k_2-1}\left|\frac{dg_1}{de_1(t)}\right| |h_2|, \tag{20.8}$$

there exists $\dot{V}_{h_2 g_1} < 0$. Hence, trajectory $(e_1(t), e_2(t))$ is drawn to the region Ω_2. Suppose a trajectory $(e_1(t), e_2(t))$ arrives at a point $(\widetilde{e}_1, \widetilde{e}_2)$ on the boundary of Ω_2 in an infinite time, then $(\widetilde{e}_1, \widetilde{e}_2)$ is a limit point of trajectory $(e_1(t), e_2(t))$ and it is satisfied that $\dot{V}_{h_2 g_1}(\widetilde{e}_1, \widetilde{e}_2) = 0$. For the reason of that the origin is the only point on the boundary of Ω_2 making $\dot{V}_{h_2 g_1}(t) = 0$, we get $(\widetilde{e}_1, \widetilde{e}_2) = (0, 0)$. Trajectory of $(e_1(t), e_2(t))$ will directly converge to the origin or arrive at the boundary in the finite time and goes into the region Ω_2. Then, consider the situation that trajectory of $(e_1(t), e_2(t))$ has been in Ω_2 after a certain time. From the structure of Ω_2, we obtain the following inequality:

$$\begin{aligned} &\beta_1 e_1(t) - g_1(1 + k_1 \text{sign}(e_1(t))) \le e_2(t) \\ &\le \beta_1 e_1(t) + g_1(1 - k_1 \text{sign}(e_1(t))). \end{aligned} \tag{20.9}$$

Setting

$$V_1(t) = \frac{1}{2} e_1^2(t),$$

the derivative of V_1 is given as

$$\dot{V}_1(t) = e_1(t)\dot{e}_1(t) = e_1(t)(e_2(t) - \beta_1 e_1(t)) \le -(k_1-1)|e_1(t)| g_1.$$

Therefore, there exists $e_1(t) \ne 0$ which makes $\dot{V}_1(t) < 0$ and $e_1(t) \to 0$ is obtained. We have $e_2(t) \to 0$ based on the structure of Ω_2 (20.9). Furthermore, we get $e_3(t) \to 0$ from (20.5). The region Ω_3 is the self-stable region of system (20.4) by the definition and trajectories in Ω_3 will finally converge to the origin.

Lastly, consider the condition that trajectory of $(e_1(t), e_2(t), e_3(t))$ is out of Ω_3, i.e., $V_{h_3 g_2}(t) \ge 0$. Suppose $|w(t)| < W$ in (20.4). Define

$$V_{h_3 g_2} = \frac{1}{2}(h_3^2 - g_2^2).$$

The derivative of $V_{h_3 g_2}(t)$ is given as follows:

$$\begin{aligned}\dot{V}_{h_3 g_2}(t) = h_3 & \left(\frac{\partial h_3}{\partial e_3(t)}\dot{e}_3(t) + \frac{\partial h_3}{\partial e_2(t)}\dot{e}_2(t) + \frac{\partial h_3}{\partial e_1(t)}\dot{e}_1(t)\right) \\ & - g_2 \left(\frac{\partial g_2}{\partial e_2(t)}\dot{e}_2(t) + \frac{\partial g_2}{\partial e_1(t)}\dot{e}_1(t)\right).\end{aligned}$$

Setting

$$g_2 = \begin{cases} k_3|h_2|, \ |h_2| \geq g_1 \\ k_3 g_1, \ |h_2| < g_1, \end{cases} \tag{20.10}$$

where k_3 is constant. By (20.8) and (20.10), the following inequality is easily gotten:

$$k_3 > \frac{(k_1+1)^2}{k_2 - 1}\left|\frac{dg_1}{de_1(t)}\right|.$$

One has that

$$g_2 \text{sat}(h_2/g_1) = k_3 h_2. \tag{20.11}$$

Moreover, it is obtained that

$$\begin{aligned}\dot{V}_{h_3 g_2}(t) = {} & h_3^2(-\beta_1 + k_2 k_3) + h_2 h_3 \Bigg(\beta_1 k_2 k_3 - k_2^2 k_3^2 - \beta_2 f_1'(e_1(t)) \\ & + k_1 k_2 k_3 \frac{dg_1 s}{de_1(t)}\Bigg) - g_1 s h_3 \Bigg(\beta_3 \frac{|f_3(e_1(t))|}{g_1} - \beta_2 k_1 f_1'(e_1(t)) \\ & + k_1^2 k_2 k_3 \frac{dg_1 s}{de_1(t)}\Bigg) + h_3 w(t) - g_2 h_3 \frac{\partial g_2}{\partial e_2(t)} \\ & - g_2 h_2 \left(\beta_1 \frac{\partial g_2}{\partial e_2(t)} - k_2 k_3 \frac{\partial g_2}{\partial e_2(t)} + \frac{\partial g_2}{\partial e_1(t)}\right) \\ & + k_1 g_2 g_1 s \left(\beta_1 \frac{\partial g_2}{\partial e_2(t)} + \frac{\partial g_2}{\partial e_1(t)}\right).\end{aligned}$$

We know that $V_{h_3 g_2}(t) \geq 0$ is equal to $|h_3| \geq g_2$ and $|w(t)| < W$. The following inequality is obtained as

$$
\begin{aligned}
\dot{V}_{h_3 g_2}(t) \le{}& h_3^2(-\beta_1 + k_2 k_3) + h_3^2 \left|\beta_1 k_2 - k_2^2 k_3 - \frac{\beta_2}{k_3} f_2'(e_1(t))\right| \\
&+ h_3^2 k_1 k_2 \left|\frac{d g_1 s}{d e_1(t)}\right| + \frac{h_3^2}{k_3}\left|-\beta_3 \frac{|f_3(e_1(t))|}{g_1} + \beta_2 k_1 f_2'(e_1(t))\right| \\
&+ |h_3| W + h_3^2 k_1^2 k_2 \left|\frac{d g_1 s}{d e_1(t)}\right| + \frac{k_1 h_3^2}{k_3}\left|\frac{\partial g_2}{\partial e_1(t)} + \beta_1 \frac{\partial g_2}{\partial e_2(t)}\right| \\
&+ h_3^2 \left|\frac{\partial g_2}{\partial e_2(t)}\right| + \frac{h_3^2}{k_3}\left|\frac{\partial g_2}{\partial e_1(t)} + \beta_1 \frac{\partial g_2}{\partial e_2(t)} - k_2 k_3 \frac{\partial g_2}{\partial e_2(t)}\right|.
\end{aligned}
$$

Through adjusting parameters β_2 and β_3, we get

$$
h_3^2 \left|\beta_1 k_2 - k_2^2 k_3 - \frac{\beta_2}{k_3} f_2'(e_1(t))\right| = 0
$$

and

$$
\frac{h_3^2}{k_3}\left|-\beta_3 \frac{|f_3(e_1(t))|}{g_1} + \beta_2 k_1 f_2'(e_1(t))\right| = 0.
$$

Furthermore, we have

$$
\begin{aligned}
\dot{V}_{h_3 g_2}(t) \le{}& h_3^2(-\beta_1 + k_2 k_3) + h_3^2 \left|\frac{d g_1 \operatorname{sign}(e_1(t))}{d e_1(t)}\right| k_1 k_2 (1 + k_1) \\
&+ |h_3| W + h_3^2 \left|\frac{\partial g_2}{\partial e_2(t)}\right| + \frac{h_3^2}{k_3}\left|\frac{\partial g_2}{\partial e_1(t)} + \beta_1 \frac{\partial g_2}{\partial e_2(t)} - k_2 k_3 \frac{\partial g_2}{\partial e_2(t)}\right| \\
&+ \frac{k_1 h_3^2}{k_3}\left|\frac{\partial g_2}{\partial e_1(t)} + \beta_1 \frac{\partial g_2}{\partial e_2(t)}\right|.
\end{aligned}
$$

When $(e_1(t), e_2(t), e_3(t)) \neq (0, 0, 0)$, there exists an appropriate parameter β_1 which makes $\dot{V}_{h_3 g_2}(t) < 0$. Hence, trajectory $(e_1(t), e_2(t), e_3(t))$ is be attracted by the region Ω_3. Suppose there exists a trajectory $(e_1(t), e_2(t), e_3(t))$ arriving at a point $(\widetilde{e}_1, \widetilde{e}_2, \widetilde{e}_3)$ on the boundary of Ω_3 with unlimited time, then $(\widetilde{e}_1, \widetilde{e}_2, \widetilde{e}_3)$ is a limit point of trajectory $(e_1(t), e_2(t), e_3(t))$, and $\dot{V}_{h_3 g_2}(\widetilde{e}_1, \widetilde{e}_2, \widetilde{e}_3) = 0$. Since, the origin is the only point on the boundary of Ω_3 making $\dot{V}_{h_3 g_2}(t) = 0$, we get $(\widetilde{e}_1, \widetilde{e}_2, \widetilde{e}_3) = (0, 0, 0)$. Trajectory $(e_1(t), e_2(t), e_3(t))$ will directly converge to the origin or arrive at the boundary in the finite time and go into the region Ω_3. Therefore, if the proper parameters β_1, β_2, and β_3 are satisfied, the observed states $z_1(t)$, $z_2(t)$, and $z_3(t)$ converge to $x_1(t)$, $x_2(t)$, and $x_3(t)$.

20.2.3 Nonlinear Error Feedback Controller

Nonlinear control and error control are paid more and more attention, since better control effect is obtained in practical application. Error signals between the tracking differentiator and the extended state observer are defined as follows:

$$\begin{cases} r_1(t) = v_1(t) - z_1(t) \\ r_2(t) = v_2(t) - z_2(t), \end{cases} \tag{20.12}$$

where $v_1(t)$ and $v_2(t)$ are the given position and velocity, respectively, $z_1(t)$ and $z_2(t)$ are the observations of corresponding output. The nonlinear state error feedback controller $u(t)$ is given as

$$u = \alpha_1 fal(r_1(t), \sigma_3, \delta_1) + \alpha_2 fal(r_2(t), \sigma_4, \delta_2) - z_3(t)/b_0, \tag{20.13}$$

where α_1 and α_2 are two gains of the nonlinear controller, $z_3(t)/b_0$ is compensation for system (19.2). The errors between the given and the output is presented

$$\begin{cases} s_1(t) = v_1(t) - x_1(t) \\ s_2(t) = v_2(t) - x_2(t), \end{cases} \tag{20.14}$$

Furthermore, the error system is shown as

$$\begin{cases} \dot{s}_1(t) = s_2(t) \\ \dot{s}_2(t) = v_3(t) - f(x_1(t), x_2(t)) - b_0 u(t), \end{cases} \tag{20.15}$$

where $v_3(t)$ is the derivative of $v_2(t)$, and it is continuous and bounded. For notational simplicity, denote $fal(r_1(t), \sigma_3, \delta_1)$ by $fal(r_1(t))$, $fal(r_2(t), \sigma_4, \delta_2)$ by $fal(r_2(t))$. From (20.12) and (20.14), it is easy to get the following expression:

$$\begin{cases} r_1(t) = s_1(t) - e_1(t) \\ r_2(t) = s_2(t) - e_2(t). \end{cases} \tag{20.16}$$

Moreover, we have

$$\dot{s}_2(t) = v_3(t) - b_0\alpha_1 fal(s_1(t) - e_1(t)) - b_0\alpha_2 fal(s_2(t) - e_2(t)) + e_3(t),$$

where $e_3(t) = z_3(t) - f(x_1(t), x_2(t), t)$.

Theorem 20.5 *Consider the closed-loop system (20.15) with the error feedback controller (20.13). By choosing appropriate nonlinear controller gains α_1 and α_2 in controller (20.13), the closed-loop system (20.15) is stable. That is, output $x_1(t)$ and $x_2(t)$ converge to input $v_1(t)$ and $v_2(t)$, respectively.*

Proof Construct a Lyapunov function as follows:

$$
\begin{aligned}
V(t) = &\frac{1}{2}(s_1(t) - e_1(t))^2 + \frac{1}{2}(s_2(t) - e_2(t))^2 \\
&+(s_1(t) - e_1(t))(s_2(t) - e_2(t)).
\end{aligned} \tag{20.17}
$$

The derivative of $V(t)$ is given as

$$
\begin{aligned}
\dot{V}(t) = &-(s_2(t) - e_2(t))(b_0\alpha_1 fal(s_1(t) - e_1(t)) + b_0\alpha_2 fal(s_2(t) - e_2(t))) \\
&-(s_1(t) - e_1(t))(b_0\alpha_1 fal(s_1(t) - e_1(t)) + b_0\alpha_2 fal(s_2(t) - e_2(t))) \\
&+(s_1(t) - e_1(t))(s_2(t) - e_2(t)) + \beta_1 e_1(t)(s_1(t) - e_1(t)) \\
&+(s_2(t) - e_2(t))^2 + (s_1(t) - e_1(t) + s_2(t) - e_2(t))(v_3(t) \\
&+\beta_2 f_1(e_1(t))) + \beta_1 e_1(t)(s_2(t) - e_2(t)).
\end{aligned}
$$

Furthermore, set

$$
\begin{aligned}
&(s_1(t) - e_1(t) + s_2(t) - e_2(t))(fal(s_1(t) - e_1(t)) + fal(s_2(t) - e_2(t))) \\
&\quad = \Phi(s_1(t) - e_1(t), s_2(t) - e_2(t)).
\end{aligned}
$$

Letting $\alpha_1 = \alpha^*$ and $\alpha_2 = \alpha^*$ for simplify analysis, we have

$$
\begin{aligned}
\dot{V}(t) = &-b_0\alpha^* \cdot \Phi(s_1(t) - e_1(t), s_2(t) - e_2(t)) \\
&+(s_1(t) - e_1(t))(s_2(t) - e_2(t)) + \beta_1 e_1(t)(s_1(t) - e_1(t)) \\
&+(s_2(t) - e_2(t))^2 + (s_1(t) - e_1(t) + s_2(t) - e_2(t))(v_3 + \beta_2 f_1(e_1(t))) \\
&+\beta_1 e_1(t)(s_2(t) - e_2(t)).
\end{aligned}
$$

One has that

$$
\begin{aligned}
\dot{V}(t) = &-b_0\alpha^* \cdot \Phi(s_1(t) - e_1(t), s_2(t) - e_2(t)) - \beta_1 e_1(t)^2 + (s_2(t) - e_2(t))^2 \\
&+(s_1(t) - e_1(t))(s_2(t) - e_2(t)) + \beta_1 e_1(t) s_1(t) + \beta_1 e_1(t)(s_2(t) - e_2(t)) \\
&+(s_1(t) - e_1(t) + s_2(t) - e_2(t))(v_3(t) + \beta_2 f_1(e_1(t))).
\end{aligned}
$$

Based on (20.16), the following equation is obtained

$$
\begin{aligned}
\Phi(s_1(t) - e_1(t), s_2(t) - e_2(t)) &= \Phi(r_1(t), r_2(t)) \\
&= (r_1(t) + r_2(t))(fal(r_1(t)) + fal(r_2(t))).
\end{aligned} \tag{20.18}
$$

The function $fal(\cdot)$ is an odd function which is monotone increasing, so we have $r_1(t) fal(r_1(t)) \geq 0$ and $r_2(t) fal(r_2(t)) \geq 0$. Furthermore, we let $\sigma_3 = \sigma_4$ and $\delta_1 = \delta_2$ such that $fal(r_1(t))$ and $fal(r_2(t))$ are the same function for analysis simplicity. In order to get the sign of expression $(r_1(t) + r_2(t))(fal(r_1(t)) + fal(r_2(t)))$, the following analysis is given:

1. If $r_1(t) \geq 0$, $r_2(t) \geq 0$, we get $fal(r_1(t)) \geq 0$, $fal(r_2(t)) \geq 0$. Then $(r_1(t) + r_2(t))(fal(r_1(t)) + fal(r_2(t))) \geq 0$.
2. If $r_1(t) < 0$, $r_2(t) < 0$, we get $fal(r_1(t)) < 0$, $fal(r_2(t)) < 0$. Then $(r_1(t) + r_2(t))(fal(r_1(t)) + fal(r_2(t))) > 0$.
3. If $r_1(t) > 0 > r_2(t)$ and $r_1(t) \geq |r_2(t)|$, we get $fal(r_1(t)) > 0$, $fal(r_2(t)) < 0$ and $fal(r_1(t)) \geq |fal(r_2(t))|$. Then, $(r_1(t) + r_2(t))(fal(r_1(t)) + fal(r_2(t))) \geq 0$.
4. If $r_1(t) > 0 > r_2(t)$ and $r_1(t) < |r_2(t)|$, we get $fal(r_1(t)) > 0$, $fal(r_2(t)) < 0$ and $fal(r_1(t)) < |fal(r_2(t))|$. Then, $(r_1(t) + r_2(t))(fal(r_1(t)) + fal(r_2(t))) > 0$.

Note that the analysis process under condition $r_2(t) > 0 > r_1(t)$ is similar with the one under $r_1(t) > 0 > r_2(t)$. Therefore, there exists $(r_1(t) + r_2(t))(fal(r_1(t)) + fal(r_2(t))) \geq 0$ if $r_2(t) > 0 > r_1(t)$ holds.

In conclusion, we get that

$$(r_1(t) + r_2(t))(fal(r_1(t)) + fal(r_2(t))) \geq 0, \tag{20.19}$$

which is equal to $\Phi(s_1(t) - e_1(t), s_2(t) - e_2(t)) \geq 0$ from (20.18). For the reason of that the convergence of ESO has been proved by the self-stable region method, we have that $e_1(t)$, $e_2(t)$, and $e_3(t)$ are bounded. Moreover, the given acceleration v_3 is bounded and both the errors $s_1(t)$ and $s_2(t)$ are bounded in practical. Letting

$$\begin{aligned} M = &|(s_2(t) - e_2(t))^2 + (s_1(t) - e_1(t))(s_2(t) - e_2(t)) + \beta_1 e_1(t) s_1(t) \\ &+ (s_1(t) - e_1(t) + s_2(t) - e_2(t))(v_3 + \beta_2 f_1(e_1(t)))| \\ &+ \beta_1 e_1(t)(s_2(t) - e_2(t)), \end{aligned}$$

it is also obtained that M is also bounded. Therefore, we have

$$\dot{V}(t) < -b_0 \alpha^* \cdot \Phi(s_1(t) - e_1(t), s_2(t) - e_2(t)) - \beta_1 e_1^2(t) + M. \tag{20.20}$$

It is hold that $\dot{V}(t) < 0$ if (20.20) holds. It is easy shown that (20.20) holds if α^* is large enough. That is, by choosing appropriate nonlinear controller gains α_1 and α_2, there exists $\dot{V}(t) < 0$ for the reason of that $\alpha_1 = \alpha_2 = \alpha^*$ has been supposed above.

Based on the above observations, we have that active disturbance rejection controller (20.13) guarantees the stability of the closed-loop system (20.15) by selecting appropriate controller parameters α_1 and α_2.

20.3 Experiments and Results

In this chapter, we study the servo position control of single pneumatic rodless cylinder which is shown in Fig. 20.2. The experiment is carried out under the conditions that the absolute pressure is 0.6 MPa, and the dead-zone of proportional valve is 4.83–5.19 V which is measured by experiments.

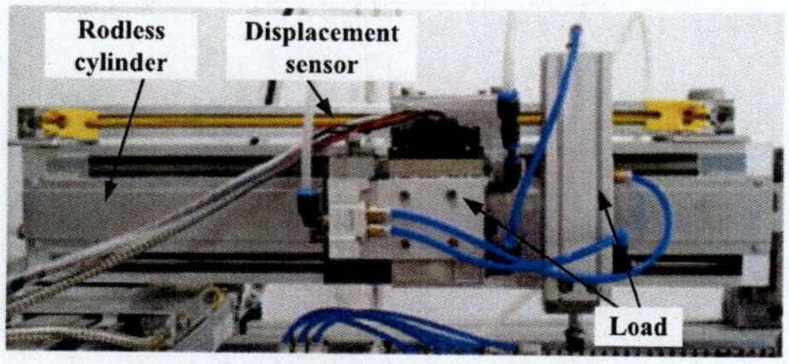

Fig. 20.2 The single pneumatic rodless cylinder

Active disturbance rejection controller and its parameters are designed as follows

$$\begin{cases} v_1(k+1) = v_1(k) + 0.1v_2(k) \\ v_2(k+1) = v_2(k) + 0.1 f_{han}(\eta(k), v_2(k), 20, 0.1) \\ z_1(k+1) = z_1(k) + 0.1(z_2(k) - 10e_1(k)) \\ z_2(k+1) = 0.1(z_3(k) - 20fal(e_1(k), 0.75, 0.1) + b_0u(k)) + z_2(k) \\ z_3(k+1) = z_3(k) + 0.1(-30fal(e_1(k), 0.5, 0.1)) \\ u(k) = \alpha_1 fal(r_1(k), 0.5, 0.01) + \alpha_2 fal(r_2(k), 0.75, 0.01) - z_3(k)/b_0. \end{cases}$$

Since the velocity sensor isn't installed on experimental platform, tracking differentiator is individually used to obtain velocity $x_2(t)(t)$ of the piston. We set $\alpha_1 = 0.7641$, $\alpha_2 = 0.5$, $b_0 = 12$ for step response and $\alpha_1 = 1.301$, $\alpha_2 = 0.5$. $b_0 = 25$ for sinusoidal experiments. The experimental results of a step signal at 200 mm and a sinusoidal signal at 0.5 Hz, 100 mm amplitude based on the designed active disturbance rejection controller have been shown in Figs. 20.3 and 20.4. Moreover, the experimental results based on proportional integral differential derivative controller have been shown in Fig. 20.5.

Figure 20.3 shows the experiment results for tracking a step signal at 200 mm. In Fig. 20.3a, the displacement output y tracks the trajectory v_1 well and the observed value z_1 is almost coincident with y. Moreover, the piston moves to the given point at about 1 s and the steady-state error is less than 0.05 mm. All the velocity state variables converge to 0 mm/s in Fig. 20.3b. Control input is the voltage signal in the pneumatic system. The change trend of the voltage increases first and then decreases, and eventually remains stable in Fig. 20.3c. When the piston moves forward, z_3 opposes motion of the piston. Figure 20.3d shows the overall change trend of z_3 decreases first and then increases. Finally, z_3 reaches a steady state.

Figure 20.4 shows the experiment results for tracking a sinusoidal trajectory at 0.5 Hz, 100 mm amplitude. The displacement output y tracks the given signal v_1 quickly and steadily in Fig. 20.4a. Considering motion of the piston in practice, the fluctuation of velocity output $x_2(t)$ is reasonable in Fig. 20.4b. Moreover, the good performance in observing is also presented by Fig. 20.4a and b. In Fig. 20.4c, control input u is relatively smooth at first and then it changes periodically with fluctuations. When the displacement output y keeps up with the given signal v_1, errors $r_1(t)$ and $r_2(t)$ become very little. Then, compensation of z_3 plays an decisive role to controller from (20.13). Therefore, the fluctuation of z_3 in Fig. 20.4d lead directly to

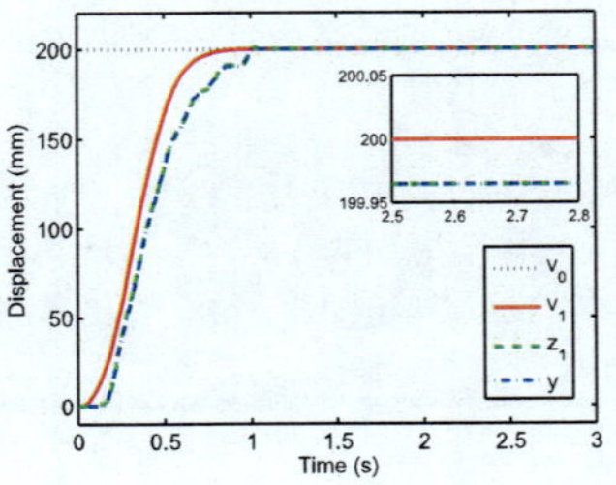

(a) Displacement.

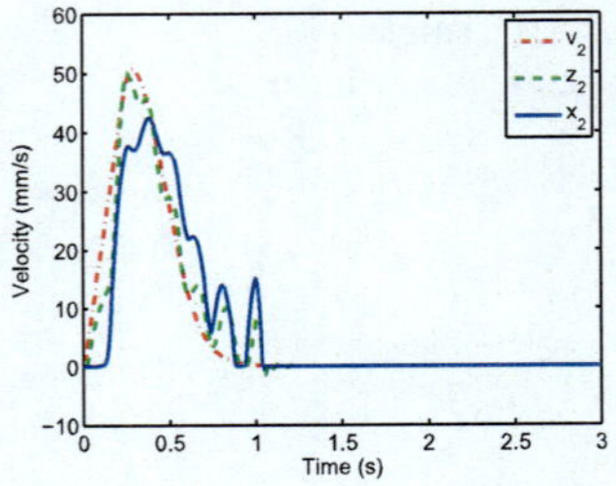

(b) Velocity.

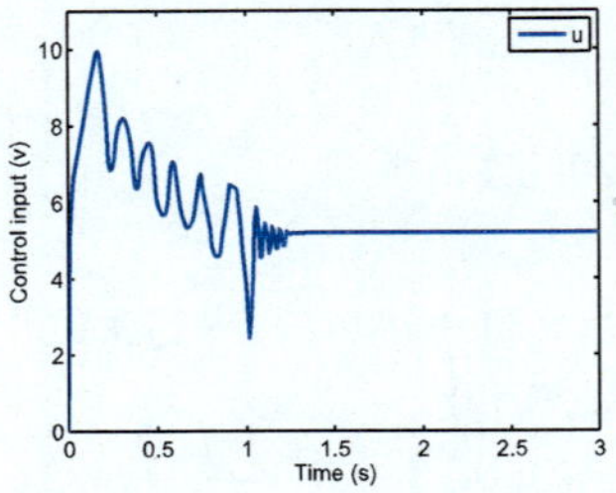

(c) Control input.

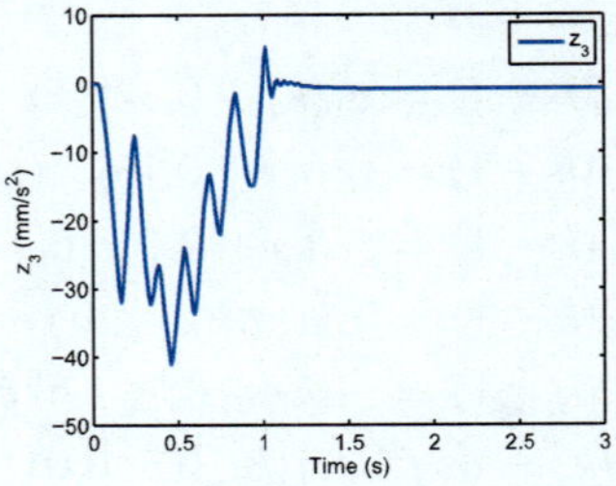

(d) Estimate of nonlinearity.

Fig. 20.3 Experimental results for tracking a step signal at 200 mm based on the active disturbance rejection control approach

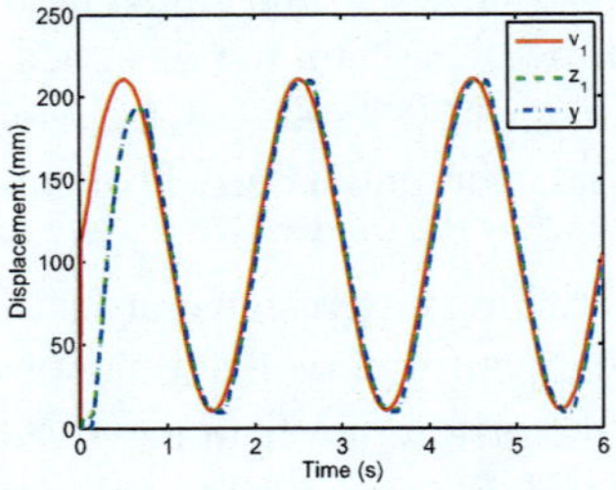

(a) Displacement.

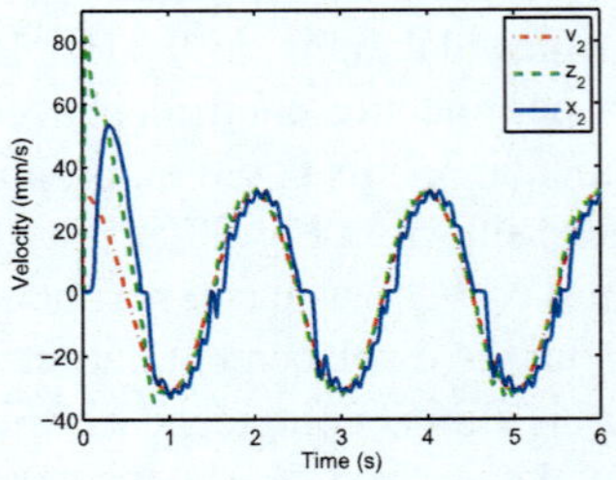

(b) Velocity.

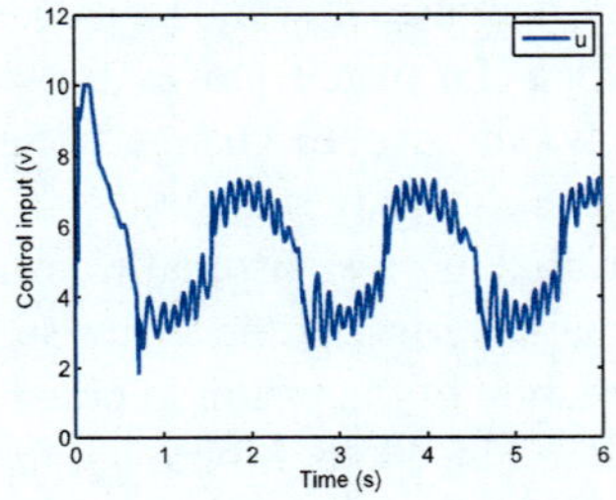

(c) Control input.

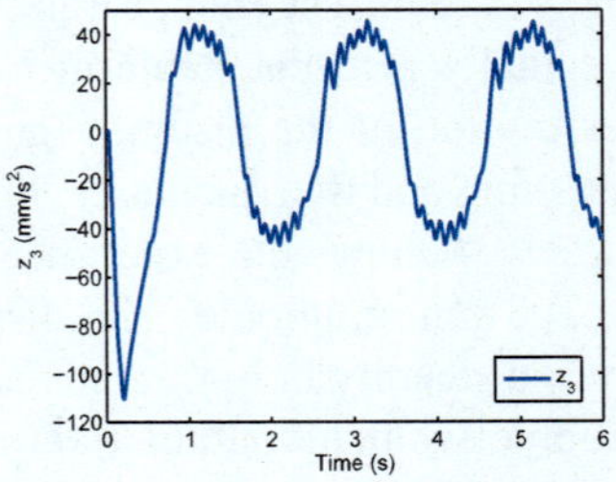

(d) Estimate of nonlinearity.

Fig. 20.4 Experimental results for tracking a sinusoidal signal at 0.5 Hz, 100 mm amplitude based on the active disturbance rejection control approach

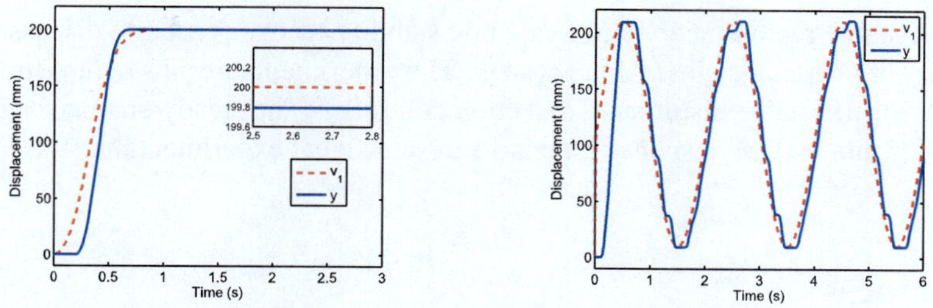

(a) Step displacement. (b) Sinusoidal displacement.

Fig. 20.5 Experimental results based on proportional integral derivative approach

the same change of u. The compensation of z_3 to nonlinear controller is vital to track a continuously changing trajectory from the sinusoidal experiment.

Figure 20.5 shows the experimental results for tracking a step signal at 200 mm and a sinusoidal signal at 0.5 Hz, 100 mm amplitude based on proportional integral derivative approach. The step response is fast in Fig. 20.5a, but the steady-state error is about 0.20 mm. The steady-state error is about 0.035 mm in Fig. 20.3a. Note that 0.035 mm $<$ 0.20 mm. Moreover, comparing with Fig. 20.4a, the sinusoidal experiment is unsatisfactory in Fig. 20.5b in which the piston can't track the given sinusoidal signal smoothly without frictional force compensation. Therefore, the control precision and tracking effect are better based on the active disturbance rejection control approach than proportional integral derivative method.

Remark 20.6 Theorem 20.4 is used to show the effectiveness of the designed ESO in (20.3). Theorem 20.5 has proven the stability of the close-loop rodless pneumatic system via nonlinear controller (20.13). Note that theoretical analysis is the precondition for doing the experiments. In fact, the theoretical analysis in our chapter is qualitative but not quantitative. Some parameters in the designed active disturbance rejection controller can be choose by referring to the theoretical analysis in Theorems 20.4 and 20.5. Therefore, the experimental results have also verified the validity of theoretical analysis.

20.4 Conclusion

In this chapter, the ADRC method for the position control of pneumatic magnetic rodless cylinder has been presented. This technique is an effective way to deal with the nonlinearity in pneumatic rodless cylinder system. ESO estimates the total nonlinearity in real time and all the observed states will be fed back to nonlinear state error feedback controller. The convergence of extended state observer is proven by the self-stable region theory and Lyapunov stability theory is used to discussed the stability of pneumatic closed-loop system. The experimental results demonstrate the validity of the active disturbance rejection controller. The steady-state error is less than 0.05 mm and the response time is 1 s in step signal experimental results.

References

1. Bone GM, Shu N (2007) Experimental comparison of position tracking control algorithms for pneumatic cylinder actuators. IEEE-ASME Trans Mechatron 12(5):557–561
2. Ramírez I (2013) Modeling and tracking control of a pneumatic servo positioning system. In: International congress of engineering mechatronics and automation, pp 1–6
3. Rahmat MF, Salim SNS, Sunar NH, Faudzi AM, Ismail ZH, Huda K (2012) Identification and non-linear control strategy for industrial pneumatic actuator. Int J Phys Sci 7(17):2565–2579
4. Morichika T, Kikkawa F, Oyama O, Yoshimitsu T (2007) Development of walking assist equipment with pneumatic cylinder. In: SICE annual conference, pp 1058–1063
5. Kirihara K, Saga N, Saito N (2008) Upper limb rehabilitation support device using a pneumatic cylinder. In: Annual conference of IEEE industrial electronics, pp 1287–1292
6. Tsai Y-C, Huang A-C (2008) Multiple-surface sliding controller design for pneumatic servo systems. Mechatronics 18(9):506–512
7. Hassan MY, Kothapalli G (2010) Comparison between neural network based PI and PID controllers. In: International multi-conference on systems, signals and devices, pp 1–6
8. Lee HK, Choi GS, Choi GH (2002) A study on tracking position control of pneumatic actuators. Mechatronics 12(6):813–831
9. Gao X, Feng Z-J (2005) Design study of an adaptive Fuzzy-PD controller for pneumatic servo system. Control Eng Pract 13(1):55–65
10. Qiu Z, Wang B, Zhang X, Han J (2013) Direct adaptive fuzzy control of a translating piezoelectric flexible manipulator driven by a pneumatic rodless cylinder. Mech Syst Signal Proc 36(2):290–316
11. Huang Y (1999) A new synthesis method for uncertain systems the self-stable region approach. Int J Syst Sci 30(1):33–38
12. Huang Y, Han J (2000) Analysis and design for the second order nonlinear continuous extended states observer. Chin Sci Bull 45(21):1938–1944

Chapter 21
Multi-controller Control

21.1 Introduction

Recently, a lot of researchers have been involved in controlling pneumatic actuators [1–4]. Some sliding mode controllers have been proposed to study the positioning control of pneumatic cylinders [5, 6]. Backstepping control based on Lyapunov methods is useful to improve positioning accuracy for a rod cylinder in [7]. Note that it is difficult to achieve high positioning accuracy and short transportation time for complicated conditions in practical industrial production. Therefore, the idea of using multiple controllers becomes popular to meet the need of industry over the last decades in [8, 9]. Various comprehensive controllers are needed to meet demand of high performance. A combination controller with both an adaptive law and a backstepping method is adopted to study tracking control for pneumatic systems without pressure sensors [10]. An adaptive robust controller containing sliding mode methods and backstepping techniques is also used to control a pneumatic cylinder to track a given trajectory [11]. The backstepping technique has been widely used for controlling pneumatic systems. Hence, to achieve ultra-fast response for a pneumatic rodless cylinder system, the backstepping technique is studied in this chapter.

In this paper, a backstepping-based controller is proposed to improve rapidity for pneumatic servo systems. A multi-controller strategy, which has been widely applied in many fields [8, 9], is introduced for pneumatic systems. Moreover, a least-square method is used to identify parameters of a proportional valve to obtain the relatively accurate model which is needed for the backstepping-based controller. An ESO is introduced to estimate nonlinear terms. Analysis based on a Lyapunov approach is presented to show convergence of the ESO. Finally, experimental results with response time 0.5 s and accuracy 0.05 mm verify the effectiveness of the proposed control strategy. The main contributions of this chapter are summarized as below:

i Pressure feedback is achieved to improve positioning speed of the pneumatic servo system by modifying experimental equipments.

ii A recursive least-square algorithm is given to obtain the accuracy mathematical model of a five-way proportional valve.

L. Zhao et al., *Pneumatic Servo Systems Analysis*, Advances in Industrial Control,
https://doi.org/10.1007/978-981-16-9515-5_21

iii A multi-controller strategy is proposed by designing a backstepping-based controller and a nonlinear error feedback controller.

21.2 Main Results

21.2.1 Schematic Diagram of Control Method

In this chapter, a multi-controller strategy is introduced to improve the positioning performance of the pneumatic rodless cylinder system. Figure 21.1 shows the control loop of the pneumatic rodless cylinder system.

In this chapter, a tracking differentiator is designed to generate the transition process of given signals and obtain the corresponding differential signal. The nonlinear function $f_{han}(\cdot)$ is used to avoid the phenomenon of high-frequency flutter in the process of digital computation effectively. Letting $\eta(t) = v_1(t) - v(t)$, the second-order tracking differentiator is devised as follows:

$$\begin{cases} v_1(k+1) = v_1(k) + hv_2(k) \\ v_2(k+1) = v_2(k) + hf_{han}(\eta, v_2(k), r, h_0), \end{cases} \tag{21.1}$$

where v represents the given signal, $v_1(t)$ is the arranging transient dynamics of $v(t)$, and $v_2(t)$ is the corresponding differential signal of $v_1(t)$. Moreover, both h_0 and r are positive constants. Note that parameter r determines the speed of transition process.

21.2.2 Extended State Observer

Extended state observers can estimate the unknown nonlinear part in real time. So analytical expression of total nonlinearity is not required in practice. According to

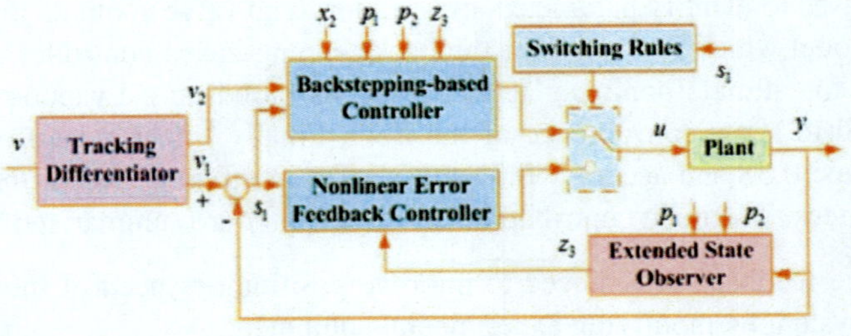

Fig. 21.1 Control loop of pneumatic rodless cylinder system

system (19.2), the nonlinear part is treated as an extended state $x_3(t)$, and $\omega(t)$ represents the derivative of $x_3(t)$. Then system (19.2) is extended to the following form:

$$\begin{cases} \dot{x}_1(t) = x_2(t) \\ \dot{x}_2(t) = x_3(t) + A(p_1(t) - p_2)/M \\ \dot{x}_3(t) = \omega(t). \end{cases} \tag{21.2}$$

Assumption 21.1 ([12]) In system (21.2), the unknown nonlinear part $x_3(t)$ is continuously differentiable and bounded. Moreover, $\omega(t)$ is also bounded in practice.

Letting $e_1(t) = z_1(t) - x_1(t)$, a third-order ESO is designed as follows:

$$\begin{cases} \dot{z}_1(t) = z_2(t) - \beta_1 e_1(t) \\ \dot{z}_2(t) = z_3(t) - \beta_2 fal(e_1(t), \sigma_1, \delta) + A(p_1(t) - p_2(t))/M \\ \dot{z}_3(t) = -\beta_3 fal(e_1(t), \sigma_2, \delta), \end{cases} \tag{21.3}$$

where $z_i(t)$ $(i = 1, 2, 3)$ are the observed values of $x_i(t)$. Moreover, β_1, β_2 and β_3 are three observer gains.

The following error system model is obtained from (21.2) and (21.3) as

$$\begin{cases} \dot{e}_1(t) = e_2(t) - \beta_1 e_1(t) \\ \dot{e}_2(t) = e_3(t) - \beta_2 f_1(t) \\ \dot{e}_3(t) = -\beta_3 f_2(t) - \omega(t), \end{cases} \tag{21.4}$$

where $e_i(t) = z_i(t) - x_i(t)$ $(i = 1, 2, 3)$, $f_1(t) = fal(e_1(t), \sigma_1, \delta)$, and $f_2(t) = fal(e_1(t), \sigma_2, \delta)$. In [12], the self-stable region method is proposed to study the convergence of a third-order ESO. However, the analysis process based on self-stable region is tedious and complicated. A new method is introduced to show the convergence of an ESO by Lyapunov stability approach in the following.

Theorem 21.1 *For system (21.4), there exist appropriate parameters β_1, β_2, and β_3 guaranteeing variables $e_1(t)$, $e_2(t)$, and $e_3(t)$ to converge to 0. That is, the extended state observer (21.3) estimates state variables of system (21.2) effectively.*

Proof Design a Lyapunov function as follows:

$$V(t) = \frac{1}{2}k_1 e_1^2(t) + \frac{1}{2}k_2 e_2^2(t) + \frac{1}{2}k_3 e_3^2(t),$$

where $k_i > 0$ $(i = 1, 2, 3)$. The derivative of $V(t)$ is obtained as

$$\begin{aligned}\dot{V}(t) &= k_1e_1(t)\dot{e}_1(t) + k_2e_2(t)\dot{e}_2(t) + k_3e_3(t)\dot{e}_3(t)\\ &= k_1e_1(t)(e_2(t) - \beta_1e_1(t)) + k_2e_2(t)(e_3(t) - \beta_2 f_1(t))\\ &\quad + k_3e_3(t)(-\beta_3 f_2(t) - \omega(t))\\ &= k_1e_1(t)e_2(t) - k_1\beta_1e_1^2(t) + k_2e_2(t)e_3(t) - k_2\beta_2e_2(t)f_1(t)\\ &\quad - k_3\beta_3e_3(t)f_2(t) - k_3e_3(t)\omega(t),\end{aligned}$$

where $\omega(t)$ is bounded in practice as in [12]. There exists a big enough parameter β_1, proper parameters β_2 and β_2 such that $\dot{V}(t) < 0$ holds. That is, the convergence of the ESO (21.3) is shown.

21.2.3 Backstepping-Based Controller

Let the following error variables be

$$\begin{cases} s_1(t) = v_1(t) - x_1(t)\\ s_2(t) = \phi_1(t) - x_2(t)\\ s_3(t) = \phi_2(t) - A(p_1(t) - p_2(t))/M,\end{cases}$$

where $\phi_1(t)$ and $\phi_2(t)$ are designed virtual control variables, and $s_i(t)$ $(i = 1, 2, 3)$ are error variables. According to (19.2), (19.3), and (19.12), the following error system is obtained as

$$\begin{cases} \dot{s}_1(t) = v_2(t) - x_2(t)\\ \dot{s}_2(t) = \dot{\phi}_1(t) - F_w(t)/M - A(p_1(t) - p_2(t))/M\\ \dot{s}_3(t) = \dot{\phi}_2(t) - Q(u(t))\Phi + \Psi,\end{cases} \tag{21.5}$$

where Φ and Ψ are described as follows:

$$\begin{cases} \Phi = \dfrac{K_0RT}{M}\left(\dfrac{g(p_1(t))}{x_0 + x_1(t)} + \dfrac{g(p_2(t))}{L_0 - x_0 - x_1(t)}\right)\\ \Psi = \dfrac{AK_0}{M}\left(\dfrac{p_1(t)}{x_0 + x_1(t)} + \dfrac{p_2(t)}{L_0 - x_0 - x_1(t)}\right)x_2(t)\end{cases}$$

and

$$Q(u(t)) = \text{sign}(u(t) - u^*)A(u(t)),$$

which is unable to obtain controller $u(t)$. Therefore, let

$$Q(u(t)) = \frac{1}{\Phi}(\Psi + \dot{\phi}_2(t) + s_2(t) + k_3 fal(s_3(t)) + k_4 fal(s_1(t))) \tag{21.6}$$

for designing the backstepping-based controller $u(t)$ of system (21.5) in this chapter. Then the backstepping-based controller $u(t)$ is calculated as follows:

$$\bar{u}(t) = Q^{-1}(u(t)), \tag{21.7}$$

where $Q^{-1}(u(t))$ is an inverse of $Q(u(t))$.

The following theorem is given to show the effectiveness of a backstepping-based controller for system (21.5).

Theorem 21.2 *There exists the backstepping-based controller (21.7) such that the error system (21.5) is stable by choosing some appropriate gains k_i $(i = 1, \ldots, 4)$. That is, state variables $s_j(t)$ $(j = 1, 2, 3)$ converge to 0 and position $x_1(t)$ tracks the reference signal $v_1(t)$ for the pneumatic rodless cylinder system (19.2).*

Proof Three steps are given on the backstepping controller for the pneumatic rodless cylinder system system (19.2) in the following.

• Step 1: Design the first Lyapunov function as

$$V_1(t) = \frac{1}{2}s_1^2(t).$$

Taking the derivative of $V_1(t)$, the following equation is obtained:

$$\dot{V}_1(t) = s_1(t)\dot{s}_1(t) = s_1(t)(v_2(t) - x_2(t)).$$

Letting $fal(s_1(t)) = fal(s_1(t), \sigma_3, \delta_1)$, and

$$\phi_1(t) = v_2(t) + k_1(t)fal(s_1(t)),$$

there exists

$$\dot{V}_1(t) = s_1(t)(\phi_1(t) - k_1 fal(s_1(t)) - x_2(t)) = -k_1 s_1(t) fal(s_1(t)) + s_1(t)s_2(t).$$

When $s_1(t) \neq 0$, there is $s_1(t)fal(s_1(t)) > 0$. If $s_2(t) = 0$, then there exists a positive parameter k_1 such that $\dot{V}_1(t) < 0$ holds. State variable $s_1(t)$ converges to 0 for the error system (21.5).

• Step 2: Define the second Lyapunov function as

$$V_2(t) = V_1(t) + \frac{1}{2}s_2^2(t).$$

Differentiating $V_2(t)$, the following equation is obtained:

$$\begin{aligned}\dot{V}_2(t) &= \dot{V}_1(t) + s_2(t)\dot{s}_2(t)\\ &= -k_1 s_1(t) fal(s_1(t)) + s_1(t)s_2(t)\\ &\quad + s_2(t)\left(\dot{\phi}_1 - \frac{F_w(t)}{M} - \frac{A}{M}(p_1(t) - p_2(t))\right).\end{aligned}$$

Since nonlinear terms are unavailable, the estimated value $z_3(t)$ of nonlinear terms is used in the virtual controller ϕ_2. Let $fal(s_2(t)) = fal(s_2(t), \sigma_4, \delta_2)$, and

$$\phi_2(t) = \dot{\phi}_1(t) - z_3(t) + s_1(t) + k_2 fal(s_2(t)).$$

Note that variable $\dot{\phi}_1$ is obtained by the tracking differentiator (21.1). The following expression is obtained as

$$\begin{aligned}\dot{V}_2(t) &= -k_1 s_1(t) fal(s_1(t)) + s_1(t)s_2(t) + s_2(t)\Big(\phi_2(t) + z_3(t)\\ &\quad - s_1(t) - k_2 fal(s_2(t)) - \frac{F_w(t)}{M} - \frac{A}{M}(p_1(t) - p_2(t))\Big)\\ &= -k_1 s_1(t) fal(s_1(t)) - k_2 s_2(t) fal(s_2(t)) + s_2(t)s_3(t)\\ &\quad + s_2(t)\left(z_3(t) - \frac{F_w(t)}{M}\right).\end{aligned}$$

Suppose that Δ is a bounded positive number and

$$\left| z_3(t) - \frac{F_w(t)}{M} \right| < \Delta.$$

One has that

$$\dot{V}_2 < -k_1 s_1(t) fal(s_1(t)) - k_2 s_2(t) fal(s_2(t)) + s_2(t)s_3(t) + | s_2(t) | \Delta.$$

If $s_3(t) = 0$, then there exist two parameters k_1 and k_2 such that $(s_1(t), s_2(t))$ converges to $(0, 0)$.

- Step 3: Design the third Lyapunov function as

$$V_3(t) = V_2(t) + \frac{1}{2}s_3^2(t).$$

The derivative of $V_3(t)$ is shown as follows:

$$\begin{aligned}\dot{V}_3(t) &= \dot{V}_2(t) + s_3(t)\dot{s}_3(t)\\ &= -k_1 s_1 fal(s_1(t)) - k_2 s_2(t) fal(s_2(t)) + s_2(t)s_3(t)\\ &\quad + s_2(t)\left(z_3(t) - \frac{F_w(t)}{M}\right) + s_3(t)(\dot{\phi}_2 - Q(u(t))\Phi + \Psi).\end{aligned}$$

Design $Q(u(t))$ as

$$Q(u(t)) = \frac{1}{\Phi}(\Psi + \dot{\phi}_2 + s_2(t) + k_3 fal(s_3(t)) + k_4 fal(s_1(t))),$$

where $fal(s_3(t)) = fal(s_3(t), \sigma_5, \delta_3)$. Variable $\dot{\phi}_2$ is also obtained by the tracking differentiator (21.1). There exists

$$\begin{aligned}
\dot{V}_3(t) = & -k_1 s_1(t) fal(s_1(t)) - k_2 s_2(t) fal(s_2(t)) - k_3 s_3(t) fal(s_3(t)) \\
& -k_4 fal(s_1(t)) s_3(t) + s_2(t)\left(z_3(t) - \frac{F_w(t)}{M}\right) \\
< & -k_1 s_1 fal(s_1(t)) - k_2 s_2(t) fal(s_2(t)) - k_3 s_3(t) fal(s_3(t)) \\
& -k_4 fal(s_1(t)) s_3(t) + \mid s_2(t) \mid \Delta \\
< & -k_1 s_1(t) fal(s_1(t)) + \frac{1}{4} k_4^2 fal^2(s_1(t)) - k_3 s_3(t) fal(s_3(t)) + s_3^2(t) \\
& -\frac{1}{4} k_4^2 fal^2(s_1(t)) - s_3^2(t) - k_4 fal(s_1(t)) s_3(t) - k_2 s_2(t) fal(s_2(t)) \\
& + \mid s_2(t) \mid \Delta \\
< & -k_1 s_1(t) fal(s_1(t)) + \frac{1}{4} k_4^2 fal^2(s_1(t)) - k_3 s_3(t) fal(s_3(t)) + s_3^2(t) \\
& -\left(\frac{1}{2} k_4 fal(s_1(t)) + s_3(t)\right)^2 - k_2 \mid s_2(t) \mid fal(s_2(t)) + \mid s_2(t) \mid \Delta \\
< & -k_1 s_1(t) fal(s_1(t)) + \frac{1}{4} k_4^2 fal^2(s_1(t)) \\
< & -k_3 s_3(t) fal(s_3(t)) + s_3^2(t) - \left(\frac{1}{2} k_4 fal(s_1(t)) + s_3(t)\right)^2 \\
& - \mid s_2(t) \mid (k_2 fal(s_2(t)) - \Delta).
\end{aligned}$$

When $(s_1(t), s_2(t), s_3(t)) \neq (0, 0, 0)$, $s_1(t) fal(s_1(t)) > 0$, and $s_3(t) fal(s_1(t)) > 0$ hold as shown in Fig. 21.5. There exists a big enough adjustable parameter k_1 and a smaller adjustable parameter k_4, such that $-k_1 s_1(t) fal(s_1(t)) + \frac{1}{4} k_4^2 fal^2(s_1(t)) < 0$ holds. Similarly, there exist two big enough adjustable parameters k_3 and k_2, such that $-k_3 s_3(t) fal(s_3(t)) + s_3^2(t) < 0$ and $- \mid s_2(t) \mid (k_2 fal(s_2(t)) - \Delta) < 0$. Therefore, there exist appropriate adjustable parameters k_i, $i = 1, 2, 3, 4$, such that $\dot{V}_3(t) < 0$ holds for any $(s_1(t), s_2(t), s_3(t)) \neq (0, 0, 0)$. This completes the proof.

21.2.4 Multi-controller Strategy

In [12], an active disturbance rejection controller is designed to realize high positioning accuracy for a pneumatic servo system. According to [13], a simplified nonlinear error controller is designed as

$$u(t) = \alpha_1 fal(s_1(t)) - \alpha_2 z_3(t), \tag{21.8}$$

where both α_1 and α_2 are adjustable parameters.

In order to improve the convergence rate, a multi-controller strategy is expressed in this chapter as follows:

$$u(t) = \begin{cases} \text{Controller (21.7)}, \ |s_1(t)| > s_1^* \\ \text{Controller (21.8)}, \ |s_1(t)| \le s_1^*, \end{cases} \tag{21.9}$$

where s^* is the switching term. That is, both the backstepping-based controller and the nonlinear error feedback controller are used in this chapter. The error $s_1(t)$ between input signal and output position is considered as the basis of switching. In addition, an ESO is designed to estimate nonlinearity in system to compensate. Owing to the tracking differentiator, a differential signal is easily obtained.

21.3 Experiments and Results

In this section, the physical map of a single pneumatic rodless cylinder is shown in Fig. 21.2. Set air supply pressure 0.6 MPa and atmospheric pressure 0.1 MPa. Firstly, a step signal with 0.2 m is given as the reference input. The step experimental results are shown in Figs. 21.3, 21.4, 21.5 and 21.6. In addition, the main adjustable parameters are listed in Table 21.1.

Owing to the accurate system model, there aren't fluctuations for estimated value $z_3(t)$ in Fig. 21.3. Limited to the physical structure of the five-way valve, a small amount of compressed air flows in rodless cylinder such that the pressures $p_1(t)$ and p_2 are changing slowly Fig. 21.4. Control value converges to the region of null voltage smoothly in Fig. 21.5. The experimental results of position is shown in Fig. 21.6. Reference input $v_0(t)$ is the step signal with 0.2 m. Signal $v_1(t)$ is obtained by a tracking differentiator. Output $y(t)$ and the observed value $z_1(t)$ are almost coincident. Both accuracy 0.05 mm and rapidity response 0.5 s are achieved according to the curve $y(t)$.

Besides the step response, a sine signal with 1 Hz, 0.1m amplitude is given to study the trajectory tracking control of the rodless cylinder in Fig. 21.7. Obviously, output $y(t)$ tracks given signal $v_1(t)$ well, and the observed value z_1 is almost coincident

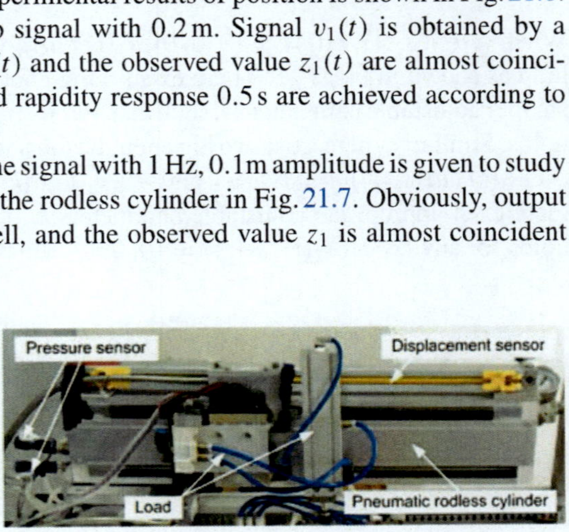

Fig. 21.2 Single pneumatic rodless cylinder

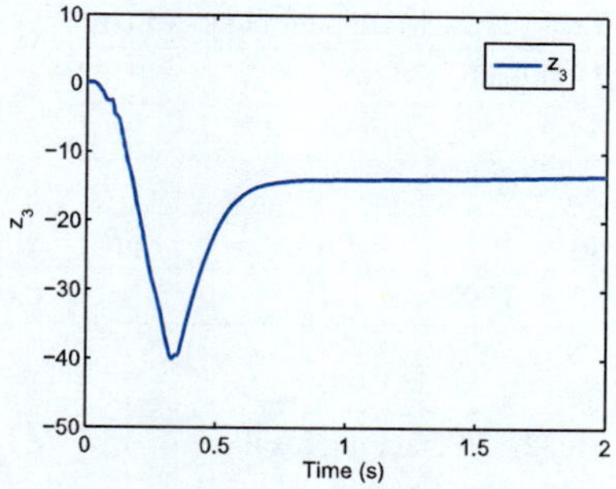

Fig. 21.3 Estimated value of pneumatic rodless cylinder system

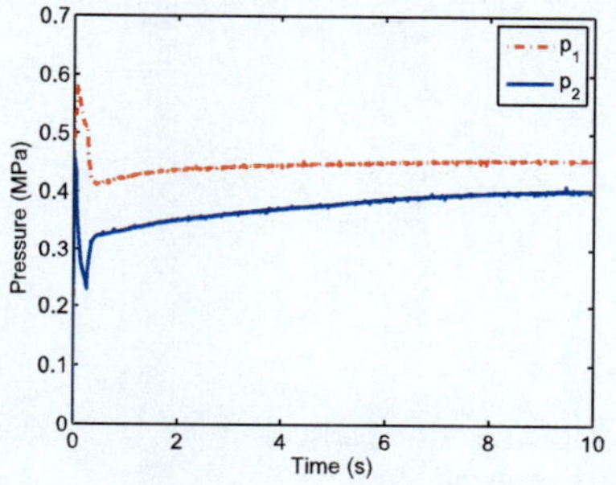

Fig. 21.4 Pressures of pneumatic rodless cylinder system

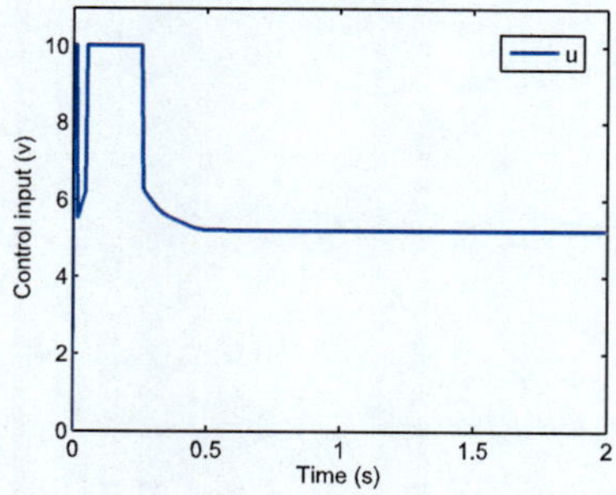

Fig. 21.5 Control value of pneumatic rodless cylinder system

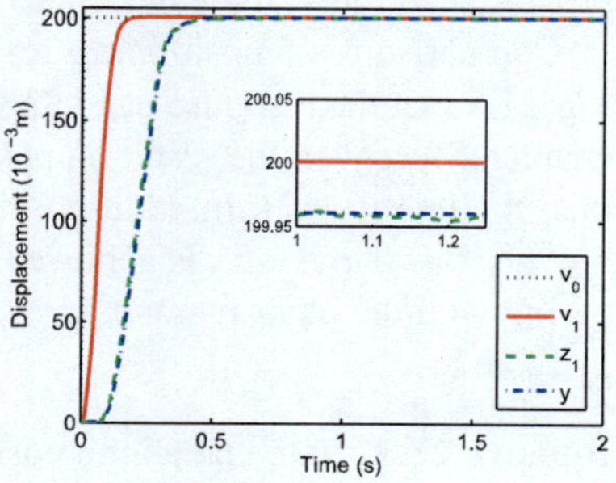

Fig. 21.6 Positioning results for pneumatic rodless cylinder system

Table 21.1 Main parameters of controller

Parameters		
$k_1 = 1$	$\beta_1 = 200$	$\alpha_1 = 0.1712$
$k_2 = 0.5$	$\beta_2 = 600$	$\alpha_2 = 0.001$
$k_3 = 10$	$\beta_3 = 1000$	$p^* = 0.29$
$k_4 = 1000$	$s_1^* = 0.062\,\mathrm{m}$	

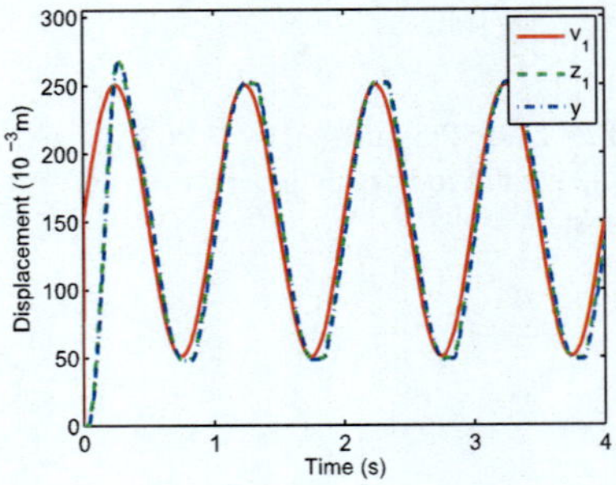

Fig. 21.7 Tracking result for pneumatic rodless cylinder system

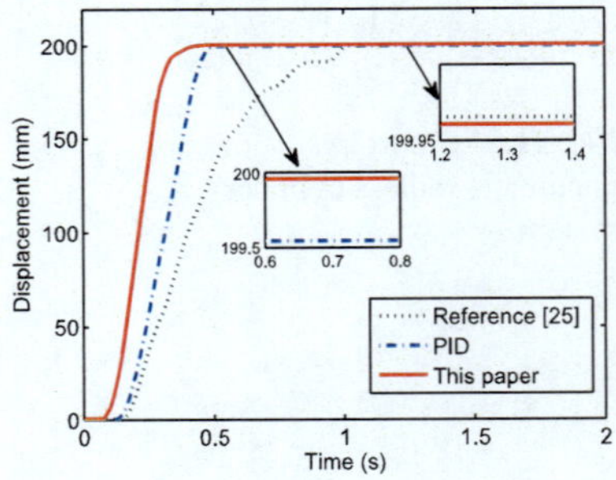

Fig. 21.8 Comparisons of experimental results

with $y(t)$. Therefore, Fig. 21.7 shows a good performance in trajectory tracking which proves the validity of the designed control strategy.

Comparisons of positioning results are shown in Fig. 21.8. It is shown from Fig. 21.8 that the response delay has been improved by the proposed method in this chapter. Moreover, the result on positioning changes more smoothly in this chapter than the one in [12]. In addition, the better performances with accuracy 0.05 mm and response time 0.5 s is achieved by the designed multi-controller strategy comparing with the other two methods. Furthermore, quantitative comparisons is given in Table 21.2.

Remark 21.3 Note that a high positioning accuracy 0.05 mm has been achieved in [12]. However, output in [13] has fluctuation phenomenon, and the response time 1 s is too long. Though, the result based on PID has advantage in response time 0.5 s, the positioning accuracy 0.5 mm isn't desired. The output in this chapter realizes good performances in high positioning accuracy 0.05 mm and rapidity 0.5 s.

Table 21.2 Comparisons results

	Response time (s)	Accuracy (mm)
PID controller	0.5	0.5
Controller in [12]	1	0.05
Controller in this chapter	0.5	0.05

Remark 21.4 The performance of tracking sine signal shown in Fig. 21.7 is determined by the multi-controller strategy which is designed in this chapter. The frequency of sine signal has been improved in Fig. 21.7 with 1 Hz, but the result in [12] is with 0.5 Hz. The principal reason for experimenting of tracking a sine signal is to further show that the multi-controller strategy is applicable for varying signals.

21.4 Conclusion

In this chapter, a nonlinear backstepping controller has been applied to improve response speed of the pneumatic rodless cylinder system. Meanwhile, another nonlinear controller is also used to keep high positioning accuracy. Therefore, a multi-controller strategy has been proposed to realize the switch between two different controllers. An ESO has been designed to estimate nonlinear terms and the estimated value is compensated in controller design. Experimental results verify the effectiveness of the designed control strategy.

References

1. Bone GM, Shu N (2007) Experimental comparison of position tracking control algorithms for pneumatic cylinder actuators. IEEE-ASME Trans Mechatron 12(5):557–561
2. Wang J, Pu J, Moore P (1999) A practical control strategy for servo-pneumatic actuator systems. Control Eng Pract 7(12):1483–1488
3. Saleem A, Taha B, Tutunji T, Al-Qaisia A (2015) Identification and cascade control of servo-pneumatic system using particle swarm optimization. Simul Model Pract Theory 52:164–179
4. Sobczyk MR, Gervini VI, Perondi EA, Cunha MAB (2016) A continuous version of the LuGre friction model applied to the adaptive control of a pneumatic servo system. J Frankl Inst 353(13):3021–3039
5. Paul AK, Mishra JE, Radke MG (1994) Reduced order sliding mode control for pneumatic actuator. IEEE Trans Control Syst Technol 2(3):271–276
6. Surgenor BW, Vaughan ND (1997) Continuous sliding mode control of a pneumatic actuator. J Dyn Syst Meas Control 119(3):578–581
7. Rao Z, Bone GM (2008) Nonlinear modeling and control of servo pneumatic actuators. IEEE Trans Control Syst Technol 16(3):562–569
8. Paul A, Akar M, Safonov MG, Mitra U (2005) Adaptive power control for wireless networks using multiple controllers and switching. IEEE Trans Neural Netw 16(5):1212–1218

9. Seron MM, De Dona JA (2010) Actuator fault tolerant multi-controller scheme using set separation based diagnosis. Int J Control 83(11):2328–2339
10. Ren H-P, Huang C (2013) Adaptive backstepping control of pneumatic servo system. In: IEEE international symposium on industrial electronics, pp 1–6
11. Nuchkrua T, Leephakpreeda T (2013) Fuzzy self-tuning PID control of hydrogen-driven pneumatic artificial muscle actuator. J Bionic Eng 10(3):329–340
12. Qiu Z, Wang B, Zhang X, Han J (2013) Direct adaptive fuzzy control of a translating piezoelectric flexible manipulator driven by a pneumatic rodless cylinder. Mechan Syst Signal Proc 36(2):290–316
13. Zheng Q, Dong L, Lee DH, Gao Z (2009) Active disturbance rejection control for MEMS gyroscopes. IEEE Trans Control Syst Technol 17(6):1432–1438

Chapter 22
Linear Feedback Control

22.1 Introduction

Pneumatic rodless cylinder systems are extensively applied into a variety of industry fields, such as food packaging, automobile manufacturing industry, manipulation, mobile robotic systems, and other automation systems [1–4]. Many approaches have been proposed to improve the position control performance of pneumatic cylinder systems [5, 6]. However, actuator saturation is often overlooked in controller design of the pneumatic cylinder systems in previous works. The actuator saturation is inevitable in many practical control systems due to the limitation of actuators or inherent physical constraints of systems. The actuator saturation has a great negative effect on the control performance of a system, even giving rise to undesirable inaccuracy and leading instability [7–9]. In [10], a switching-based adaptive control scheme has been proposed to cope with actuator saturation in nonlinear teleoperation systems. Adaptive control of single input uncertain nonlinear systems in the presence of input saturation and unknown external disturbance based on backstepping approaches has been introduced in [11]. However, owing to the difficulty of the problem on actuator saturation, few of available results are presented by taking saturation into account in the design and analysis of pneumatic rodless cylinder systems. Considering the good tracking and robust performance of ADRC algorithm, a strictly invariant set accompanied by the actuator saturation has to be estimated in the ADRC control scheme, which is an effective method for the pneumatic rodless cylinder system control. This work is important and challenging in both theory and practice, which motivate us to research.

In this chapter, positioning tracking control of a pneumatic rodless cylinder system with actuator saturation is investigated via an ADRC. A linear ESO is designed to estimate and compensate strong friction force and other nonlinearities in the pneumatic rodless cylinder system. An actuator saturation linear feedback control law is developed to further improve the tracking performance. Furthermore, a linear matrix inequality-based optimization algorithm is employed to estimate a strictly invariance set for the closed-loop system. Experiment results with response time 0.5 s and

L. Zhao et al., *Pneumatic Servo Systems Analysis*, Advances in Industrial Control,
https://doi.org/10.1007/978-981-16-9515-5_22

accuracy 0.005 mm for a 200 mm step signal demonstrate the effectiveness of the proposed control strategy. The main contributions of this paper are summarized as follows:

i An ADRC with actuator saturation is designed for the pneumatic rodless cylinder to deal with the internal uncertainties and the external disturbances.
ii A LMI-based framework is established for the ADRC to enlarge the estimate of the domain of attraction of the pneumatic rodless cylinder system.
iii Experiment results illustrate that the proposed controller significantly improves the positioning accuracy by considering the actuator saturation.

22.2 Main Results

22.2.1 Schematic Diagram of Control Method

In order to improve positioning accuracy of the pneumatic rodless cylinder system with nonlinearities and actuator saturation, active disturbance rejection control is proposed in this chapter. The block diagram of the active disturbance rejection controller for the pneumatic rodless cylinder system subject to actuator saturation is shown in Fig. 22.1. The following second-order differentiator is described as follows:

$$\dot{v}_1(t) = v_2(t) \tag{22.1}$$

$$\dot{v}_2(t) = fhan(v_1(t) - v_0(t), v_2(t), r, h_0), \tag{22.2}$$

where $v_0(t)$ is the given signal, $v_1(t)$ is the tracking signal of $v_0(t)$, $v_2(t)$ is the differential signal of $v_1(t)$, r and h_0 are two adjustable parameters. The speed of the transition process can be regulated by adjusting In this chapter, the tracking error $e_d(t)$ is a given signal of the tracking differentiator, i.e., $e_d(t) = x_1(t)$, $\dot{e}_d(t) = x_2(t)$. That is, $v_1(t)$ is the tracking signal of $x_1(t)$, $v_2(t)$ is the differential signal of $x_2(t)$. An approximation error of the tracking differentiator is omitted as in [12]. Therefore, $v_1(t)$ and $v_2(t)$ are used to replace $x_1(t)$ and $x_2(t)$ in controller design, respectively.

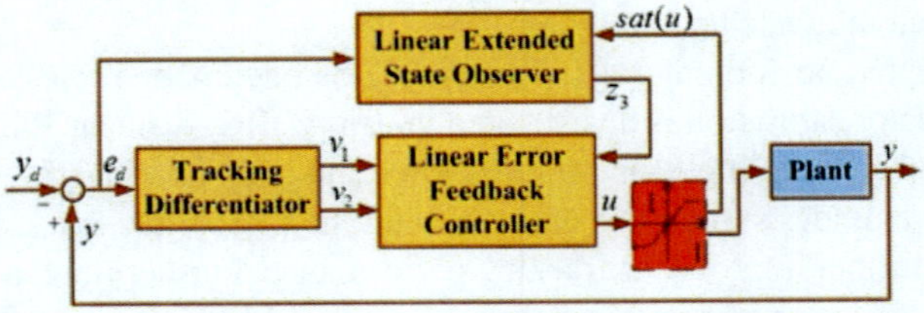

Fig. 22.1 The hardware block diagram of pneumatic rodless cylinder system

22.2.2 Linear Extended State Observer

The ESO is designed to deal with uncertainties of the pneumatic rodless cylinder system (19.2). The unknown nonlinear dynamics $f(x_1(t), x_2(t))$ mainly contains friction force, which is continuously differentiable and bounded. It is treated as an extended state $x_3(t)$, i.e., $f(x_1(t), x_2(t)) = x_3(t)$. Then the pneumatic rodless cylinder system (19.2) is rewritten as follows:

$$\dot{x}_1(t) = x_2(t) \tag{22.3}$$
$$\dot{x}_2(t) = x_3(t) + b_0 u_m \text{sat}(u(t)) \tag{22.4}$$
$$\dot{x}_3(t) = h(t), \tag{22.5}$$

where $h(t)$ is the derivative of $x_3(t)$. Note that $h(t)$ is bounded in practice. If $(x_1(t), x_2(t)) = (0, 0)$, then $h(t) = 0$.

The ESO for system (22.3)–(22.5) is represented as Follows:

$$\dot{z}_1(t) = z_2(t) + \alpha_1(x_1(t) - z_1(t)) \tag{22.6}$$
$$\dot{z}_2(t) = z_3(t) + \alpha_2(x_1(t) - z_1(t)) + b_0 u_m \text{sat}(u(t)) \tag{22.7}$$
$$\dot{z}_3(t) = \alpha_3(x_1(t) - z_1(t)), \tag{22.8}$$

where

$$\alpha_1 = \frac{l_1}{\varepsilon}, \ \alpha_2 = \frac{l_2}{\varepsilon^2}, \ \alpha_3 = \frac{l_3}{\varepsilon^3}$$

and ε is a small positive constant. Moreover, $z_1(t)$, $z_2(t)$, and $z_3(t)$ are the observations of $x_1(t)$, $x_2(t)$, and $x_3(t)$, respectively, $L = [l_1, l_2, l_3]^T$ are observer gains which are selected such that the characteristic polynominal $s^3 + l_1 s^2 + l_2 s + l_3$ is Hurwitz.

For $i = 1, 2, 3$, let

$$e_i(t) = x_i(t) - z_i(t) \tag{22.9}$$
$$\xi_i(t) = \frac{e_i(t)}{\varepsilon^{3-i}}. \tag{22.10}$$

Associated with system (22.3)–(22.5) and ESO (22.6)–(22.8), the error system is obtained that

$$\dot{\xi}_1(t) = \frac{1}{\varepsilon}\xi_2(t) - \frac{l_1}{\varepsilon}\xi_1(t) \tag{22.11}$$
$$\dot{\xi}_2(t) = \frac{1}{\varepsilon}\xi_3(t) - \frac{l_2}{\varepsilon}\xi_1(t) \tag{22.12}$$
$$\dot{\xi}_3(t) = h(t) - \frac{l_3}{\varepsilon}\xi_1(t). \tag{22.13}$$

The error system (22.11)–(22.13) are rewritten as follows:

$$\dot{\xi}(t) = \frac{1}{\varepsilon}(\Lambda + LB_3)\xi(t) + B_4 h(t), \tag{22.14}$$

where $\xi(t) = [\xi_1(t), \xi_2(t), \xi_3(t)]^T$,

$$\Lambda = \begin{bmatrix} 0\,1\,0 \\ 0\,0\,1 \\ 0\,0\,0 \end{bmatrix}, \ B_3 = \begin{bmatrix} -1\,0\,0 \end{bmatrix}, \ B_4 = \begin{bmatrix} 0 \\ 0 \\ 1 \end{bmatrix}.$$

Assumption 22.1 ([13]) For $\forall \xi(t) = [\xi_1(t), \xi_2(t), \xi_3(t)]^T \in R^3$, there exist constants β_{11}, β_{12}, β_{13}, β_{14} and ϑ, and positive definite, radially unbounded, and continuous differentiable functions $V_1, W_1 : R^3 \to R$ such that

$$\beta_{11}\|\xi(t)\|^2 \leq V_1(\xi(t)) \leq \beta_{12}\|\xi(t)\|^2 \tag{22.15}$$

$$\beta_{13}\|\xi(t)\|^2 \leq W_1(\xi(t)) \leq \beta_{14}\|\xi(t)\|^2 \tag{22.16}$$

$$\sum_{i=1}^{2}(\xi_{i+1}(t) - l_i\xi_1(t))\frac{\partial V_1(\xi(t))}{\partial \xi_i(t)} - l_3\xi_1(t)\frac{\partial V_1(\xi(t))}{\partial \xi_3(t)} \leq -W_1(\xi(t)) \tag{22.17}$$

$$\left|\frac{\partial V_1(\xi(t))}{\partial \xi_3(t)}\right| \leq \vartheta\|\xi(t)\|. \tag{22.18}$$

Theorem 22.2 *Consider the error system formed of (22.14) and the ESO (22.6)–(22.8). There exists a positive constant $\tau > 0$ with $t > \tau$ such that*

$$\lim_{\varepsilon \to 0} |x_i(t) - z_i(t)| = 0, \ i = 1, 2, 3.$$

Then the error system (22.14) is convergent. That is, the ESO (22.6)–(22.8) designed in this chapter is effective.

Proof First, the derivative of the x_3 is satisfied with

$$|h(t)| = \left| x_2(t)\frac{\partial f(x_1(t), x_2(t))}{\partial x_1(t)} + (x_3(t) + b_0 u_m \mathrm{sat}(u(t))) \cdot \frac{\partial f(x_1(t), x_2(t))}{\partial x_2(t)} \right|. \leq N \tag{22.19}$$

There is a positive constant $N > 0$, such that $|h(t)| \leq N$ for all $t \geq 0$.

Since the ESO gain matrix L is designed such that $\Lambda + LB_3$ is Hurwitz, the conditions in Assumption 22.1 are naturally guaranteed. Let P be the positive definite matrix solution of the following Lyapunov Equation:

$$P(\Lambda + LB_3) + (\Lambda + LB_3)^T P = -I. \tag{22.20}$$

Let the Lyapunov functions $V_1(\xi(t)), W_1(\xi(t)) : R^3 \to R$ be given as follows:

$$V_1(\xi(t)) = \xi^T(t)P\xi(t),\ \ W_1(\xi(t)) = \|\xi(t)\|^2. \tag{22.21}$$

One has that

$$\beta_{\min}(P)\|\xi(t)\|^2 \le V_1(\xi(t)) \le \beta_{\max}(P)\|\xi(t)\|^2 \tag{22.22}$$

$$\sum_{i=1}^{2}(\xi_{i+1}(t) - l_i\xi_1(t))\frac{\partial V_1(\xi(t))}{\partial \xi_i(t)} - l_3\xi_1(t)\frac{\partial V_1(\xi(t))}{\partial \xi_3(t)}$$
$$= -\xi^T(t)\xi(t) = -\|\xi(t)\|^2 = -W_1(\xi(t)) \tag{22.23}$$

$$\left|\frac{\partial V_1(\xi(t))}{\partial \xi_3(t)}\right| \le \left\|\frac{\partial V_1(\xi(t))}{\partial \xi(t)}\right\| = \|2\xi(t)^T P\|$$
$$\le 2\|P\|\|\xi(t)\| = 2\beta_{\max}(P)\|\xi(t)\|. \tag{22.24}$$

By conditions (22.22)–(22.24), the derivative of $V_1(\xi(t))$ with respect to t is computed as

$$\begin{aligned}\frac{dV_1(\xi(t))}{dt} &= \frac{1}{\varepsilon}\left\{\sum_{i=1}^{2}(\xi_{i+1}(t) - l_i\xi_1(t))\frac{\partial V_1(\xi(t))}{\partial \xi_i(t)}\right.\\ &\quad \left.+(\varepsilon h(t) - l_3\xi_1(t))\frac{\partial V_1(\xi(t))}{\partial \xi_3(t)}\right\}\\ &\le -\frac{1}{\varepsilon}W_1(\xi(t)) + |h|\cdot\left|\frac{\partial V_1(\xi(t))}{\partial \xi_3(t)}\right|\\ &\le -\frac{1}{\varepsilon}\|\xi(t)\|^2 + 2N\beta_{\max}(P)\|\xi(t)\|\\ &\le -\frac{V_1(\xi(t))}{\beta_{\max}(P)\varepsilon} + 2N\beta_{\max}(P)\frac{\sqrt{V_1(\xi(t))}}{\sqrt{\beta_{\min}(P)}}.\end{aligned} \tag{22.25}$$

For

$$\frac{dV_1(\xi(t))}{dt} = 2\sqrt{V_1(\xi(t))}\frac{d\sqrt{V_1(\xi(t))}}{dt},$$

the inequality (22.25) is rewritten as follows:

$$\frac{d\sqrt{V_1(\xi(t))}}{dt} \le -\frac{1}{2\beta_{\max}(P)\varepsilon}\sqrt{V_1(\xi(t))} + \frac{N\beta_{\max}(P)}{\sqrt{\beta_{\min}(P)}}.$$

Consider (22.22), one has that

$$
\begin{aligned}
\|\xi(t)\| &\le \frac{\sqrt{V_1(\xi(t))}}{\sqrt{\beta_{\min}(P)}} \\
&\le \left(\frac{\sqrt{V_1(\xi(0))}}{\sqrt{\beta_{\min}(P)}} - \frac{2N\beta_{\max}^2(P)\varepsilon}{\beta_{\min}(P)}\right)\cdot \\
&\quad \exp\left(-\frac{t}{2\beta_{\max}(P)\varepsilon}\right) + \frac{2N\beta_{\max}^2(P)\varepsilon}{\beta_{\min}(P)}.
\end{aligned}
$$

Together with (22.9), it follows that

$$
\begin{aligned}
|e_i(t)| &= \varepsilon^{3-i}\left|\xi_i\left(\frac{t}{\varepsilon}\right)\right| \le \varepsilon^{3-i}\left\|\xi\left(\frac{t}{\varepsilon}\right)\right\| \\
&\le \varepsilon^{3-i}\left[\left(\frac{\sqrt{V_1(\xi(0))}}{\sqrt{\beta_{\min}(P)}} - \frac{2N\beta_{\max}^2(P)\varepsilon}{\beta_{\min}(P)}\right)\cdot\right. \\
&\quad \left.\exp\left(-\frac{t}{2\beta_{\max}(P)\varepsilon^2}\right) + \frac{2N\beta_{\max}^2(P)\varepsilon}{\beta_{\min}(P)}.\right]
\end{aligned}
\tag{22.26}
$$

It is shown that

$$
\lim_{\varepsilon\to 0} |e_i(t)| = 0,\ i = 1, 2, 3.
$$

is uniform in $t \in [\tau, \infty)$.

22.2.3 Linear Error Feedback Controller

The ADRC control law with a state feedback gain K is given by

$$
u(t) = \frac{1}{b_0}\left(Kx(t) - \frac{z_3(t)}{u_m}\right), \tag{22.27}
$$

where $K = [k_1, k_2]$, $x(t) = [x_1(t), x_2(t)]^T$.

System (19.2) is written as the following equivalent state space form:

$$
\dot{x}(t) = Ax(t) + u_m B_1 \mathrm{sat}(u(t)) + B_2 f(x_1(t), x_2(t)), \tag{22.28}
$$

where

$$
A = \begin{bmatrix} 0 & 1 \\ 0 & 0 \end{bmatrix},\ B_1 = \begin{bmatrix} 0 \\ b_0 \end{bmatrix},\ B_2 = \begin{bmatrix} 0 \\ 1 \end{bmatrix}
$$

Disturbance rejection with guaranteed the domain of attraction is mainly studied in this subsection. Some lemmas are shown in the following.

Lemma 22.3 ([14]) *For a positive number γ, one has that*

$$2D^T S \leq \frac{1}{\gamma} D^T D + \gamma S^T S, \ \forall D, S \in R^n.$$

Let $Q \in R^{n \times n}$ be a positive definite matrix. For $\rho > 0$, denote

$$\Psi(Q, \rho) = \{x(t) \in R^n : x^T Q x(t) \leq \rho\}. \tag{22.29}$$

For a matrix $H \in R^{m \times n}$, denote the lth row of H as h_l and define

$$\Theta(H) = \{x \in R^n : |h_l x(t)| \leq 1, l \in [1, m]\}. \tag{22.30}$$

Let E be the set of $m \times m$ diagonal matrices with elements being either 1 or 0. Suppose that each element of E is labeled as E_i, $i \in G = [1, 2^m]$, and denote $E_i^- = I - E_i$. Furthermore, set $\sum_{i=1}^{2^m} \eta_i = 1$ with $0 \leq \eta_i \leq 1$, it follows that

$$\sum_{i=1}^{2^m} \eta_i (E_i + E_i^-) = I.$$

The convex hull is recalled to represent the saturated feedback controller (22.27).

Lemma 22.4 ([8]) *Let $u, v \in R^m$, $u = [u_1, u_2, \ldots, u_m]$, $v = [v_1, v_2, \ldots, v_m]$. If $|v_j| < 1$ for all $j \in [1, m]$, then*

$$sat(u(t)) \in co\{E_i u(t) + E_i^- v, i \in G\}.$$

Let $K, H \in R^{m \times n}$ be given. Note that $|h_j x| \leq 1$ for all $j \in [1, m]$. By Lemma 22.4, it is obtained that

$$\text{sat}(Kx) \in \text{co}\{E_i Kx + E_i^- Hx, i \in G\}.$$

Consequently, it is easy to get

$$\text{sat}(Kx) = \sum_{i=1}^{2^m} \eta_i (E_i Kx + E_i^- Hx) \tag{22.31}$$

According to (22.31), for any $x(t) \in \Theta(H)$, the saturated feedback controller (22.27) for the single-input and single-output dynamic system (22.28) with $m = 1, n = 2, E = \{0, 1\}$ is expressed as follows:

$$\begin{aligned}\text{sat}(u(t)) &= \text{sat}[b_0^{-1}Kx(t) - (b_0 u_m)^{-1} z_3(t)] \\ &= \sum_{i=1}^{2} \eta_i [E_i (b_0^{-1} K x(t) - (b_0 u_m)^{-1} z_3(t)) \\ &\quad + E_i^- (Hx(t) - (b_0 u_m)^{-1} z_3(t))] \\ &= \sum_{i=1}^{2} \eta_i (b_0^{-1} E_i K + E_i^- H) x(t) - (b_0 u_m)^{-1} z_3(t). \end{aligned} \tag{22.32}$$

Therefore, based on the saturation controller (22.32) and the estimate error (22.9), for any $x(t) \in \Theta(H)$, the closed-loop system (22.28) is described by

$$\dot{x}(t) = \hat{A}x(t) + B_2 e_3(t), \tag{22.33}$$

where $\hat{A} = A + \sum_{i=1}^{2} \eta_i u_m B_1 (b_0^{-1} E_i K + E_i^- H)$. According to inequation (22.26) in Theorem 1, the estimate error $e_3(t)$ is bounded, i.e., $|e_3(t)| \le C$ where $C = \sqrt{V_1(\xi(0))/\beta_{\min}(P)}$. Consequently, the closed-loop system (22.33) is rewritten as follows:

$$\dot{x}(t) = \bar{A}x(t) + \hat{B}_2 \omega(t) \tag{22.34}$$

with $\hat{B}_2 = CB_2$ and $\omega^2(t) \le 1$.

Lemma 22.5 ([8]) *A set in R^n is said to be invariant if all the trajectories starting from it will remain in it regardless of disturbances. Let $V(x(t)) = x^T(t)Qx(t)$. If for all $x(t) \in \partial\Psi(Q, \rho)$ and all $\omega(t)$, $\omega^T(t)\omega(t) \le 1$, the relation $\dot{V}(x(t)) < 0$ holds, then the ellipsoid $\Psi(Q, \rho)$ is said to be a strictly invariant.*

The following theorem shows that a strictly invariant set is estimated for the closed-loop system (22.34).

Theorem 22.6 *Consider the closed-loop system formed of (22.34). Given two ellipsoids $\Psi(Q, \rho_1)$ and $\Psi(Q, 1)$, $0 < \rho_1 < 1$, if there exist matrices $H_1, H_2 \in R^{1\times 2}$ and a positive γ such that*

$$\begin{aligned} &\mathcal{H}\{QA + u_m Q B_1 (b_0^{-1} E_i K + E_i^- H_1)\} \\ &\quad + \frac{1}{\gamma} Q \hat{B}_2 \hat{B}_2^T Q + \frac{\gamma}{\rho_1} Q < 0, \ \forall i \in [1, 2] \end{aligned} \tag{22.35}$$

$$\begin{aligned} &\mathcal{H}\{QA + u_m Q B_1 (b_0^{-1} E_i K + E_i^- H_2)\} \\ &\quad + \frac{1}{\gamma} Q \hat{B}_2 \hat{B}_2^T Q + \gamma Q < 0, \ \forall i \in [1, 2] \end{aligned} \tag{22.36}$$

and $\Psi(Q, \rho_1) \subset \Theta(H_1)$, $\Psi(Q, 1) \subset \Theta(H_2)$, with the notation $\mathcal{H}\{*\} = * + *^T$, then the ellipsoid $\Psi(Q, \rho)$, $\rho \in [\rho_1, 1]$ is a strictly invariant set.

Proof A candidate Lyapunov function is chosen as

$$V(x(t)) = x^T(t)Qx(t), \tag{22.37}$$

The derivative of $V(x(t))$ along with (22.34) is given as

$$\begin{aligned}\dot{V}(x(t)) &= 2x^T(t)Q\dot{x}(t) = 2x^T(t)Q\left[\hat{A}x(t) + \hat{B}_2\omega(t)\right] \\ &= 2x^T(t)Q\hat{A}x(t) + 2x^T(t)Q\hat{B}_2\omega(t).\end{aligned}$$

According to Lemma 22.3, there exists $\gamma > 0$ such that

$$\begin{aligned}2x^T(t)Q\hat{B}_2\omega(t) &\le \frac{1}{\gamma}x^T(t)Q\hat{B}_2\hat{B}_2^TQx(t) + \gamma\omega^2(t) \\ &\le \frac{1}{\gamma}x^T(t)Q\hat{B}_2\hat{B}_2^{\,T}Qx(t) + \gamma.\end{aligned}$$

By Lemma 22.4, one has that

$$2x^T(t)Q\hat{A}x(t) \le \max\nolimits_{i\in G}\left\{2x^T(t)Q[A + u_mB_1(b_0^{-1}E_iK + E_i^-H_1)]x(t)\right\}.$$

Hence, there exists

$$\dot{V}(x(t)) \le \max_{i\in G}\{x^T(t)\Omega x(t)\} + \frac{1}{\gamma}x^T(t)Q\hat{B}_2\hat{B}_2^TQx(t) + \gamma$$

with

$$\Omega = \mathcal{H}\{QA + u_mQB_1(b_0^{-1}E_iK + E_i^-H_1)\}.$$

It follows from (22.35) that

$$\dot{V}(x(t)) < -\frac{\gamma}{\rho_1}x^T(t)Qx(t) + \gamma$$

holds for all $x(t) \in \Psi(Q, \rho_1)$. Note that $\dot{V}(x(t)) < 0$ holds on the boundary of $x^T(t)Qx(t) = \rho_1$ with $\Psi(Q, \rho_1)$. Therefore, $\Psi(Q, \rho_1)$ is a strictly invariant set according to Lemma 22.5. Similarly, it follows from (22.36) that $\Psi(Q, 1)$ is also a strictly invariant set for $x(t) \in \Psi(Q, 1)$. The conditions $\Psi(Q, \rho_1) \subset \Theta(H_1)$ and $\Psi(Q, 1) \subset \Theta(H_2)$ are equivalent to

$$\begin{bmatrix}\rho_1^{-1} & H_1 \\ H_1^T & Q\end{bmatrix} \ge 0 \text{ and } \begin{bmatrix}1 & H_2 \\ H_2^T & Q\end{bmatrix} \ge 0,$$

respectively. Since $\rho \in [\rho_1, 1]$, there exists a $\kappa \in [0, 1]$ such that

$$\frac{1}{\rho} = \kappa \frac{1}{\rho_1} + (1 - \kappa).$$

Letting $H = \kappa H_1 + (1 - \kappa) H_2$, it follows that

$$\begin{bmatrix} \rho & H \\ H^T & Q \end{bmatrix} \geq 0$$

Based on (22.35) and (22.36), by convexity, one has that

$$\begin{aligned} &\mathcal{H}\{QA + u_m QB_1(b_0^{-1}E_i K + E_i^- H)\} \\ &+\frac{1}{\gamma} Q\hat{B}_2\hat{B}_2^T Q + \frac{\gamma}{\rho} Q < 0, \ \forall i \in [1, 2] \end{aligned} \tag{22.38}$$

which implies that $\Psi(Q, \rho)$ is a strictly invariant set for the closed-loop system (22.33).

In order to obtain the disturbance rejection with guaranteed the domain of attraction, two invariant ellipsoids $\Psi(Q, \rho_1)$ and $\Psi(Q, 1)$ are constructed satisfying conditions (22.35) and (22.36) in Theorem 22.6, respectively. Moreover, two shape reference sets $\mathcal{X}_\infty = \Psi(R_1, 1)$ and $\mathcal{X}_0 = \Psi(R_2, 1)$ are given such that $\mathcal{X}_0 \subset \Psi(Q, 1)$ and $\Psi(Q, \rho_1) \subset \mu\mathcal{X}_\infty$ hold with minimized μ. Note that μ is a measure of the degree of disturbance rejection. Let $\vartheta = \mu^2$, $M = Q^{-1}$, $Y = KM$, $Z_1 = H_1 M$ and $Z_2 = H_2 M$ as in [8]. By fixing ρ_1 and γ, the optimization constraints is formulated as

$$\inf_{M>0, Y, Z_1, Z_2} \ \vartheta \tag{22.39}$$

$$\begin{aligned} \text{s.t.} \ &(a) \ M \geq R_2^{-1}, \ \rho_1 M \leq \vartheta R_1^{-1}, \\ &(b) \ \mathcal{H}\{AM + u_m QB_1(b_0^{-1}E_i Y + E_i^- Z_1)\} \\ &\quad +\gamma^{-1}\hat{B}_2\hat{B}_2^T + \rho_1\gamma^{-1}M < 0, \ \forall i \in [1, 2], \\ &(c) \ \mathcal{H}\{AM + u_m QB_1(b_0^{-1}E_i Y + E_i^- Z_2)\} \\ &\quad +\gamma^{-1}\hat{B}_2\hat{B}_2^T + \gamma M < 0, \ \forall i \in [1, 2], \\ &(d) \begin{bmatrix} \rho_1^{-1} & Z_1 \\ Z_1^T & M \end{bmatrix} \geq 0, \\ &(e) \begin{bmatrix} 1 & Z_2 \\ Z_2^T & M \end{bmatrix} \geq 0 \end{aligned}$$

Remark 22.7 The control strategy proposed in this chapter is applicable to all the linear positioning issues of rodless cylinders in pneumatic servo systems. It has strong robustness using an ESO to estimate and compensate total uncertainties. The stability analysis of this controller has been shown in Theorem 22.6.

Remark 22.8 The actuator saturation is taken into account in the position control design by using a polytopic form. It is shown that the control precision is further improved by employing a ADRC-based actuator saturation linear feedback control law. Moreover, a modified ESO is designed in this chapter. In comparison with traditional form, the proposed ESO is more simple and easy to be applied in a pneumatic rodless cylinder system. This will result in fast operation speed, as well as better response time. Therefore, the response time and positioning accuracy are both improved by the proposed method in this chapter.

22.3 Experiments and Results

In this section, the position control of single rodless pneumatic cylinder with actuator saturation is investigated which is shown in Fig. 22.2. The objective of this chapter is to achieve the accurate position control of rodless pneumatic cylinder with actuator saturation via ADRC control strategy. To demonstrate the effectiveness of this control method, a step signal with magnitude of 100 mm is given as the reference signal of the rodless cylinder. The ADRC controller parameters are determined based on the Theorems 22.2 and 22.6. The parameters of the pneumatic rodless cylinder system and the parameters of ADRC are listed in Table 22.1. The control sampling time is 0.01 s. The absolute pressure of air supply is 0.6 MPa.

Strictly invariant sets for the close-loop system (22.28) are obtained based on the given gains K of the ADRC controller. By setting $\rho_1 = 0.01$, $\gamma = 0.005$, the convex optimization problem (22.39) is solved. The solution is $\vartheta = 6.766496 \times 10^{-3}$ and

$$Q = 10^{-4} \times \begin{bmatrix} 0.2882 & 0.1111 \\ 0.1111 & 0.2890 \end{bmatrix}$$

The invariant ellipsoids of the close-loop system (22.28) are shown in Fig. 22.3. The large ellipsoid is $\Psi(Q, 1)$, the small one is $\Psi(Q, \rho_1)$. That is, all the trajectories of the rodless cylinder system (22.28) starting from within the large ellipsoid $\Psi(Q, 1)$ will enter the small ellipsoid $\Psi(Q, \rho_1)$ and remain in it.

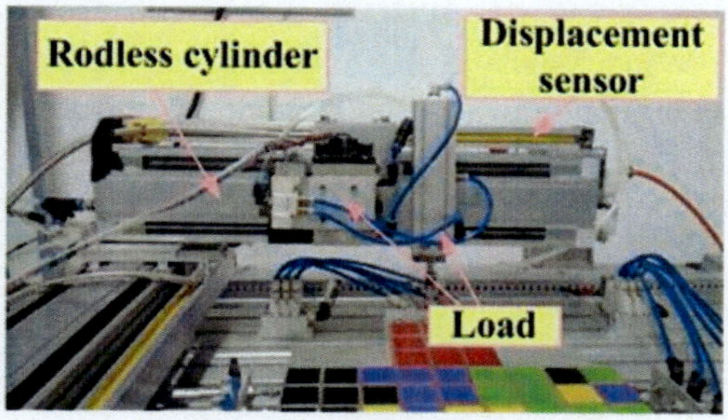

Fig. 22.2 Single pneumatic rodless cylinder

Table 22.1 Parameters of pneumatic rodless cylinder system and linear-ADRC

Air supply pressure	0.6 Mpa	
Input voltage	0–10 V	
Rodless cylinder	$T = 300\,\text{K}$	$A = 314\,\text{mm}^2$
	$R = 287\,\text{J/(k}\cdot\text{ mol)}$	$k = 1.4$
TD	$r_0 = 25$	$h_0 = 0.8$
Linear-ESO	$l_1 = 5$	$l_2 = 5$
	$l_3 = 0.4$	$\varepsilon = 0.1$
Controller	$k_1 = 0.5$	$k_2 = 0.3$
	$b_0 = 45$	$u_m = 5\,\text{V}$

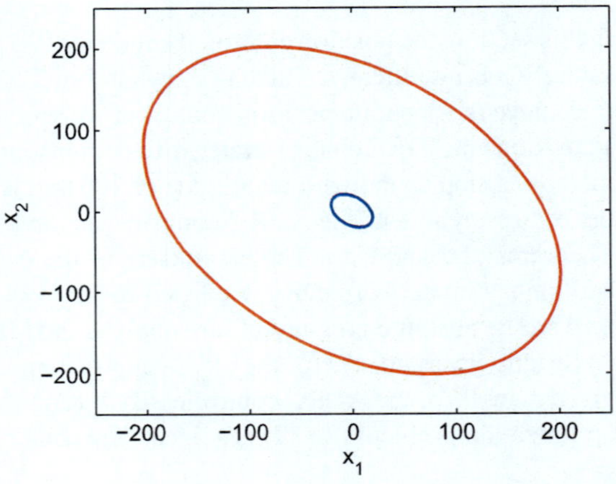

Fig. 22.3 The invariant ellipsoids for pneumatic rodless cylinder system

The positioning results of the pneumatic rodless cylinder system with actuator saturation are shown in Fig. 22.4a. The actual output displacement y tracks the reference signal y_d accurately and quickly without any overshoot. The steady-state error within ± 0.005 mm, which is equal to the sensor resolution. The response time is 0.5 s. The positioning result of the pneumatic rodless cylinder system without actuator saturation as a comparison experiment is shown in Fig. 22.4b. There is a large overshoot for the rodless cylinder without actuator saturation in the step response. The control voltage of the proportional directional control valve is shown in Fig. 22.4c. The input signal is the voltage signal. Note that the control voltage increases to the saturation limit 10 V, and eventually remains to 5 V which is the null voltage of the proportional directional control valve. Thereby, the input voltage without actuator saturation obviously exceeds the saturation limit of the proportional directional control valve as shown in Fig. 22.4c. Comparisons of positioning results on tracking the step signal are shown in Fig. 22.4d. The better performances with accuracy 0.005 mm and response time 0.5 s are achieved by controller subject to actuator saturation in this chapter comparing with the methods in [15, 16] and PID. Furthermore, quantitative comparisons on the three control strategies is given in Table 22.2.

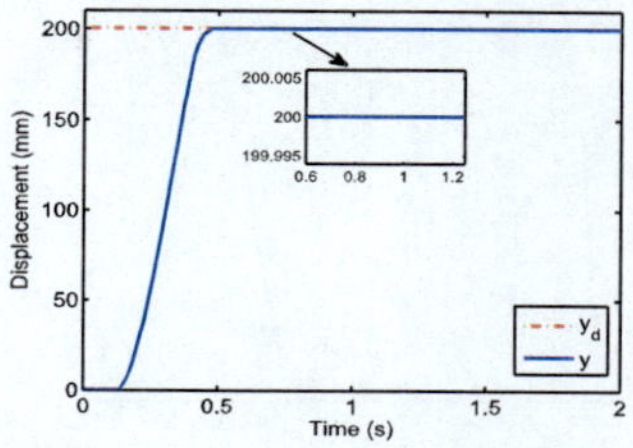

(a) Step response based on ADRC with actuator saturation.

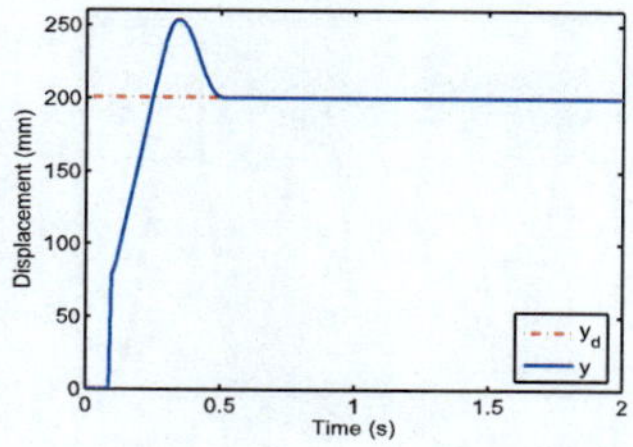

(b) Step response based on ADRC without actuator saturation.

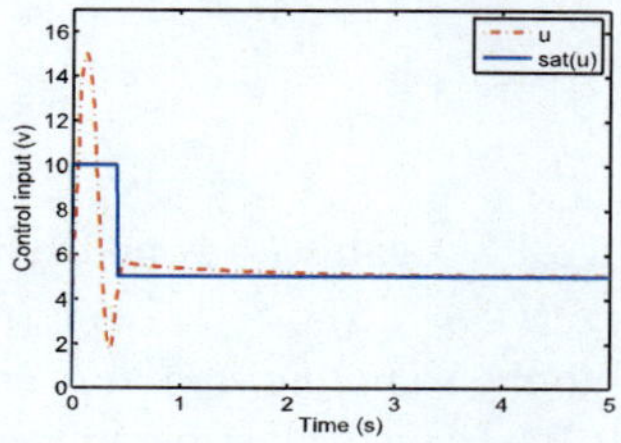

(c) Control voltage for step response.

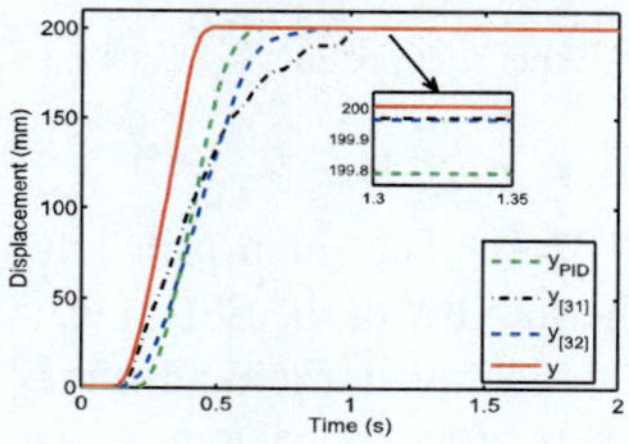

(d) Comparisons of step responses.

Fig. 22.4 Experiment results for tracking a step signal with 200 mm

Table 22.2 Comparing results

	Response time (s)	Steady-state error (mm)
PID	0.68	0.215
Ref. [15]	1	0.05
Ref. [16]	0.8	0.05
This chapter	0.5	0.005

In order to study the trajectory tracking control of the rodless cylinder with actuator saturation, a reference signal of amplitude 100 mm and frequency 0.5 Hz is considered. The tracking performance is shown in Fig. 22.5a. The displacement output y tracks the reference signal y_d much more quickly and smoothly than those in [15, 16]. The tracking performance of a sinusoidal signal with PID is shown in Fig. 22.5b.

Remark 22.9 An active disturbance rejection position control scheme has been presented for a rodless cylinder in servo systems without pressure states in [15]. In [16], a multi-controller strategy is proposed by designing a backstepping-based controller and a nonlinear error feedback controller. However, actuator saturation which is inevitable in pneumatic cylinders has not been taken into consideration in [15, 16]. In this chapter, actuator saturation is taken into consideration in the design of the active disturbance rejection controller. Experiment results reveal that the proposed controller significantly improves the positioning accuracy and the steady-state position error is within 0.005 mm for a step signal.

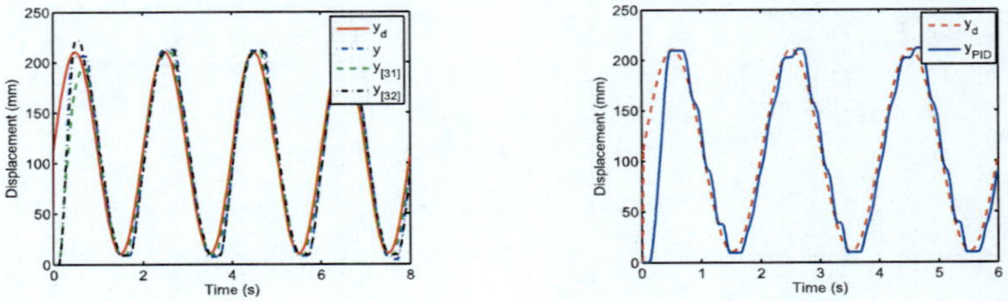

(a) Comparisons of sinusoidal tracking results.

(b) Sinusoidal tracking result based on PID.

Fig. 22.5 Experiment results for tracking a sinusoidal signal

Remark 22.10 The main principle of selecting the ESO gains l_1, l_2 and l_3 is to ensure the stability of the ESO, i.e., $z_1(t) \to x_1(t), z_2(t) \to x_2(t), z_3(t) \to x_3(t)$. In general, the parameter l_1 is inversely proportional to the sampling step h, l_2 and l_3 are inversely proportional to h^2 and h^3, respectively. Based on the law, parameters l_1, l_2 and l_3 are adjusted by real-time tracking and adjusting of the observer in practice. The controller gains k_1 and k_2 have definitude physical meaning. As a PD controller, k_1 is the proportionality coefficient, k_2 is the differential gain. That is, the regulation of the ADRC controller gain is similar to the PD controller.

22.4 Conclusion

This chapter presents a novel control strategy for the pneumatic rodless cylinder system subject to actuator saturation. A linear ESO-based control law is proposed for the accurate position control problem. An LMI-based optimization algorithm has been put forward to estimate a strictly invariance set of the closed-loop system. Experiment results validate the effectiveness of the proposed method. In particular, the active disturbance rejection controller significantly improves the positioning accuracy. The work is important and key for the control of pneumatic right angle composite motion system.

References

1. Pandian SR, Takemura F, Hayakawa Y, Kawamura S (2002) Pressure observer-controller design for pneumatic cylinder actuators. IEEE-ASME Trans Mechatron 7(4):490–499
2. Abry F, Brun X, Sesmat S, Bideaux E, Ducat C (2016) Electropneumatic cylinder backstepping position controller design with real-time closed-loop stiffness and damping tuning. IEEE Trans Control Syst Technol 24(2):541–552
3. Specker T, Buchholz M, Dietmayer K (2014) A new approach of dynamic friction modelling for simulation and observation. In: International federation of automatic control, pp 4523–4528

4. Yao J, Deng W, Jiao Z (2015) Adaptive control of hydraulic actuators with lugre model-based friction compensation. IEEE Trans Ind Electron 62(10):6469–6477
5. Rao Z, Bone GM (2008) Nonlinear modeling and control of servo pneumatic actuators. IEEE Trans Control Syst Technol 16(3):562–569
6. Taheri B, Case D, Richer E (2014) Force and stiffness backstepping-sliding mode controller for pneumatic cylinders. IEEE-ASME Trans Mechatron 19(6):1799–1809
7. Hu T, Lin Z, Chen BM (2002) An analysis and design method for linear systems subject to actuator saturation and disturbance. Automatica 38(2):351–359
8. Hu T, Lin Z, Chen BM (2002) Analysis and design for discrete-time linear systems subject to actuator saturation. Syst Control Lett 45(2):97–112
9. Chen BM, Lee TH, Peng K, Venkataramanan V (2003) Composite nonlinear feedback control for linear systems with input saturation: theory and an application. IEEE Trans Autom Control 48(3):427–439
10. Zhai D, Xia Y (2016) Adaptive control for teleoperation system with varying time delays and input saturation constraints. IEEE Trans Ind Electron 63(11):6921–6929
11. Wen C, Zhou J, Liu Z, Su H (2011) Robust adaptive control of uncertain nonlinear systems in the presence of input saturation and external disturbance. IEEE Trans Autom Control 56(7):1672–1678
12. Xia Y, Shi P, Liu G-P, Rees D, Han J (2007) Active disturbance rejection control for uncertain multivariable systems with time-delay. IET Control Theory Appl 1(1):75–81
13. Guo B-Z, Zhao Z-L (2011) On the convergence of an extended state observer for nonlinear systems with uncertainty. Syst Control Lett 60(6):420–430
14. Petersen IR (1987) A stabilization algorithm for a class of uncertain linear systems. Syst Control Lett 8(4):351–357
15. Zhao L, Yang Y, Xia Y, Liu Z (2015) Active disturbance rejection position control for a magnetic rodless pneumatic cylinder. IEEE Trans Ind Electron 62(9):5835–5846
16. Zhao L, Xia Y, Yang Y, Liu Z (2017) Multicontroller positioning strategy for a pneumatic servo system via pressure feedback. IEEE Trans Ind Electron 64(6):4800–4809

Index

L. Zhao et al., *Pneumatic Servo Systems Analysis*, Advances in Industrial Control,
https://doi.org/10.1007/978-981-16-9515-5